ADVANCES IN APPLIED COMBINATORICS

ADVANCES IN APPLIED COMBINATORICS

Edited by:

Stefano Spezia

ARCLER PRESS

www.arclerpress.com

Advances in Applied Combinatorics

Stefano Spezia

Arcler Press
2010 Winston Park Drive,
2nd Floor
Oakville, ON L6H 5R7
Canada
www.arclerpress.com
Tel: 001-289-291-7705
001-905-616-2116
Fax: 001-289-291-7601
Email: orders@arclereducation.com

ISBN: 978-1-77407-352-0 (Hardcover)

DECLARATION

ABOUT THE EDITOR

Stefano Spezia is Ph.D. holder in Applied Physics at the University of Palermo since April 2012. His major research experience is in noise-induced effects in nonlinear systems, especially in the fields of modeling of complex biological systems and simulation of semiconductor spintronic devices. Associate member of the Italian Physical Society and European Physical Society.

TABLE OF CONTENTS

SECTION 3: DERANGEMENTS AND THE EULER'S TOTIENT

SECTION 4: PARTITIONS AND GENERATING FUNCTIONS

SECTION 5: LINEAR RECURRENCES AND THE FIBONACCI NUMBERS

SECTION 6: GRAPH ALGORITHMS

SECTION 7: PERMUTATION GROUPS

LIST OF CONTRIBUTORS

Thabet Abdeljawad
Department of Mathematics, Faculty of Art and Sciencs, Çankaya University, Balgat, 06530 Ankara, Turkey

Dumitru Baleanu
Department of Mathematics, Faculty of Art and Sciencs, Çankaya University, Balgat, 06530 Ankara, Turkey
Institute of Space Sciences, 76900 Magurele-Bucharest, Romania
Department of Chemical and Materials Engineering, Faculty of Engineering, King Abdulaziz University, Jeddah 21589, Saudi Arabia

Fahd Jarad
Department of Mathematics, Faculty of Art and Sciencs, Çankaya University, Balgat, 06530 Ankara, Turkey

Ravi P. Agarwal
Department of Mathematics, Texas A & M University, 700 University Boulevard, Kingsville, TX, USA

Jianwei Zhou
Department of Mathematics, Linyi University, Linyi 276005, China

Anthony Sofo
Victoria University College, Victoria University, Melbourne City, VIC 8001, Australia

Ulrich Abel
Department MND, Technische Hochschule Mittelhessen, Wilhelm-Leuschner-Straße 13, 61169 Friedberg, Germany

Vijay Gupta
Department of Mathematics, Netaji Subhas Institute of Technology, Sector 3 Dwarka, New Delhi 110078, India

Mircea Ivan
Department of Mathematics, Technical University of Cluj-Napoca, Str. Memorandumului nr. 28, 400114 Cluj-Napoca, Romania

A. Vijayalekshmi
S. T. Hindu College, Nagercoil, India

J. Virgin Alangara Sheeba
S. T. Hindu College, Nagercoil, India

David Leach
Department of Mathematics, University of West Georgia, 1601 Maple Street, Carrollton, GA 30118, USA

Alexandra Gavryushkina
Department of Computer Science, The University of Auckland, Auckland, New Zealand

David Welch
Department of Computer Science, The University of Auckland, Auckland, New Zealand

Alexei J Drummond
Department of Computer Science, The University of Auckland, Auckland, New Zealand
Allan Wilson Centre for Molecular Ecology and Evolution, University of Auckland, Auckland, New Zealand

Taekyun Kim
Department of Mathematics, College of Science, Tianjin Polytechnic University, Tianjin, China
Department of Mathematics, Kwangwoon University, Seoul, Republic of Korea

Dae San Kim
Department of Mathematics, Sogang University, Seoul, Republic of Korea

Gwan-Woo Jang
Department of Mathematics, Kwangwoon University, Seoul, Republic of Korea

Jongkyum Kwon
Department of Mathematics Education and ERI, Gyeongsang National University, Jinju, Republic of Korea.

Andrew V. Sills
Department of Mathematical Sciences, Georgia Southern University, Statesboro, GA, 30460-8093, USA

Francesca Arrigo
Department of Mathematics and Statistics, University of Strathclyde, Glasgow, G1 1XH, UK

Peter Grindrod
Mathematical Institute, University of Oxford, Andrew Wiles Building, Radcliffe Observatory Quarter, Woodstock Road, Oxford, OX2 6GG, UK

Desmond J.Higham
Department of Mathematics and Statistics, University of Strathclyde, Glasgow, G1 1XH, UK

Vanni Noferini
Department of Mathematical Sciences, University of Essex, Wivenhoe Park, Colchester, CO4 3SQ, UK

Tian-Xiao He
Department of Mathematics and Computer Science, Illinois Wesleyan University, Bloomington, IL 61702, USA

Peter J.-S. Shiue
Department of Mathematical Sciences, University of Nevada Las Vegas, Las Vegas, NV 89154, USA

Andrew YZ Wang
School of Mathematical Sciences, University of Electronic Science and Technology of China, Chengdu, 611731, P.R. China

Peibo Wen
School of Mathematical Sciences, University of Electronic Science and Technology of China, Chengdu, 611731, P.R. China

Bartłomiej Bosek
Theoretical Computer Science Department, Faculty of Mathematics and Computer Science, Jagiellonian University, Krakow, Poland ´

Dariusz Leniowski
Institute of Computer Science, University of Warsaw, Warsaw, Poland

Piotr Sankowski
Institute of Computer Science, University of Warsaw, Warsaw, Poland

Anna Zych-Pawlewicz
Institute of Computer Science, University of Warsaw, Warsaw, Poland

Zemin Liu
Zhejiang University, China

Vincent W. Zheng
Advanced Digital Sciences Center, Singapore,

Zhou Zhao
Zhejiang University, China

Hongxia Yang
Alibaba Group, China

Kevin Chen-Chuan Chang
University of Illinois at Urbana-Champaign, USA

Minghui Wu
Zhejiang University City College, China

Jing Ying
Zhejiang University, China

Matheswaran Saravanan
Department of Computer Science and Engineering, VMKV Engineering College, Salem, Tamil Nadu 636308, India

Muthusamy Madheswaran
Department of Electronics and Communication Engineering, Mahendra

Engineering College, Namakkal, Tamil Nadu 637503, India

Timothy Woodcock
Department of Mathematics, Stonehill College Easton, Massachusetts, USA 02357

Vladimir Kornyak
Laboratory of Information Technologies, Joint Institute for Nuclear Research, 141980 Dubna, Moscow Region, Russia

LIST OF ABBREVIATIONS

BCDCP	Base Station Controlled Dynamic Clustering Protocol
CTPEDCDA	Cluster-based and Tree-based Power Efficient Data Collection and Aggregation
CH	Cluster Head
DMSTRP	Dynamic Minimal Spanning Tree Routing Protocol
EESR	Energy efficient spanning tree
MPP	Meta-Path Proximity
MST	Minimum Spanning Tree
MDS	Multidimensional scaling
MCHRP	Multiple Cluster Heads Routing Protocol
PRA	Path Ranking Algorithm
SN	Sensor Node
SAP	Shortest augmenting path
SPE	Subgraph-augmented path embedding
WMST	Weighted Minimum Spanning Tree
WSN	Wireless sensor network

PREFACE

Combinatorics is the science of combinations. It is a very important topic in the field of Discrete Mathematics. Among many other things, it helps us to formulate methods for enumerating a wide range of objects that fulfil a certain feature of interest in a given field. In particular, combinatorics has links to physical sciences, data processing, probability, statistics, numerical analysis, information and coding theory, and various other domains.

Section 1 of *Advances in Applied Combinatorics* book begins with the introduction of the binomial coefficients, the permutations and the combinatorial proofs. Among them, it discusses of fractional calculus by making use of the binomial theorem, of spectral norms of circulant matrices whose entries are binomial coefficients combined with either Fibonacci numbers or Lucas numbers, and of optimal algorithms for sorting a signed permutation by short operations. In the end, Section 1 presents a combinatorial enumeration argument for proving the applicability of stability control in epidemics complex networks.

Section 2 focuses on graph theory and partially ordered sets (posets). In particular, it initially provides a study about the total dominator chromatic number of paths, cycles and ladder graph. Then, it treats of the modular Leech trees and of recursive algorithms for phylogenetic tree counting. Lastly, it presents a new proof of Dilworth's theorem based upon the min-flow/max-cut property in flow networks, and its applications to h-partite graphs with multiple partial orders.

Sections 3 treats initially of some identities of derangement polynomials also by using umbral calculus. In the end, it focuses on the generalized Euler's totient, its connections to other totients and with counting formulae.

Section 4 deals with partitions and generating functions. Among them, it presents a convergent series formula which generalizes the Hardy-Ramanujan-Rademacher formula for the number of integer partitions. Lastly, it provides a study of a closed-form exponential generating function associated with non-backtracking walks around a graph.

Sections 5 focuses on sequences of numbers and polynomials defined by 2nd order linear recurrence, and on the Fibonacci numbers and the partial finite sums of their reciprocals.

Section 6 reviews some of the most important graph algorithms. In detail, it treats initially of the shortest augmenting path (SAP) algorithm that is one of the fundamental approaches to the maximum matching and maximum flow problems. In the end, it deals with minimum spanning trees (MST) with applications both to phylogenetic studies and to wireless sensor networks.

Finally, the last Section 7 discusses the permutations groups. In particular, it provides a brief survey of primitive groups of prime power degree. In the end, it presents a quantum formalism that does not involve any concepts associated with actual infinities, because formulated in constructive finite terms by using a unitary representation of a finite group.

SECTION 1:
BINOMIAL COEFFICIENTS, PERMUTATIONS AND COMBINATORIAL PROOFS

1

Fractional Sums and Differences with Binomial Coefficients

Thabet Abdeljawad[1], Dumitru Baleanu[1,2,3] , Fahd Jarad[1] , and Ravi P. Agarwal[4]

[1]Department of Mathematics, Faculty of Art and Sciencs, Çankaya University, Balgat, 06530 Ankara, Turkey

[2]Institute of Space Sciences, 76900 Magurele-Bucharest, Romania

[3]Department of Chemical and Materials Engineering, Faculty of Engineering, King Abdulaziz University, Jeddah 21589, Saudi Arabia

[4]Department of Mathematics, Texas A & M University, 700 University Boulevard, Kingsville, TX, USA

ABSTRACT

In fractional calculus, there are two approaches to obtain fractional derivatives. The first approach is by iterating the integral and then defining a fractional order by using Cauchy formula to obtain Riemann fractional

Citation: Thabet Abdeljawad, Dumitru Baleanu, Fahd Jarad, and Ravi P. Agarwal, "Fractional Sums and Differences with Binomial Coefficients," Discrete Dynamics in Nature and Society, vol. 2013, Article ID 104173, 6 pages, 2013. https://doi.org/10.1155/2013/104173.

Definition 2 (see [25–28]). (i) For a natural number m, the m rising (ascending) factorial of t is defined by

$$t^{\overline{m}} = \prod_{k=0}^{m-1} (t+k), \qquad t^{\overline{0}} = 1. \tag{5}$$

(ii) For any real number, the α rising function is defined by

$$t^{\overline{\alpha}} = \frac{\Gamma(t+\alpha)}{\Gamma(t)}, \quad t \in \mathbb{R} - \{\ldots, -2, -1, 0\}, \quad 0^{\overline{\alpha}} = 0. \tag{6}$$

Regarding the rising factorial function, we observe the following:

(i) $$\nabla\left(t^{\overline{\alpha}}\right) = \alpha t^{\overline{\alpha-1}}, \tag{7}$$

(ii) $$\left(t^{\overline{\alpha}}\right) = (t+\alpha-1)^{(\alpha)}, \tag{8}$$

(iii) $$\Delta_t (s-\rho(t))^{\overline{\alpha}} = -\alpha (s-\rho(t))^{\overline{\alpha-1}}. \tag{9}$$

Notation:

(i) For a real $\alpha>0$, we set $n = [\alpha] + 1$, where $[\alpha]$ is the greatest integer less than α.

(ii) For real numbers a and b, we denote $\mathbb{N}_a = \{a, a+1, \ldots\}$ and $_b\mathbb{N} = \{b, b-1, \ldots\}$.

(iii) For $n \in \mathbb{N}$ and real a, we denote

$$_{\ominus}\Delta^n f(t) \triangleq (-1)^n \Delta^n f(t). \tag{10}$$

(iv) For $n \in \mathbb{N}$ and real b, we denote

$$\nabla_{\ominus}^n f(t) \triangleq (-1)^n \nabla^n f(t). \tag{11}$$

The following definition and the properties followed can be found in [29] and the references therein.

Definition 3 (see [29]). Let $\sigma(t) = t + 1$ and $\rho(t) = t - 1$ be the forward and backward jumping operators, respectively. Then

(i) the (delta) left fractional sum of order $\alpha>0$ (starting from a) is defined by

$$\Delta_a^{-\alpha} f(t) = \frac{1}{\Gamma(\alpha)} \sum_{s=a}^{t-\alpha} (t-\sigma(s))^{(\alpha-1)} f(s), \quad t \in \mathbb{N}_{a+\alpha} \tag{12}$$

(ii) The (delta) right fractional sum of order $\alpha>0$ (ending at b) is defined by

1

Fractional Sums and Differences with Binomial Coefficients

Thabet Abdeljawad[1], Dumitru Baleanu[1,2,3], Fahd Jarad[1], and Ravi P. Agarwal[4]

[1]Department of Mathematics, Faculty of Art and Sciencs, Çankaya University, Balgat, 06530 Ankara, Turkey

[2]Institute of Space Sciences, 76900 Magurele-Bucharest, Romania

[3]Department of Chemical and Materials Engineering, Faculty of Engineering, King Abdulaziz University, Jeddah 21589, Saudi Arabia

[4]Department of Mathematics, Texas A & M University, 700 University Boulevard, Kingsville, TX, USA

ABSTRACT

In fractional calculus, there are two approaches to obtain fractional derivatives. The first approach is by iterating the integral and then defining a fractional order by using Cauchy formula to obtain Riemann fractional

Citation: Thabet Abdeljawad, Dumitru Baleanu, Fahd Jarad, and Ravi P. Agarwal, "Fractional Sums and Differences with Binomial Coefficients," Discrete Dynamics in Nature and Society, vol. 2013, Article ID 104173, 6 pages, 2013. https://doi.org/10.1155/2013/104173.

integrals and derivatives. The second approach is by iterating the derivative and then defining a fractional order by making use of the binomial theorem to obtain Grünwald-Letnikov fractional derivatives. In this paper we formulate the delta and nabla discrete versions for left and right fractional integrals and derivatives representing the second approach. Then, we use the discrete version of the Q-operator and some discrete fractional dual identities to prove that the presented fractional differences and sums coincide with the discrete Riemann ones describing the first approach.

1. Introduction and Preliminaries

Fractional calculus (FC) is developing very fast in both theoretical and applied aspects. As a result, FC is used intensively and successfully in the last few decades to describe the anomalous processes which appear in complex systems [1–6]. Very recently, important results in the field of fractional calculus and its applications were reported (see e.g., [7–10] and the references therein). The complexity of the real world phenomena is a great source of inspiration for the researchers to invent new fractional tools which will be able to dig much dipper into the mysteries of the mother nature. Historically the FC passed through different periods of evolutions, and it started to face very recently a new provocation: how to formulate properly its discrete counterpart [11–24]. At this stage, we have to stress on the fact that in the classical discrete equations their roots are based on the functional difference equations, therefore, the natural question is to find the generalization of these equations to the fractional case. In other words, we will end up with generalizations of the basic operators occurring in standard difference equations. As it was expected, there were several attempts to do this generalization as well as to apply this new techniques to investigate the dynamics of some complex processes. In recent years, the discrete counterpart of the fractional Riemann-Liouville, Caputo, was investigated mainly thinking how to apply techniques from the time scales calculus to the expressions of the fractional operators. Despite of the beauty of the obtained results, one simple question arises: can we obtain the same results from a new point of view which is more simpler and more intuitive? Having all above mentioned thinks in mind we are going to use the binomial theorem in order to get Grünwald-Letnikov fractional derivatives. After that, we proved that the results obtained coincide with the ones obtained by the discretization of the Riemann-Liouville operator. In this manner, we believe that it becomes more clear what the fractional difference equations bring new in description of the related complex phenomena described. For a natural number n, the fractional polynomial is defined by

$$t^{(n)} = \prod_{j=0}^{n-1} (t-j) = \frac{\Gamma(t+1)}{\Gamma(t+1-n)}, \tag{1}$$

where Γ denotes the special gamma function and the product is zero when $t+1-j=0$ for some j. More generally, for arbitrary α, define

$$t^{(\alpha)} = \frac{\Gamma(t+1)}{\Gamma(t+1-\alpha)}, \tag{2}$$

where the convention of that division at pole yields zero. Given that the forward and backward difference operators are defined by

$$\Delta f(t) = f(t+1) - f(t), \qquad \nabla f(t) = f(t) - f(t-1), \tag{3}$$

respectively, we define iteratively the operators $\Delta^m = \Delta(\Delta^{m-1})$ and $\nabla^m = \nabla(\nabla^{m-1})$, where m is a natural number.

Here are some properties of the factorial function.

Lemma 1 (see [13]). Assume the following factorial functions are well defined.

i. $\Delta t^{(\alpha)} = \alpha t^{(\alpha-1)}$.

ii. $(t-\mu)t^{(\mu)} = t^{(\mu+1)}$, *where* $\mu \in \mathbb{R}$.

iii. $\mu^{(\mu)} = \Gamma(\mu+1)$.

iv. *If* $t \leq r$, *then* $t^{(\alpha)} \leq r^{(\alpha)}$ *for any* $\alpha > r$.

v. *If* $0 < \alpha < 1$, *then* $t^{(\alpha\nu)} \geq (t^{(\nu)})^{\alpha}$.

vi. $t^{(\alpha+\beta)} = (t-\beta)^{(\alpha)} t^{(\beta)}$.

Also, for our purposes we list down the following two properties, the proofs of which are straightforward:

$$\nabla_s (s-t)^{(\alpha-1)} = (\alpha-1)(\rho(s)-t)^{(\alpha-2)}$$

$$\nabla_t (\rho(s)-t)^{(\alpha-1)} = -(\alpha-1)(\rho(s)-t)^{(\alpha-2)}. \tag{4}$$

For the sake of the nabla fractional calculus, we have the following definition.

Definition 2 (see [25–28]). (i) For a natural number m, the m rising (ascending) factorial of t is defined by

$$t^{\overline{m}} = \prod_{k=0}^{m-1} (t+k), \qquad t^{\overline{0}} = 1. \tag{5}$$

(ii) For any real number, the α rising function is defined by

$$t^{\overline{\alpha}} = \frac{\Gamma(t+\alpha)}{\Gamma(t)}, \quad t \in \mathbb{R} - \{\dots,-2,-1,0\}, \quad 0^{\alpha} = 0. \tag{6}$$

Regarding the rising factorial function, we observe the following:

(i) $\nabla\left(t^{\overline{\alpha}}\right) = \alpha t^{\overline{\alpha-1}},$ (7)

(ii) $\left(t^{\overline{\alpha}}\right) = (t+\alpha-1)^{(\alpha)},$ (8)

(iii) $\Delta_t (s-\rho(t))^{\overline{\alpha}} = -\alpha (s-\rho(t))^{\overline{\alpha-1}}.$ (9)

Notation:

(i) For a real $\alpha>0$, we set $n = [\alpha] + 1$, where $[\alpha]$ is the greatest integer less than α.

(ii) For real numbers a and b, we denote $\mathbb{N}_a = \{a, a+1, \dots\}$ and ${}_b\mathbb{N} = \{b, b-1, \dots\}$.

(iii) For $n \in \mathbb{N}$ and real a, we denote

$${}_{\ominus}\Delta^n f(t) \triangleq (-1)^n \Delta^n f(t). \tag{10}$$

(iv) For $n \in \mathbb{N}$ and real b, we denote

$$\nabla_{\ominus}^n f(t) \triangleq (-1)^n \nabla^n f(t). \tag{11}$$

The following definition and the properties followed can be found in [29] and the references therein.

Definition 3 (see [29]). Let $\sigma(t) = t + 1$ and $\rho(t) = t - 1$ be the forward and backward jumping operators, respectively. Then

(i) the (delta) left fractional sum of order $\alpha>0$ (starting from a) is defined by

$$\Delta_a^{-\alpha} f(t) = \frac{1}{\Gamma(\alpha)} \sum_{s=a}^{t-\alpha} (t-\sigma(s))^{(\alpha-1)} f(s), \quad t \in \mathbb{N}_{a+\alpha} \tag{12}$$

(ii) The (delta) right fractional sum of order $\alpha>0$ (ending at b) is defined by

$$_b\Delta^{-\alpha} f(t) = \frac{1}{\Gamma(\alpha)} \sum_{s=t+\alpha}^{b} (s - \sigma(t))^{(\alpha-1)} f(s)$$

$$= \frac{1}{\Gamma(\alpha)} \sum_{s=t+\alpha}^{b} (\rho(s) - t)^{(\alpha-1)} f(s), \quad t \in {}_{b-\alpha}\mathbb{N}. \tag{13}$$

(iii) The (nabla) left fractional sum of order $\alpha>0$ (starting from a) is defined by

$$\nabla_a^{-\alpha} f(t) = \frac{1}{\Gamma(\alpha)} \sum_{s=a+1}^{t} (t - \rho(s))^{\overline{\alpha-1}} f(s), \quad t \in \mathbb{N}_{a+1}. \tag{14}$$

(iv) The (nabla) right fractional sum of order $\alpha>0$ (ending at b) is defined by

$$_b\nabla^{-\alpha} f(t) = \frac{1}{\Gamma(\alpha)} \sum_{s=t}^{b-1} (s - \rho(t))^{\overline{\alpha-1}} f(s)$$

$$= \frac{1}{\Gamma(\alpha)} \sum_{s=t}^{b-1} (\sigma(s) - t)^{\overline{\alpha-1}} f(s), \quad t \in {}_{b-1}\mathbb{N}. \tag{15}$$

Regarding the delta left fractional sum, we observe the following:

- $\Delta_a^{-\alpha}$ maps functions defined on $\mathbb{N}_a$ to functions defined on $\mathbb{N}_{a+\alpha}$.
- $u(t) = \Delta_a^{-n} f(t), n \in \mathbb{N}$, satisfies the initial value problem:

$$\Delta^n u(t) = f(t), \quad t \in N_a,$$

$$u(a + j - 1) = 0, \quad j = 1, 2, \ldots, n. \tag{16}$$

(iii) The Cauchy function $(t - \sigma(s))^{(n-1)}/(n - 1)!$ vanishes at $s = t - (n - 1), \ldots, t - 1$.

Regarding the delta right fractional sum, we observe the following:

- $_b\Delta^{-\alpha}$ maps functions defined on $_b\mathbb{N}$ to functions defined on $_{b-\alpha}\mathbb{N}$.
- $u(t) = {}_b\Delta^{-n} f(t), n \in \mathbb{N}$, satisfies the initial value problem:

$$\nabla_\ominus^n u(t) = f(t), \quad t \in {}_bN,$$

$$u(b - j + 1) = 0, \quad j = 1, 2, \ldots, n. \tag{17}$$

(iii) The Cauchy function $(\rho(s) - t)^{(n-1)}/(n-1)!$ vanishes at $s = t+1, t+2, \ldots, t+(n-1)$.

Regarding the nabla left fractional sum, we observe the following:

(i) $\nabla_a^{-\alpha}$ maps functions defined on $\mathbb{N}_a$ to functions defined on $\mathbb{N}_a$.

(ii) $\nabla_a^{-n} f(t)$ satisfies the nth-order discrete initial value problem:

$$\nabla^n y(t) = f(t), \qquad \nabla^i y(a) = 0, \quad i = 0, 1, \ldots, n-1. \tag{18}$$

(iii) The Cauchy function $(t - \rho(s))^{\overline{n-1}}/\Gamma(n)$ satisfies $\nabla^n y(t) = 0$.

Regarding the nabla right fractional sum we observe the following:

(i) ${}_b\Delta^{-\alpha}$ maps functions defined on ${}_b\mathbb{N}$ to functions defined on ${}_b\mathbb{N}$.

(ii) $\nabla_a^{-n} f(t)$ satisfies the nth-order discrete initial value problem:

$$_{\ominus}\Delta^n y(t) = f(t), \qquad _{\ominus}\Delta^i y(b) = 0, \quad i = 0, 1, \ldots, n-1. \tag{19}$$

The proof can be done inductively. Namely, assuming it is true for n, we have

$$_{\ominus}\Delta^{n+1}\, {}_b\nabla^{-(n+1)} f(t) = {}_{\ominus}\Delta^n \left[-\Delta\, {}_b\nabla^{-(n+1)} f(t)\right]. \tag{20}$$

By the help of (9), it follows that

$$_{\ominus}\Delta^{n+1}\, {}_b\nabla^{-(n+1)} f(t) = {}_{\ominus}\Delta^n\, {}_b\nabla^{-n} f(t) = f(t). \tag{21}$$

The other part is clear by using the convention that $\sum_{k=s+1}^{s} = 0$.

(iii) The Cauchy function $(s - \rho(t))^{\overline{n-1}}/\Gamma(n)$ satisfies $_{\ominus}\Delta^n y(t) = 0$.

Definition 4. (i) [12] The (delta) left fractional difference of order $\alpha>0$ (starting from a) is defined by

$$\Delta_a^\alpha f(t) = \Delta^n \Delta_a^{-(n-\alpha)} f(t) = \frac{\Delta^n}{\Gamma(n-\alpha)} \sum_{s=a}^{t-(n-\alpha)} (t - \sigma(s))^{(n-\alpha-1)} f(s),$$

$$t \in \mathbb{N}_{a+(n-\alpha)}. \tag{22}$$

(ii) [19]The (delta) right fractional difference of order $\alpha > 0$ (ending at b) is defined by

$$
\begin{aligned}
{}_b\Delta^{\alpha} f(t) &= \nabla^n_{\ominus}\, {}_b\Delta^{-(n-\alpha)} f(t) \\
&= \frac{(-1)^n \nabla^n}{\Gamma(n-\alpha)} \sum_{s=t+(n-\alpha)}^{b} (s-\sigma(t))^{(n-\alpha-1)} f(s), \\
& \qquad t \in {}_{b-(n-\alpha)}\mathbb{N}. \qquad (23)
\end{aligned}
$$

(iii) [20] The (nabla) left fractional difference of order $\alpha > 0$ (starting from a) is defined by

$$
\begin{aligned}
\nabla^{\alpha}_{a} f(t) &= \nabla^n \nabla_a^{-(n-\alpha)} f(t) \\
&= \frac{\nabla^n}{\Gamma(n-\alpha)} \sum_{s=a+1}^{t} (t-\rho(s))^{\overline{n-\alpha-1}} f(s), \\
& \qquad t \in \mathbb{N}_{a+1}. \qquad (24)
\end{aligned}
$$

(iv) [29, 30] The (nabla) right fractional difference of order α>0 (ending at b) is defined by

$$
\begin{aligned}
{}_b\nabla^{\alpha} f(t) &= {}_{\ominus}\Delta^n\, {}_b\nabla^{-(n-\alpha)} f(t) \\
&= \frac{(-1)^n \Delta^n}{\Gamma(n-\alpha)} \sum_{s=t}^{b-1} (s-\rho(t))^{\overline{n-\alpha-1}} f(s), \\
& \qquad t \in {}_{b-1}\mathbb{N}. \qquad (25)
\end{aligned}
$$

Regarding the domains of the fractional type differences we observe the following.

- The delta left fractional difference Δ^{α}_{a} maps functions defined on $\mathbb{N}_a$ to functions defined on $\mathbb{N}_{a+(n-\alpha)}$.
- The delta right fractional difference ${}_b\Delta^{\alpha}$ maps functions defined on ${}_b\mathbb{N}$ to functions defined on ${}_{b-(n-\alpha)}\mathbb{N}$.
- The nabla left fractional difference Δ^{α}_{a} maps functions defined on $\mathbb{N}_a$ to functions defined on $\mathbb{N}_{a+n}$.

- The nabla right fractional difference ${}_b\Delta^{\alpha}$ maps functions defined on ${}_b\mathbb{N}$ to functions defined on ${}_{b-n}\mathbb{N}$.

Lemma 5 (see [15]). Let $0 \leq n-1 < \alpha \leq n$, and let $y(t)$ be defined on $\mathbb{N}_a$. Then the following statements are valid:

- $(\Delta_a^{\alpha})y(t-\alpha) = \nabla_{a-1}^{\alpha}y(t)\ for\ t \in \mathbb{N}_{n+a},$
- $(\Delta_a^{-\alpha})y(t+\alpha) = \nabla_{a-1}^{-\alpha}y(t)\ for\ t \in \mathbb{N}_a.$

Lemma 6 (see [29]). Let $y(t)$ be defined on ${}_{b+1}\mathbb{N}$. Then the following statements are valid:

- $({}_b\Delta^{\alpha})y(t+\alpha) = {}_{b+1}\nabla^{\alpha}y(t)\ for\ t \in {}_{b-n}\mathbb{N},$
- $({}_b\Delta^{-\alpha})y(t-\alpha) = {}_{b+1}\nabla^{-\alpha}y(t)\ for\ t \in {}_b\mathbb{N}.$

If $f(s)$ is defined on $N_a \cap {}_bN$ and $a \equiv b(\text{mod}1)$ then $(Qf)(s) = f(a+b-s)$. The Q-operator generates a dual identity by which the left type and the right type fractional sums and differences are related. Using the change of variable $u=a+b-s$, in [18] it was shown that

$$\Delta_a^{-\alpha}Qf(t) = Q\,{}_b\Delta^{-\alpha}f(t), \tag{26}$$

and, hence,

$$\Delta_a^{\alpha}Qf(t) = (Q\,{}_b\Delta^{\alpha}f)(t). \tag{27}$$

The proof of (27) follows by (26) and by noting that

$$-Q\nabla f(t) = \Delta Qf(t). \tag{28}$$

Similarly, in the nabla case we have

$$\nabla_a^{-\alpha}Qf(t) = Q\,{}_b\nabla^{-\alpha}f(t), \tag{29}$$

and, hence,

$$\nabla_a^{\alpha}Qf(t) = (Q\,{}_b\nabla^{\alpha}f)(t). \tag{30}$$

The proof of (30) follows by (29) and that

$$-Q\Delta f(t) = \nabla Qf(t). \tag{31}$$

For more details about the discrete version of the Q-operator we refer to [29].

From the difference calculus or time scale calculus, for a natural n and a sequence f, we recall

$$\Delta^n f(t) = \sum_{k=0}^{n} (-1)^k \binom{n}{k} f(t+n-k),$$

$$\nabla^n f(t) = \sum_{k=0}^{n} (-1)^k \binom{n}{k} f(t-n+k). \quad (32)$$

THE FRACTIONAL DIFFERENCES AND SUMS WITH BINOMIAL COEFFICIENTS

We first give the definition of fractional order of (32) in the left and right sense.

Definition 7. The (binomial) delta left fractional difference and sum of order $\alpha>0$ for a function f defined on $\mathbb{N}_a$ are defined by

(a) $$B\Delta_a^{\alpha} = \sum_{k=0}^{\alpha+t-a} (-1)^k \binom{\alpha}{k} f(t+\alpha-k), \quad t \in \mathbb{N}_{a+n-\alpha}, \quad (33)$$

(b) $$B\Delta_a^{-\alpha} = \sum_{k=0}^{\alpha+t-a} (-1)^k \binom{-\alpha}{k} f(t-\alpha-k), \quad t \in \mathbb{N}_{a+\alpha}, \quad (34)$$

where $(-1)^k \binom{-\alpha}{k} = \binom{\alpha+k-1}{k}$

Definition 8. The (binomial) nabla left fractional difference and sum of order $\alpha>0$ for a function f defined on $\mathbb{N}_a$, are defined by

(a) $$B\nabla_a^{\alpha} = \sum_{k=0}^{t-a-1} (-1)^k \binom{\alpha}{k} f(t-k), \quad t \in \mathbb{N}_{a+n}, \quad (35)$$

(b) $$B\nabla_a^{-\alpha} = \sum_{k=0}^{t-a-1} (-1)^k \binom{-\alpha}{k} f(t-k), \quad t \in \mathbb{N}_{a}. \quad (36)$$

Analogously, in the right case we can define the following.

Definition 9. The (binomial) delta right fractional difference and sum of order $\alpha>0$ for a function f defined on ${}_b\mathbb{N}$ are defined by

(a) $$_b\Delta B^{\alpha} = \sum_{k=0}^{\alpha+b-t} (-1)^k \binom{\alpha}{k} f(t-\alpha+k), \quad t \in b-n+\alpha\mathbb{N}, \quad (37)$$

(b) $$_b\Delta B^{-\alpha} = \sum_{k=0}^{-\alpha+b-t} (-1)^k \binom{-\alpha}{k} f(t+\alpha+k), \quad t \in b-\alpha\mathbb{N}. \tag{38}$$

Definition 10. The (binomial) nabla right fractional difference and sum of order α>0 for a function f defined on $_b\mathbb{N}$ are defined by

(a) $$_b\nabla B^{\alpha} = \sum_{k=0}^{b-t-1} (-1)^k \binom{\alpha}{k} f(t-k), \quad t \in {}_{b-n}\mathbb{N}, \tag{39}$$

(b) $$_b\nabla B^{-\alpha} = \sum_{k=0}^{b-t-1} (-1)^k \binom{-\alpha}{k} f(t-k), \quad t \in {}_b\mathbb{N}. \tag{40}$$

We next proceed to show that the Riemann fractional differences and sums coincide with the binomial ones defined above. We will use the dual identities in Lemma 5 and Lemma 6, and the action of the discrete version of the Qoperator to follow easy proofs and verifications. In [20], the author used a delta Leibniz's rule to obtain the following alternative definition for Riemann delta left fractional differences:

$$\Delta_a^{\alpha} f(t) = \frac{1}{\Gamma(-\alpha)} \sum_{s=a}^{t+\alpha} (t-\sigma(s))^{(-\alpha-1)} f(s),$$

$$\alpha \notin \mathbb{N},\ t \in \mathbb{N}_{a+n-\alpha}, \tag{41}$$

then proceeded with long calculations and showed, actually, that

$$\Delta_a^{\alpha} f(t) = B\Delta_a^{\alpha} f(t), \qquad \Delta_a^{-\alpha} f(t) = B\Delta_a^{-\alpha} f(t). \tag{42}$$

Theorem 11. Let f be defined on suitable domains and α>0. Then,

(1) $$\Delta_a^{\alpha} f(t) = B\Delta_a^{\alpha} f(t), \qquad \Delta_a^{-\alpha} f(t) = B\Delta_a^{-\alpha} f(t), \tag{43}$$

(2) $$_b\Delta^{\alpha} f(t) = {}_b\Delta B^{\alpha} f(t), \qquad {}_b\Delta^{-\alpha} f(t) = \Delta B_a^{-\alpha} f(t), \tag{44}$$

(3) $$\nabla_a^{\alpha} f(t) = B\nabla_a^{\alpha} f(t), \qquad \nabla_a^{-\alpha} f(t) = B\nabla_a^{-\alpha} f(t), \tag{45}$$

(4) $$_b\nabla^{\alpha} f(t) = {}_b\nabla B^{\alpha} f(t), \qquad {}_b\nabla^{-\alpha} f(t) = {}_b\nabla B^{-\alpha} f(t). \tag{46}$$

Proof. (1) follows by (42).

(2) By the discrete Q-operator action we have

$$
\begin{aligned}
{}_b\Delta^{\alpha} f(t) &= Q\Delta_a(Qf)(t) \\
&= Q\sum_{k=0}^{\alpha+t-a}(-1)^k\binom{\alpha}{k}(Qf)(t+\alpha-k) \\
&= {}_b\Delta B^{\alpha} f(t).
\end{aligned} \tag{47}
$$

The fractional sum part is also done in a similar way by using the Q-operator.

(3) By the dual identity in Lemma 5 (i) and (42), we have

$$\nabla_a^{\alpha} f(t) = \Delta_{a+1}^{\alpha} f(t+\alpha) = B\Delta_{a+1}^{\alpha} f(t+\alpha) = B\nabla_a^{\alpha} f(t). \tag{48}$$

The fractional sum part can be proved similarly by using Lemma 5 (ii) and (42).

(4) The proof can be achieved by either (2) and Lemma 6 or, alternatively, by (3) and the discrete Q-operator.

Remark 12. In analogous to (41), the authors in [31] used a nabla Leibniz's rule to prove that

$$\nabla_a^{\alpha} f(t) = \frac{1}{\Gamma(-\alpha)}\sum_{s=a+1}^{t}(t-\rho(s))^{\overline{-\alpha-1}} f(s). \tag{49}$$

In [30], the authors used a delta Leibniz's Rule to prove the following formula for nabla right fractional differences:

$${}_b\nabla^{\alpha} f(t) = \frac{1}{\Gamma(-\alpha)}\sum_{s=t}^{b-1}(s-\rho(t))^{\overline{-\alpha-1}} f(s). \tag{50}$$

Similarly, we can use a nabla Leibniz's rule to prove the following formula for the delta right fractional differences:

$${}_b\Delta^{\alpha} f(t) = \frac{1}{\Gamma(-\alpha)}\sum_{s=t-\alpha}^{b}(s-\sigma(t))^{(-\alpha-1)} f(s). \tag{51}$$

We here remark that the proofs of the last three parts of Theorem 11 can be done alternatively by proceeding as in [20] starting from (49), (50), and (51). Also, it is worth mentioning that mixing both delta and nabla operators in defining delta and nabla right Riemann fractional differences was essential in proceeding, through the dual identities and the discrete Qoperator or delta and nabla type Leibniz's rules, to obtain the main results in this paper [29].

CONCLUSION

The impact of fractional calculus in both pure and applied branches of science and engineering started to increase substantially. The main idea of iterating an operator and then generalizing to any order (real or complex) started to be used in the last decade to obtain appropriate discretization for the fractional operators. We mention, from the theory of time scales view point, that how to obtain the fractional operators was a natural question and it was not correlated to the wellknown Grunwald-Letnikov approach. We believe that the ¨ discretizations obtained recently in the literature for the fractional operators are different from the one reported within Grunwald-Letnikov method. Bearing all of these thinks in ¨ mind we proved that the discrete operators via binomial theorem will lead to the same results as the ones by using the discretization of the Riemann-Liouville operators via time scales techniques. The discrete version of the impressive dual tool Q-operator has been used to prove the equivalency.

REFERENCES

1. S. G. Samko, A. A. Kilbas, and O. I. Marichev, Marichev, Fractional Integrals and Derivatives: Theory and Applications, Gordon and Breach, Yverdon, Switzerland, 1993.
2. I. Podlubny, Fractional Differential Equations, vol. 198, Academic Press., San Diego, Calif, USA, 1999.
3. A. A. Kilbas, H. M. Srivastava, and J. J. Trujillo, Theory and Application of Fractional Differential Equations, vol. 204 of NorthHolland Mathematics Studies, Elsevier Science, Amsterdam,The Netherland, 2006.
4. B. J. West, M. Bologna, and P. Grigolini, Physics of Fractal Operators, Springer, New York, NY, USA, 2003.
5. R. L. Magin, Fractional Calculus in Bioengineering, Begell House, West Redding, Conn, USA, 2006.
6. D. Baleanu, K. Diethelm, E. Scalas, and J. J. Trujillo, Fractional Calculus Models and Numerical Methods, vol. 3 of Series on Complexity, Nonlinearity and Chaos, World Scientific, Hackensack, NJ, USA, 2012.
7. M. D. Ortigueira, "Fractional central differences and derivatives," Journal of Vibration and Control, vol. 14, no. 9-10, pp. 1255–1266, 2008.
8. P. Lino and G. Maione, "Tuning PI(.) fractional order controllers for position control of DC-servomotors," in Proceedings of the IEEE International Symposium on Industrial Electronics (ISIE '10), pp. 359–363, July 2010.
9. J. A. T. Machado, C. M. Pinto, and A. M. Lopes, "Power law and entropy analysis of catastrophic phenomena," Mathematical Problems in Engineering, vol. 2013, Article ID 562320, 10 pages, 2013.
10. C. F. Lorenzo, T. T. Hartley, and R. Malti, "Application of the principal fractional metatrigonometric functions for the solution of linear commensurate-order time-invariant fractional differential equations," Philosophical Transactions of the Royal Society A, vol. 371, no. 1990, Article ID 20120151, 2013.
11. H. L. Gray and N. F. Zhang, "On a new definition of the fractional difference," Mathematics of Computation, vol. 50, no. 182, pp. 513–529, 1988.
12. K. S. Miller and B. Ross, "Fractional difference calculus," in

Proceedings of the International Symposium on Univalent Functions, Fractional Calculus and Their Applications, pp. 139–152, Nihon University, Koriyama, Japan, 1989.

13. F. M. Atıcı and P. W. Eloe, "A transform method in discrete fractional calculus," International Journal of Difference Equations, vol. 2, no. 2, pp. 165–176, 2007.
14. F. M. Atıcı and P. W. Eloe, "Initial value problems in discrete fractional calculus," Proceedings of the American Mathematical Society, vol. 137, no. 3, pp. 981–989, 2009.
15. F. M. Atıcı and P. W. Eloe, "Discrete fractional calculus with the nabla operator," Electronic Journal of Qualitative Theory of Differential Equations, no. 3, pp. 1–12, 2009.
16. F. M. Atıcı and S. S¸engul, "Modeling with fractional difference ¨ equations," Journal of Mathematical Analysis and Applications, vol. 369, no. 1, pp. 1–9, 2010.
17. F. M. Atıcı and P. W. Eloe, "Gronwall's inequality on discrete fractional calculus," Computerand Mathematics with Applications, vol. 64, no. 10, pp. 3193–3200, 2012.
18. T. Abdeljawad, "On Riemann and Caputo fractional differences," Computers and Mathematics with Applications, vol. 62, no. 3, pp. 1602–1611, 2011.
19. T. Abdeljawad and D. Baleanu, "Fractional differences and integration by parts," Journal of Computational Analysis and Applications, vol. 13, no. 3, pp. 574–582, 2011.
20. M. Holm, The theory of discrete fractional calculus development and application dissertation., University of Nebraska, Lincoln, Neb, USA, 2011.
21. G. A. Anastassiou, "Principles of delta fractional calculus on time scales and inequalities," Mathematical and Computer Modelling, vol. 52, no. 3-4, pp. 556–566, 2010.
22. G. A. Anastassiou, "Nabla discrete fractional calculus and nabla inequalities," Mathematical and Computer Modelling, vol. 51, no. 5-6, pp. 562–571, 2010.
23. G. A. Anastassiou, "Foundations of nabla fractional calculus on time scales and inequalities," Computers and Mathematics with Applications, vol. 59, no. 12, pp. 3750–3762, 2010.
24. N. R. O. Bastos, R. A. C. Ferreira, and D. F. M. Torres, "Discretetime

fractional variational problems," Signal Processing, vol. 91, no. 3, pp. 513–524, 2011.

25. M. Bohner and A. Peterson, Advances in Dynamic Equations on Time Scales, Birkhauser, Boston, Mass, USA, 2003. ¨

26. G. Boros and V. Moll, Ircsistible Integrals, Symbols, Analysis and Expreiments in the Evaluation of Integrals, Cambridge University Press, Cambridge, UK, 2004.

27. R. L. Graham, D. E. Knuth, and O. Patashnik, Concrete Mathematics: A Foundation for Copmuter Science, Addison-Wesley, Reading, Mass, USA, 2nd edition, 1994.

28. J. Spanier and K. B. Oldham, "The pochhammer polynomials $(x)n$," in An Atlas of Functions, pp. 149–156, Hemisphere, Washington, DC, USA, 1987.

29. T. Abdeljawad, "Dual identities in fractional difference calculus within Riemann," Advances in Difference Equations, vol. 2013, article 36, 2013.

30. T. Abdeljawad and F. M. Atıcı, "On the definitions of nabla fractional operators," Abstract and Applied Analysis, vol. 2012, Article ID 406757, 13 pages, 2012.

31. K. Ahrendt, L. Castle, M. Holm, and K. Yochman, "Laplace transforms for the nabla-difference operator and a fractional variation of parameters formula," Communications in Applied Analysis. In press.

2

The Identical Estimates of Spectral Norms for Circulant Matrices with Binomial Coefficients Combined with Fibonacci Numbers and Lucas Numbers Entries

Jianwei Zhou

Department of Mathematics, Linyi University, Linyi 276005, China

ABSTRACT

Improved estimates for spectral norms of circulant matrices are investigated, and the entries are binomial coefficients combined with either Fibonacci numbers or Lucas numbers. Employing the properties of given circulant matrices, this paper improves the inequalities for their spectral norms, and gets corresponding identities of spectral norms. Moreover, by some well-known identities, the explicit identities for spectral norms are obtained. Some numerical tests are listed to verify the results.

Citation: Jianwei Zhou, "The Identical Estimates of Spectral Norms for Circulant Matrices with Binomial Coefficients Combined with Fibonacci Numbers and Lucas Numbers Entries," Journal of Function Spaces, vol. 2014, Article ID 672398, 5 pages, 2014. https://doi.org/10.1155/2014/672398.

INTRODUCTION

Circulant matrices have connection to physics, signal and image processing, probability, statistics, numerical analysis, algebraic coding theory, and many other areas. There are lots of examples from statistical signal processing and information theory that illustrate the application of the circulant matrices, which emphasize how the asymptotic eigenvalue distribution theorem allows one to evaluate results for processes (for the details please refer to [1–3] and the reference therein). Meanwhile a real circulant stochastic process can be described with autocovariance matrices, which are subjected to a cyclical permutation. With the help of autocovariance circulant matrices, it is easy to provide derivations of some results that are central to the analysis of statistical periodograms and empirical spectral density functions (see [4]).

In past decades, the estimates for spectral norms of matrices have been investigated in lots of literatures. Moreover, the determinants and inverses of circulant matrices are stated in many articles. The norms of circulant matrices play an important role in analysing the process of statistics, numerical analysis, and many other problems (for more details, please refer to [3, 5–10] and the reference therein). Bryc and Sethuraman [11] investigated the maximum eigenvalue for circulant matrices. Solak [7] obtained lower and upper bounds for the spectral norm of circulant matrices, where the entries are classical Fibonacci numbers. İpek [8] establishes spectral norms of circulant matrices with Fibonacci and Lucas numbers. Furthermore, circulant matrices take up an important status in stochastic calculus, Meckes [12, 13] gave some results on the spectral norm of a special random Toeplitz matrix and random circulant matrices, Mehta [14] made a deep discussion on random circulant matrices.

The outline of this paper is as follows. In Section 2, we state some preliminaries and recall some well-known results. In Section 3, we focus on the identities of estimations for spectral norms. In Section 4, we present various numerical examples to exhibit the accuracy and efficiency of our results. Finally, we summarise this paper and illustrate our future work.

PRELIMINARIES

The Fibonacci and Lucas sequences $\{F_n\}$ and $\{L_n\}$ are defined by the recurrence relations:

$$F_n = F_{n-1} + F_{n-2}, \qquad L_n = L_{n-1} + L_{n-2}, \tag{1}$$

with $F_0 = 0$, $F_1 = 1$, $L_0 = 2$, and $L_1 = 1$, respectively.

Obviously, the Fibonacci and Lucas sequences are listed in the following sequence:

$$\begin{array}{lllllllllll} n & 0 & 1 & 2 & 3 & 4 & 5 & 6 & 7 & \cdots \\ F_n & 0 & 1 & 1 & 2 & 3 & 5 & 8 & 13 & \cdots \\ L_n & 2 & 1 & 3 & 4 & 7 & 11 & 18 & 29 & \cdots, \end{array} \tag{2}$$

and their corresponding Binet forms are (see [15])

$$F_n = \frac{1}{\sqrt{5}}\left[\left(\frac{1+\sqrt{5}}{2}\right)^n - \cos(\pi n)\left(\frac{1+\sqrt{5}}{2}\right)^{-n}\right],$$

$$L_n = \left(\frac{1+\sqrt{5}}{2}\right)^n + \cos(\pi n)\left(\frac{1+\sqrt{5}}{2}\right)^{-n}. \tag{3}$$

Now, we recall that, for

$$B = \begin{pmatrix} F_0 & F_1 & \cdots & F_{n-1} \\ F_{n-1} & F_0 & \cdots & F_{n-2} \\ F_{n-2} & F_{n-1} & \cdots & F_{n-3} \\ \vdots & \vdots & \ddots & \vdots \\ F_1 & F_2 & \cdots & F_0 \end{pmatrix}_{n\times n}, \tag{4}$$

there hold the following estimates:

$$\sqrt{F_nF_{n-1} + 4\left(F_{n-1}^2 + F_{n-1}F_{n-2} + \frac{1-(-1)^n}{2}\right)} \le \|B\|_2,$$

$$\|B\|_2$$

$$\le \sqrt{F_nF_{n-1} + 4\left(F_{n-1}^2 + F_{n-1}F_{n-2} + \frac{1-(-1)^n}{2}\right)}$$

$$\times \sqrt{1 + F_nF_{n-1} + 4\left(F_{n-1}^2 + F_{n-1}F_{n-2} - \frac{3+(-1)^n}{2}\right)}. \tag{5}$$

For the details please refer to [7].

There are lots of identities for Fibonacci numbers and Lucas numbers combined with Binomial coefficients (for more details please refer to [8, 16–18] and the reference therein). In this paper, we focus on the following identities:

$$\sum_{i=0}^{n}\binom{n}{i}F_i = F_{2n}, \qquad \sum_{i=0}^{n}\binom{n}{i}L_i = L_{2n},$$

$$\sum_{i=0}^{n}\binom{n}{i}F_{i+k} = F_{2n+k}, \qquad \sum_{i=0}^{n}\binom{n}{i}L_{i+k} = L_{2n+k}. \tag{6}$$

Furthermore, for all $k \in \mathbb{Z}^+$, there hold the following identities:

$$\sum_{i=0}^{n}\binom{n}{i}F_{2i+k} = \begin{cases} 5^{n/2}F_{n+k}, & (n \text{ even}), \\ 5^{(n-1)/2}L_{n+k}, & (n \text{ odd}), \end{cases} \tag{7}$$

$$\sum_{i=0}^{n}\binom{n}{i}L_{2i+k} = \begin{cases} 5^{n/2}L_{n+k} & (n \text{ even}), \\ 5^{(n+1)/2}F_{n+k} & (n \text{ odd}). \end{cases} \tag{8}$$

Definition 1 (see [19]). A circulant matrix is an $n\times n$ complex matrix with the following form:

$$A = \begin{pmatrix} a_0 & a_1 & \cdots & a_{n-1} \\ a_{n-1} & a_0 & \cdots & a_{n-2} \\ a_{n-2} & a_{n-1} & \cdots & a_{n-3} \\ \vdots & \vdots & \ddots & \vdots \\ a_1 & a_2 & \cdots & a_0 \end{pmatrix}_{n\times n}. \tag{9}$$

The first row of A is $(a_0, a_1,\ldots,_{-1})$ and its $(j + 1)$th row is obtained by giving its jth row a right circular shift by one positions.

Definition 2 (see [3]). The spectral norm $\|\cdot\|_2$ of a matrix A with complex entries is the square root of the largest eigenvalue of the positive semidefinite matrix A^*A:

$$\|A\|_2 = \sqrt{\lambda_{\max}(A^*A)}, \tag{10}$$

where A^* denotes the conjugate transpose of A. Therefore if A is an $n\times n$ real symmetric matrix or A is a normal matrix, then

$$\|A\|_2 = \max_{1\le i\le n}|\lambda_i|, \tag{11}$$

where $\lambda_1, \lambda_2,\ldots,\lambda_n$ are the eigenvalues of A.

THE IDENTITIES OF ESTIMATIONS FOR SPECTRAL NORMS

We give the main theorems of this paper in the following parts.

Theorem 3. Let B_1 be as the matrix in (9), and let the first row of B_1 be $(\binom{n}{0}F_0, \binom{n}{1}F_1, \ldots, \binom{n}{n}F_n)$. Then one has

$$\|B_1\|_2 = F_{2n}. \qquad (12)$$

Proof. Combining with Definition 2, the spectral radius of B_1 is equal to its spectral norm, where we used the fact that B_1 is normal. Moreover, by the irreducible and entrywise nonnegative properties, we deduce that $\|B_1\|2$ is equal to its Perron value. Denote by V = $(1, 1, \ldots, 1)^T$ an n-dimensional column vector. There holds

$$B_1 v = \left(\sum_{i=0}^{n}\binom{n}{i}F_i\right)v. \qquad (13)$$

Obviously, $\sum_{i=0}^{n}\binom{n}{i}F_i$ is an eigenvalue of B_1 associated with the positive eigenvector V, which is the Perron value of B_1. Employing the first identity in (6), we have

$$\|B_1\|_2 = F_{2n}. \qquad (14)$$

This completes the proof.

With the same approach, we obtain the following corollary.

Table 1: Spectral norms of B_i (i = 1, 2, 3, 4) and k=1

n	1	2	3	4	5	6	7	8
$\|B_1\|_2$	1	3	8	21	55	144	377	987
F_{2n}	1	3	8	21	55	144	377	987
$\|B_2\|_2$	3	7	18	47	123	322	843	2207
L_{2n}	3	7	18	47	123	322	843	2207
$\|B_3\|_2$	2	5	13	34	89	233	610	1597
F_{2n+1}	2	5	13	34	89	233	610	1597
$\|B_4\|_2$	4	11	29	76	199	521	1364	3571
L_{2n+1}	4	11	29	76	199	521	1364	3571

Corollary 4: Let B_2 be as the matrix in (9), and let the first row of B_2 be $(\binom{n}{0}L_0, \binom{n}{1}L_1, \ldots, \binom{n}{n}L_n)$. Then one has the following identity:

$$\|B_2\|_2 = L_{2n}. \qquad (15)$$

Theorem 5. Let B_3 be with the form as (9). For all $k \in \mathbb{Z}^+$, if the first row of B_3 is $(\binom{n}{0}F_k, \binom{n}{1}F_{1+k}, \ldots, \binom{n}{n}F_{n+k})$, then one obtains

$$\|B_3\|_2 = F_{2n+k}. \qquad (16)$$

Proof. Following the same techniques of the above theorem and combining with the fact that B_3 is irreducible and entrywise nonnegative, we declare that the spectral norm of B_3 is equal to its Perron value. Let $V^T = (1, 1, \ldots, 1)_{1\times n}$. Then

$$B_3 v = \left(\sum_{i=0}^{n} \binom{n}{i} F_{i+k} \right) v. \qquad (17)$$

Obviously, we declare that $\sum_{i=0}^{n} \binom{n}{i} F_{i+k}$ is an eigenvalue of B_3 associated with V. With simple analysis, we obtain that $\sum_{i=0}^{n} \binom{n}{i} F_{i+k}$ is equal to the Perron value of B_3. Combining with the third identity of binomial coefficients and Fibonacci numbers in (6), we obtain

$$\|B_3\|_2 = F_{2n+k}, \qquad (18)$$

which completes the proof.

Similarly, there holds the following corollary.

Corollary 6. Let B_4 be as the matrix in (9). For all $k \in \mathbb{Z}^+$, the first row of B_4 is $(\binom{n}{0} L_k, \binom{n}{1} L_{1+k}, \ldots, \binom{n}{n} L_{n+k})$; then

$$\|B_4\|_2 = L_{2n+k}. \qquad (19)$$

Now, we are at the point to recall the following lemma to verify the identities of spectral norms with other approaches.

Lemma 7 (see [3]). Let A be a nonnegative matrix. If the column sums of A are equal, then

$$\rho(A) = \|A\|_1, \qquad (20)$$

where $\rho(A) = \max\{|\lambda| : \lambda \text{ is an eigenValue of matrix } A\}$ and $\|\cdot\|_1$ denotes the maximum column sum matrix norm.

Theorem 8: Let B_5 be with the form as (9) and let the first row of B_5 be $(\binom{n}{0} F_k, \binom{n}{1} F_{2+k}, \ldots, \binom{n}{n} F_{2n+k})$. Then one deduces the following identity:

$$\|B_5\|_2 = \begin{cases} 5^{n/2} F_{n+k} & (n \ even), \\ 5^{(n-1)/2} L_{n+k} & (n \ odd), \end{cases} \quad (21)$$

Where $k \in \mathbb{Z}^+$

Proof. Obviously, the circulant matrix B_5 is normal; with the results of Definition 2, we declare that the spectral radius of B_5 is equal to (B_5); that is, $||B_5||2 = \rho(B_5)$. Furthermore, applying entrywise nonnegative properties and column sum of B_5 are certain constant K_{col}, which is described in (7). By Lemma 7, we obtain

$$\rho(B_5) = K_{col}. \quad (22)$$

Employing the identities of Fibonacci numbers and Binomial coefficients in (7), we have

$$\|B_5\|_2 = \begin{cases} 5^{n/2} F_{n+k} & (n \text{ even}), \\ 5^{(n-1)/2} L_{n+k} & (n \text{ odd}). \end{cases} \quad (23)$$

This completes the proof.

Furthermore, we give the following corollary without proofs, which can be proved with the same approaches as the above theorem.

Corollary 9. Let $B6$ be as the matrix in (9). For all $k \in \mathbb{Z}^+$, the first row of $B6$ is $(\binom{n}{0} L_k, \binom{n}{1} L_{2+k}, \ldots, \binom{n}{n} L_{2n+k})$; then we have the following identity:

$$\|B_6\|_2 = \begin{cases} 5^{n/2} L_{n+k} & (n \ even), \\ 5^{(n+1)/2} F_{n+k} & (n \ odd). \end{cases} \quad (24)$$

NUMERICAL EXAMPLES

In this section, we give some examples to verify our identities in the above theorems and corollaries.

Example 10. In this example, we give the numerical results for B_i ($i = 1, 2, 3, 4$) in Table 1.

Table 2: Spectral norms of B_6 and B_6.

n	1	2	3	4	5	6	7	8
$\|B_5\|_2$	3	10	35	125	450	1625	5875	21250
$5^{n/2}F_{n+1}$	\	10	\	125	\	1625	\	21250
$5^{(n-1)/2}L_{n+1}$	3	\	35	\	450	\	5875	\
$\|B_6\|_2$	5	20	100	275	1000	3625	13125	47500
$5^{(n+1)/2}F_{n+1}$	5	\	100	\	1000	\	13125	\
$5^{n/2}L_{n+1}$	\	20	\	275	\	3625	\	47500

Example 11. For simplicity, let $k=1$. We give the numerical results for B_5 and B_6 in Table 2.

With the data in Tables1 and 2, we declare that the identity for the spectral norm of B_i $(i = 1, \ldots, 6)$ holds.

CONCLUSION

This paper had discussed the identical estimates of spectral norms for some circulant matrices, which are listed by explicit formulations. In the future, we are going to investigate the determinants, inverses of circulant matrices with certain entries, and, inspired by [6], we will investigate the properties of g-circulant matrices. Particularly worth mentioning is the fact that, for the g-circulant matrix, we had some numerical results to prove the fact that the same identical estimates hold precisely, and we will concern on the theoretical confirmation in part of the future work.

ACKNOWLEDGMENTS

The author thanks Professor Z. L. Jiang for valuable discussions and suggestions and wishes to express sincere thanks to referees for their useful suggestions and comments.This work is partly supported by National Natural Science Foundation of China (Grant no. 11201212), Promotive Research Fund for Excellent Young and Middle-Aged Scientists of Shandong Province (Grant no. BS2012DX004), and the AMEP of Linyi University.

REFERENCES

1. W.-S. Chou, B.-S. Du, and P. J.-S. Shiue, "A note on circulant transition matrices in Markov chains," Linear Algebra and Its Applications, vol. 429, no. 7, pp. 1699–1704, 2008.
2. R. M. Gray and L. D. Davisson, An Introduction to Statistical Signal Processing, Cambridge University Press, London, UK, 2005.
3. R. A. Horn and C. R. Johnson, Matrix Analysis, Cambridge University Press, Cambridge, UK, 1985.
4. D. S. G. Pollock, "Circulant matrices and time-series analysis," Working Paper No. 442, Queen Mary &Westfield College, 2000.
5. A. Bose, R. S. Hazra, and K. Saha, "Spectral norm of circulanttype matrices," Journal of Theoretical Probability, vol. 24, no. 2, pp. 479–516, 2011.
6. E. Ngondiep, S. Serra-Capizzano, and D. Sesana, "Spectral features and asymptotic properties for g-circulants and g- Toeplitz sequences," SIAM Journal on Matrix Analysis and Applications, vol. 31, no. 4, pp. 1663–1687, 2010.
7. S. Solak, "On the norms of circulant matrices with the Fibonacci and Lucas numbers," Applied Mathematics and Computation, vol. 160, no. 1, pp. 125–132, 2005.
8. A. ˙Ipek, "On the spectral norms of circulant matrices with classical Fibonacci and Lucas numbers entries," Applied Mathematics and Computation, vol. 217, no. 12, pp. 6011–6012, 2011.
9. J. W. Zhou and Z. L. Jiang, "Spectral norms of circulant-type matrices with Binomial coefficients and Harmonic numbers," International Journal of Computational Methods, vol. 11, no. 5, Article ID 1350076, 14 pages, 2014.
10. J. W. Zhou and Z. L. Jiang, "Spectral norms of circulant and Skew-circulant matrices with Binomial coefficients entries," in Proceedings of the 9th International Symposium on Linear Drives for Industry Applications, vol. 271 of Lecture Notes in Electrical Engineering, pp. 219–224, Springer, Berlin, Germany, 2014.
11. W. Bryc and S. Sethuraman, "A remark on the maximum eigenvalue for circulant matrices," in High Dimensional Probability V: The Luminy Volume, vol. 5, pp. 179–184, Institute of Mathematical Statistics Collections, Beachwood, Ohio, USA, 2009.
12. M. W. Meckes, "On the spectral norm of a random Toeplitz matrix,"

Electronic Communications in Probability, vol. 12, pp. 315–325, 2007.

13. M. W. Meckes, "Some results on random circulant matrices," in High Dimensional Probability V: The Luminy Volume, vol. 5, pp. 213–223, Institute of Mathematical Statistics Collections, Beachwood, Ohio, USA, 2009.
14. M. L. Mehta, Random Matrices, vol. 142 of Pure and Applied Mathematics, Elsevier/Academic Press, Amsterdam, The Netherlands, 3rd edition, 2004.
15. E. G. Kocer, N. Tuglu, and A. Stakhov, "On the m-extension of the Fibonacci and Lucas p-numbers," Chaos, Solitons and Fractals, vol. 40, no. 4, pp. 1890–1906, 2009.
16. M. Akbulak and D. Bozkurt, "On the norms of Toeplitz matrices involving Fibonacci and Lucas numbers," Hacettepe Journal of Mathematics and Statistics, vol. 37, no. 2, pp. 89–95, 2008.
17. M. Benoumhani, "A sequence of Binomial coefficients related to Lucas and Fibonacci numbers," Journal of Integer Sequences, vol. 6, no. 2, pp. 1–10, 2003.
18. R. Melham, "Sums involving Fibonacci and Pell numbers," Portugaliae Mathematica, vol. 56, no. 3, pp. 309–317, 1999.
19. W. T. Stallings and T. L. Boullion, "The pseudoinverse of an r-circulant matrix," Proceedings of the American Mathematical Society, vol. 34, no. 2, pp. 385–388, 1972.

3

Harmonic Numbers and Cubed Binomial Coefficients

Anthony Sofo

Victoria University College, Victoria University, Melbourne City, VIC 8001, Australia

ABSTRACT

Euler related results on the sum of the ratio of harmonic numbers and cubed binomial coefficients are investigated in this paper. Integral and closed-form representation of sums are developed in terms of zeta and polygamma functions. The given representations are new.

INTRODUCTION

The well-known Riemann zeta function is defined as

$$\zeta(z) = \sum_{r=1}^{\infty} \frac{1}{r^z}, \quad \text{Re}(z) > 1. \tag{1.1}$$

The generalized harmonic numbers of order α are given by

Citation: Anthony Sofo, "Harmonic Numbers and Cubed Binomial Coefficients," International Journal of Combinatorics, vol. 2011, Article ID 208260, 14 pages, 2011. https://doi.org/10.1155/2011/208260

$$H_n^{(\alpha)} = \sum_{r=1}^{n} \frac{1}{r^\alpha} \quad \text{for } (\alpha, n) \in \mathbb{N} \times \mathbb{N},\ \mathbb{N} := \{1,2,3,\ldots\}, \tag{1.2}$$

and for α=1,

$$H_n^{(1)} = H_n = \int_0^1 \frac{1-t^n}{1-t} dt = \sum_{r=1}^{n} \frac{1}{r} = \gamma + \psi(n+1), \tag{1.3}$$

where γ denotes the Euler-Mascheroni constant defined by

$$\gamma = \lim_{n\to\infty} \left(\sum_{r=1}^{n} \frac{1}{r} - \log(n) \right) = -\psi(1) \approx 0.5772156649\ldots, \tag{1.4}$$

and where $\psi(z)$ denotes the Psi, or digamma function defined by

$$\psi(z) = \frac{d}{dz} \log \Gamma(z) = \frac{\Gamma'(z)}{\Gamma(z)} = \sum_{n=0}^{\infty} \left(\frac{1}{n+1} - \frac{1}{n+z} \right) - \gamma, \tag{1.5}$$

and the Gamma function $\Gamma(z) = \int_0^\infty u^{z-1} e^{-u} du$, for $\mathbb{R}(z) > 0$.

Variant Euler sums of the form

$$\sum_{n=1}^{\infty} \frac{\left(H_{2n}^{(1)} - (1/2)H_n^{(1)}\right) x^{2n}}{n^p} \quad \text{for } p = 1 \text{ and } p = 2 \tag{1.6}$$

have been considered by Chen [1], and recently Boyadzhiev [2] evaluated various binomial identities involving power sums with harmonic numbers $H_n^{(1)}$. Other remarkable harmonic number identities known to Euler are

$$\sum_{n=1}^{\infty} \frac{H_n^{(1)}}{(n+1)^2} = \zeta(3), \qquad \sum_{n=1}^{\infty} \frac{H_n^{(1)}}{n^3} = \frac{5}{4}\zeta(4), \tag{1.7}$$

there is also a recurrence formula

$$(2n+1)\zeta(2n) = 2\sum_{r=1}^{n-1} \zeta(2r)\zeta(2n-2r), \tag{1.8}$$

which shows that in particular, for n=2,5ζ(4)=2(ζ(2))^2 and more generally that $\zeta(2n)$ is a rational multiple of $(\zeta(n))^2$. Another elegant recursion known to Euler [3] is

$$2\sum_{n=1}^{\infty} \frac{H_n^{(1)}}{n^q} = (q+2)\zeta(q+1) - \sum_{r=1}^{q-2} \zeta(r+1)\zeta(q-r). \tag{1.9}$$

Further work in the summation of harmonic numbers and binomial coefficients has also been done by Flajolet and Salvy [4] and Basu [5]. In this paper it is intended to add, in a small way, some results related to (1.7) and to extend the result of Cloitre, as reported in [6], $\sum_{n=1}^{\infty}(H_n^{(1)}/\binom{n+k}{k}) = (k/(k-1)^2)$ for $k > 1$. Specifically, we investigate integral representations and closed form representations for sums of harmonic numbers and cubed binomial coefficients. The works of [7–13] also investigate various representations of binomial sums and zeta functions in simpler form by the use of the Beta function and other techniques. Some of the material in this paper was inspired by the work of Mansour, [8], where he used, in part, the Beta function to obtain very general results for finite binomial sums.

INTEGRAL REPRESENTATIONS AND IDENTITIES

The following Lemma, given by Sofo [11], is stated without proof and deals with the derivative of a reciprocal binomial coefficient.

Lemma 2.1. Let a be a positive real number, $z \geq 0$, is a positive integer and let $Q(an,z) = \binom{an+z}{z}^{-1}$ be an analytic function of z. Then,

$$Q'(an,z) = \frac{dQ}{dz} = \begin{cases} -Q(an,z)\left[\psi(z+1+an) - \psi(z+1)\right] & \text{for } z > 0, \\ -H_n^{(1)}, & \text{for } z = 0 \text{ and } a = 1. \end{cases} \tag{2.1}$$

Theorem 2.2. Let $a,,,d \geq 0$ be real positive numbers, $|t| \leq 1, p \geq 0$ and let $j,k,l,m \geq 0$ be real positive numbers. Then

$$\sum_{n\geq 1} \frac{t^n \binom{n+p-1}{n-1}\left[\psi(j+1+an) - \psi(j+1)\right]}{n\binom{an+j}{j}\binom{bn+k}{k}\binom{cn+l}{l}\binom{dn+m}{m}}$$
$$= -aklmt \int\!\!\!\int\!\!\!\int\!\!\!\int_0^1 \frac{(1-x)^j(1-y)^{k-1}(1-z)^{l-1}(1-w)^{m-1}}{(1-tx^a y^b z^c w^d)^{p+1}}$$
$$\times \ln(1-x)\cdot x^{a-1} y^b z^c w^d \, dx\, dy\, dz\, dw. \tag{2.2}$$

Proof. Expand

$$\sum_{n\geq 1} \frac{t^n \binom{n+p-1}{n-1}}{n\binom{an+j}{j}\binom{bn+k}{k}\binom{cn+l}{l}\binom{dn+m}{m}}$$
$$= \sum_{n\geq 1}\left[t^n \binom{n+p-1}{n-1} \frac{an\Gamma(an)\Gamma(j+1)\Gamma(bn+1)k\cdot\Gamma(k)}{n\Gamma(an+j+1)\Gamma(bn+k+1)}\right.$$
$$\left.\times \frac{\Gamma(cn+1)l\cdot\Gamma(l)\Gamma(dn+1)m\cdot\Gamma(m)}{\Gamma(cn+l+1)\Gamma(dn+m+1)}\right]$$
$$= aklm \int_0^1 \frac{(1-x)^j}{x} \sum_{n\geq 1} x^{an} t^n \binom{n+p-1}{n-1} \times B(bn+1,k)B(cn+1,l)B(dn+1,m)dx, \tag{2.3}$$

where

$$B(\alpha,\beta)=\frac{\Gamma(\alpha)\Gamma(\beta)}{\Gamma(\alpha+\beta)}=\int_0^1(1-y)^{\alpha-1}y^{\beta-1}dy$$

$$=\int_0^1(1-y)^{\beta-1}y^{\alpha-1}dy,\quad \text{for } \alpha>0 \text{ and } \beta>0 \tag{2.4}$$

is the classical Beta function. Differentiating with respect to the parameter j, and utilizing Lemma 2.1 implies the resulting equation is as follows:

$$\sum_{n\geq1}\frac{t^n\binom{n+p-1}{n-1}(\psi(j+1+an)-\psi(j+1))}{n\binom{an+j}{j}\binom{bn+k}{k}\binom{cn+l}{l}\binom{dn+m}{m}}$$

$$=-aklm\int_0^1\frac{(1-x)^j}{x}\ln(1-x)\sum_{n\geq1}x^{an}t^n\binom{n+p-1}{n-1}$$

$$\times B(bn+1,k)B(cn+1,l)B(dn+1,m)dx,$$

$$=-aklm\int\iiint_0^1\frac{(1-x)^j\ln(1-x)}{x}(1-y)^{k-1}(1-z)^{l-1}(1-w)^{m-1}$$

$$\times\sum_{n\geq1}\binom{n+p-1}{n-1}\left(tx^ay^bz^cw^d\right)^n dx\,dy\,dz\,dw$$

$$=-aklmt\int\iiint_0^1\frac{(1-x)^j\ln(1-x)(1-y)^{k-1}(1-z)^{l-1}(1-w)^{m-1}}{(1-tx^ay^bz^cw^d)^{p+1}}$$

$$\times x^{a-1}y^bz^cw^d dx\,dy\,dz\,dw \tag{2.5}$$

for $|tx^ay^bz^cw^d|<1$.

In the following three corollaries we encounter harmonic numbers at possible rational values of the argument, of the form $H^{(\alpha)}_{(r/b)-1}$ where r=1,2,3,-k,α=1,2,3,…, and $k\in\mathbb{N}$. The polygamma function $\psi^{(\alpha)}(z)$ is defined as

$$\psi^{(\alpha)}(z)=\frac{d^{\alpha+1}}{dz^{\alpha+1}}[\log\Gamma(z)]=\frac{d^\alpha}{dz^\alpha}[\psi(z)],\quad z\neq\{0,-1,-2,-3,\dots\}. \tag{2.6}$$

To evaluate $H^{(\alpha)}_{(r/b)-1}$ we have available a relation in terms of the polygamma function $\psi^{(\alpha)}(z)$, for rational arguments z,

$$H^{(\alpha+1)}_{(r/b)-1}=\zeta(\alpha+1)+\frac{(-1)^\alpha}{\alpha!}\psi^{(\alpha)}\left(\frac{r}{b}\right), \tag{2.7}$$

where $\zeta(z)$ is the Riemann zeta function. We also define

$$H^{(1)}_{(r/b)-1} = \gamma + \psi\left(\frac{r}{b}\right), \qquad H^{(\alpha)}_{0} = 0. \tag{2.8}$$

The evaluation of the polygamma function $\psi^{(\alpha)}(r/b)$ at rational values of the argument can be explicitly done via a formula as given by Kölbig [14] (see also [15]) or Choi and Cvijović [16] in terms of the polylogarithmic or other special functions. Some specific values are given as

$$\psi^{(n)}\left(\frac{1}{2}\right) = (-1)^n n!\left(2^{n+1} - 1\right)\zeta(n+1)$$

$$H^{(5)}_{(-2/3)} = \zeta(5) - \frac{1}{24}\psi^{(4)}\left(\frac{1}{3}\right) = -\frac{2\pi^5\sqrt{3}}{9} - 120\zeta(5), \qquad H^{(2)}_{1/4} = 16 - 8G - 5\zeta(2), \tag{2.9}$$

and can be confirmed on a mathematical computer package, such as Mathematica [17].

Corollary 2.3. Let $a=1,=c=b>0,t=1,p=0,j=0$ and let $l=m=k\geq 1$ be a positive integer. Then

$$\sum_{n\geq 1} \frac{H_n^{(1)}}{n\binom{bn+k}{k}^3}$$

$$= -k^3 \int\iiint_0^1 \frac{\left[(1-y)(1-z)(1-w)\right]^{k-1}}{\left(1 - x(yzw)^b\right)} \ln(1-x)\cdot(yzw)^b\, dx\, dy\, dz\, dw \tag{2.10}$$

$$= \sum_{r=1}^{k}(-1)^{r+1}\binom{k}{r}^3$$

$$\times\Bigg[-\frac{r^2}{4b^2}\zeta(4) - \left(\frac{r^2}{b^2}H^{(1)}_{(r/b)-1} + \frac{r}{b} + \frac{r^2}{b}XR(k)\right)\zeta(3)$$

$$+\left(1 - \frac{r}{b}H^{(1)}_{(r/b)-1} - \frac{r^2}{b^2}H^{(2)}_{(r/b)-1} + r\left\{1 - \frac{r}{b}H^{(1)}_{(r/b)-1}\right\}XR(k) + r^2 YR(k)\right)\zeta(2)$$

$$+\frac{1}{2}\left\{\left(H^{(1)}_{(r/b)-1}\right)^2 + H^{(2)}_{(r/b)-1}\right\} + \frac{r}{b}\left\{H^{(1)}_{(r/b)-1}H^{(2)}_{(r/b)-1} + H^{(3)}_{(r/b)-1}\right\}$$

$$+\frac{r^2}{2b^2}\left\{\left(H^{(2)}_{(r/b)-1}\right)^2 + 2H^{(1)}_{(r/b)-1}H^{(3)}_{(r/b)-1} + 3H^{(4)}_{(r/b)-1}\right\}$$

$$+\left(\frac{r}{2}\left\{\left(H^{(1)}_{(r/b)-1}\right)^2 + H^{(2)}_{(r/b)-1}\right\} + \frac{r^2}{b}\left\{H^{(1)}_{(r/b)-1}H^{(2)}_{(r/b)-1} + H^{(3)}_{(r/b)-1}\right\}\right)XR(k)$$

$$+\frac{r^2}{2}\left\{\left(H^{(1)}_{(r/b)-1}\right)^2 + H^{(2)}_{(r/b)-1}\right\}YR(k)\Bigg], \tag{2.11}$$

where

$$XR(k) := -3\left(H^{(1)}_{k-r} - H^{(1)}_{r-1}\right), \tag{2.12}$$

$$YR(k) := \frac{3}{2}\left[\frac{XR^2(k)}{3} + H_{k-r}^{(2)} + H_{r-1}^{(2)}\right]. \tag{2.13}$$

Proof. Let

$$\sum_{n\geq 1}\frac{H_n^{(1)}}{n\binom{bn+k}{k}^3} = \sum_{n\geq 1}\frac{H_n^{(1)}(k!)^3}{n\prod_{r=1}^{k}(bn+r)^3}$$
$$= \sum_{n\geq 1}\frac{H_n^{(1)}(k!)^3}{n}\sum_{r=1}^{k}\left[\frac{A_r}{bn+r} + \frac{B_r}{(bn+r)^2} + \frac{C_r}{(bn+r)^3}\right], \tag{2.14}$$

where

$$C_r = \lim_{n\to(-r/b)}\left\{\frac{(bn+r)^3}{\prod_{r=1}^{k}(bn+r)^3}\right\} = (-1)^{r+1}\left(\frac{r}{k!}\binom{k}{r}\right)^3,$$

$$B_r = \lim_{n\to(-r/b)}\frac{d}{dn}\left\{\frac{(bn+r)^3}{\prod_{r=1}^{k}(bn+r)^3}\right\}$$
$$= 3(-1)^r\left(\frac{r}{k!}\binom{k}{r}\right)^3\left[H_{k-r}^{(1)} - H_{r-1}^{(1)}\right],$$

$$A_r = \frac{1}{2}\lim_{n\to(-r/b)}\frac{d^2}{dn^2}\left\{\frac{(bn+r)^3}{\prod_{r=1}^{k}(bn+r)^3}\right\}$$
$$= \frac{3}{2}(-1)^{r+1}\left(\frac{r}{k!}\binom{k}{r}\right)^3\left[3\left(H_{k-r}^{(1)} - H_{r-1}^{(1)}\right)^2 + H_{k-r}^{(2)} + H_{r-1}^{(2)}\right]. \tag{2.15}$$

Now, by interchanging sums, we have

$$\sum_{n\geq 1}\frac{H_n^{(1)}}{n\binom{bn+k}{k}^3} = \sum_{r=1}^{k}\left[A_r(k!)^3\sum_{n\geq 1}\frac{H_n^{(1)}}{n(bn+r)} + B_r(k!)^3\sum_{n\geq 1}\frac{H_n^{(1)}}{n(bn+r)^2} + C_r(k!)^3\sum_{n\geq 1}\frac{H_n^{(1)}}{n(bn+r)^3}\right]. \tag{2.16}$$

We can evaluate

$$\sum_{n\geq 1}\frac{H_n^{(1)}}{n(bn+r)} = \sum_{n\geq 1}\left(\frac{\psi(n)+(1/n)}{n(bn+r)} + \frac{\gamma}{n(bn+r)}\right) = \sum_{n\geq 1}\left(\frac{\psi(n)}{n(bn+r)} + \frac{1}{n^2(bn+r)} + \frac{\gamma}{n(bn+r)}\right)$$
$$= \sum_{n\geq 1}\left(\frac{\psi(n)}{n(bn+r)} + \frac{1}{rn^2}\right)$$
$$+ \frac{(\gamma-(b/r))}{r}\sum_{n\geq 0}\left(\frac{1}{n+1} - \frac{1}{n+(r/b)} + \left(\frac{1}{n+(r/b)} - \frac{1}{n+1+(r/b)}\right)\right)$$
$$= -\frac{b\gamma}{r^2} - \frac{\gamma^2}{2r} + \frac{3\zeta(2)}{2r} + \frac{(\psi(r/b))^2}{2r} - \frac{\psi'(r/b)}{2r} + \frac{\gamma}{r}\left(\psi\left(\frac{r}{b}\right) + \gamma + b\right), \tag{2.17}$$

here we have used the result from [18]

$$\sum_{n\geq 1}\frac{\psi(n)}{n(bn+r)} = \frac{1}{2r}\left(\left(\psi\left(\frac{r}{b}+1\right)\right)^2 - \gamma^2 + \zeta(2) - \psi'\left(\frac{r}{b}+1\right)\right). \tag{2.18}$$

Now using (2.7) and (2.8), we may write

$$\sum_{n\geq 1}\frac{H_n^{(1)}}{n(bn+r)}=\frac{\zeta(2)}{r}+\frac{1}{2r}\left\{\left(H_{(r/b)-1}^{(1)}\right)^2+H_{(r/b)-1}^{(2)}\right\}. \tag{2.19}$$

Similarly

$$\sum_{n\geq 1}\frac{H_n^{(1)}}{n(bn+r)^2}=\frac{1}{r}\sum_{n\geq 1}\frac{H_n^{(1)}}{n(bn+r)}-\frac{b}{r}\sum_{n\geq 1}\frac{H_n^{(1)}}{(bn+r)^2}$$

$$=-\frac{\zeta(3)}{br}+\frac{\zeta(2)}{r^2}-\frac{\zeta(2)H_{(r/b)-1}^{(1)}}{br}+\frac{1}{2r^2}\left\{\left(H_{(r/b)-1}^{(1)}\right)^2+H_{(r/b)-1}^{(2)}\right\}$$

$$+\frac{1}{br}\left\{H_{(r/b)-1}^{(1)}H_{(r/b)-1}^{(2)}+H_{(r/b)-1}^{(3)}\right\},$$

$$\sum_{n\geq 1}\frac{H_n^{(1)}}{n(bn+r)^3}=\sum_{n\geq 1}H_n^{(1)}\left[\frac{1}{r^2n(bn+r)}-\frac{b}{r^2(bn+r)^2}-\frac{b}{r(bn+r)^3}\right]$$

$$=-\frac{\zeta(4)}{4b^2r}-\frac{\zeta(3)}{br^2}-\frac{\zeta(3)H_{(r/b)-1}^{(1)}}{b^2r}+\frac{\zeta(2)}{r^3}-\frac{\zeta(2)H_{(r/b)-1}^{(1)}}{br^2}$$

$$-\frac{\zeta(2)H_{(r/b)-1}^{(2)}}{b^2r}+\frac{1}{2r^3}\left\{\left(H_{(r/b)-1}^{(1)}\right)^2+H_{(r/b)-1}^{(2)}\right\}$$

$$+\frac{1}{br^2}\left\{H_{(r/b)-1}^{(1)}H_{(r/b)-1}^{(2)}+H_{(r/b)-1}^{(3)}\right\}$$

$$+\frac{1}{2b^2r}\left\{\left(H_{(r/b)-1}^{(2)}\right)^2+2H_{(r/b)-1}^{(1)}H_{(r/b)-1}^{(3)}+3H_{(r/b)-1}^{(4)}\right\}. \tag{2.20}$$

Substituting (2.19), (2.20) into (2.16) where $XR(k)$ and $YR(k)$ are given by (2.12) and (2.13), respectively, on simplifying the identity (2.11) is realized.

For k=1 and b=1 the following identity is valid:

$$\sum_{n\geq 1}\frac{H_n^{(1)}}{n(n+1)^3}=\zeta(2)-\zeta(3)-\frac{\zeta(4)}{4}. \tag{2.21}$$

Theorem 2.4.

$$\sum_{n\geq 1}\frac{t^n\binom{n+p-1}{n-1}\left(\psi(j+1+an)-\psi(j+1)\right)}{n^2\binom{an+j}{j}\binom{bn+k}{k}\binom{cn+l}{l}\binom{dn+m}{m}}$$

$$=-ablmt\int\!\!\int\!\!\int\!\!\int_0^1\frac{(1-x)^j(1-y)^k(1-z)^{l-1}(1-w)^{m-1}}{(1-tx^ay^bz^cw^d)^{p+1}}$$

$$\times\ln(1-x)\cdot x^{a-1}y^{b-1}z^cw^d\,dx\,dy\,dz\,dw. \tag{2.22}$$

Proof. The proof of this theorem is very similar to that of Theorem 2.2 and will not be given here.

Corollary 2.5. Let a=1,=c=b>0,t=1,p=0,j=0, and let l=m=k≥1 be a positive integer. Then

$$\sum_{n\geq 1}\frac{H_n^{(1)}}{n^2\binom{bn+k}{k}^3}$$

$$= -bk^2\int\!\!\!\int\!\!\!\int\!\!\!\int_0^1 \frac{(1-y)^k[(1-z)(1-w)]^{k-1}}{\left(1-x(yzw)^b\right)}\ln(1-x)\cdot y^{b-1}(zw)^b dx\,dy\,dz\,dw \tag{2.23}$$

$$= \sum_{r=1}^{k}(-1)^{r+1}\binom{k}{r}^3$$

$$\times\left[\frac{r}{4b}\zeta(4) + \left(4 + \frac{r}{b}H_{(r/b)-1}^{(1)} + 3rXR(k) + 2r^2YR(k)\right)\zeta(3)\right.$$

$$+\left(-\frac{3b}{r} + 2H_{(r/b)-1}^{(1)} + \frac{r}{b}H_{(r/b)-1}^{(2)} + \left\{rH_{(r/b)-1}^{(1)} - 2b\right\}XR(k) - brYR(k)\right)\zeta(2)$$

$$-\frac{3b}{2r}\left\{\left(H_{(r/b)-1}^{(1)}\right)^2 + H_{(r/b)-1}^{(2)}\right\} - 2\left\{H_{(r/b)-1}^{(1)}H_{(r/b)-1}^{(2)} + H_{(r/b)-1}^{(3)}\right\}$$

$$-\frac{r}{2b}\left\{\left(H_{(r/b)-1}^{(2)}\right)^2 + 2H_{(r/b)-1}^{(1)}H_{(r/b)-1}^{(3)} + 3H_{(r/b)-1}^{(4)}\right\}$$

$$-\left(b\left\{\left(H_{(r/b)-1}^{(1)}\right)^2 + H_{(r/b)-1}^{(2)}\right\} + r\left\{H_{(r/b)-1}^{(1)}H_{(r/b)-1}^{(2)} + H_{(r/b)-1}^{(3)}\right\}\right)XR(k)$$

$$\left.-\frac{br}{2}\left\{\left(H_{(r/b)-1}^{(1)}\right)^2 + H_{(r/b)-1}^{(2)}\right\}YR(k)\right], \tag{2.24}$$

where $XR(k)$ is given by (2.12) and $YR(k)$ is given by (2.13).

Proof. Following similar steps to Corollary 2.3, we may write

$$\sum_{n\geq 1}\frac{H_n^{(1)}}{n^2\binom{bn+k}{k}^3}$$

$$= \sum_{r=1}^{k}\left[A_r(k!)^3\sum_{n\geq 1}\frac{H_n^{(1)}}{n^2(bn+r)} + B_r(k!)^3\sum_{n\geq 1}\frac{H_n^{(1)}}{n^2(bn+r)^2} + C_r(k!)^3\sum_{n\geq 1}\frac{H_n^{(1)}}{n^2(bn+r)^3}\right], \tag{2.25}$$

and evaluate

$$\sum_{n\geq 1}\frac{H_n^{(1)}}{n^2(bn+r)} = \sum_{n\geq 1}H_n^{(1)}\left[\frac{1}{rn^2} - \frac{b}{rn(bn+r)}\right]$$

$$= \frac{2\zeta(3)}{r} - \frac{b\zeta(2)}{r^2} - \frac{b}{2r^2}\left\{\left(H_{(r/b)-1}^{(1)}\right)^2 + H_{(r/b)-1}^{(2)}\right\},$$

$$\sum_{n\geq 1}\frac{H_n^{(1)}}{n^2(bn+r)^2} = \frac{3\zeta(3)}{r^2} - \frac{2b\zeta(2)}{r^3} + \frac{\zeta(2)H_{(r/b)-1}^{(1)}}{r^2}$$

$$-\frac{b}{r^3}\left\{\left(H_{(r/b)-1}^{(1)}\right)^2 + H_{(r/b)-1}^{(2)}\right\} - \frac{1}{r^2}\left\{H_{(r/b)-1}^{(1)}H_{(r/b)-1}^{(2)} + H_{(r/b)-1}^{(3)}\right\},$$

$$-\frac{3b}{2r^4}\left\{\left(H^{(1)}_{(r/b)-1}\right)^2+H^{(2)}_{(r/b)-1}\right\}-\frac{2}{r^3}\left\{H^{(1)}_{(r/b)-1}H^{(2)}_{(r/b)-1}+H^{(3)}_{(r/b)-1}\right\}$$

$$-\frac{1}{2br^2}\left\{\left(H^{(2)}_{(r/b)-1}\right)^2+2H^{(1)}_{(r/b)-1}H^{(3)}_{(r/b)-1}+3H^{(4)}_{(r/b)-1}\right\}. \tag{2.26}$$

By substituting (2.26) into (2.25) and collecting zeta functions, the identity (2.24) is obtained.

For k=1 and b=1 the following identity is valid:

$$\sum_{n\geq 1}\frac{H_n^{(1)}}{n^2(n+1)^3}=\frac{\zeta(4)}{4}+4\zeta(3)-3\zeta(2). \tag{2.27}$$

Theorem 2.6.

$$\sum_{n\geq 1}\frac{t^n\binom{n+p-1}{n-1}\left(\psi(j+1+an)-\psi(j+1)\right)}{n^3\binom{an+j}{j}\binom{bn+k}{k}\binom{cn+l}{l}\binom{dn+m}{m}}$$

$$=-abcmt\int\!\!\iiint_0^1\frac{(1-x)^j(1-y)^k(1-z)^l(1-w)^{m-1}}{\left(1-tx^ay^bz^cw^d\right)^{p+1}}$$

$$\times\ln(1-x)\cdot x^{a-1}y^{b-1}z^{c-1}w^d\,dx\,dy\,dz\,dw. \tag{2.28}$$

Proof. The proof of this theorem is very similar to that of Theorem 2.2 and will not be given here.

Corollary 2.7. Let a=1,=c=b>0,t=1,j=0, and let l=m=k≥1 be a positive integer. Then

$$\sum_{n\geq 1}\frac{H_n^{(1)}}{n^3\binom{bn+k}{k}^3}$$

$$=-b^2k\int\!\!\iiint_0^1\frac{\left((1-y)(1-z)\right)^k(1-w)^{k-1}}{\left(1-x(yzw)^b\right)}\ln(1-x)\cdot(yz)^{b-1}w^b\,dx\,dy\,dz\,dw \tag{2.29}$$

$$=\sum_{r=1}^{k}(-1)^{r+1}\binom{k}{r}^3$$

$$\times\left[\left(1+\frac{5}{4}XR(k)+\frac{5}{4}r^2YR(k)\right)\zeta(4)\right.$$

$$-\left(\frac{9b}{r}+H^{(1)}_{(r/b)-1}+5bXR(k)+2brYR(k)\right)\zeta(3)$$

$$+\left(\frac{6b^2}{r^2}-\frac{3bH^{(1)}_{(r/b)-1}}{r}-H^{(2)}_{(r/b)-1}-\left\{\frac{3b^2}{r}+bH^{(1)}_{(r/b)-1}\right\}XR(k)+b^2YR(k)\right)\zeta(2)$$

$$+\frac{3b^2}{r^2}\left\{\left(H^{(1)}_{(r/b)-1}\right)^2+H^{(2)}_{(r/b)-1}\right\}+\frac{3b}{r}\left\{H^{(1)}_{(r/b)-1}H^{(2)}_{(r/b)-1}+H^{(3)}_{(r/b)-1}\right\}$$

$$+\frac{1}{2}\left\{\left(H^{(2)}_{(r/b)-1}\right)^2+2H^{(1)}_{(r/b)-1}H^{(3)}_{(r/b)-1}+3H^{(4)}_{(r/b)-1}\right\}$$

$$+\left(\frac{3b^2}{2r}\left\{\left(H^{(1)}_{(r/b)-1}\right)^2+H^{(2)}_{(r/b)-1}\right\}+b\left(H^{(1)}_{(r/b)-1}H^{(2)}_{(r/b)-1}+H^{(3)}_{(r/b)-1}\right)\right)XR(k)$$

$$\left.+\frac{b^2}{2}\left\{\left(H^{(1)}_{(r/b)-1}\right)^2+H^{(2)}_{(r/b)-1}\right\}YR(k)\right], \tag{2.30}$$

where $XR(k)$ is given by (2.12) and $YR(k)$ is given by (2.13).

Proof. We follow similar steps as the previous corollary so that

$$\sum_{n\geq 1}\frac{H^{(1)}_n}{n^3\binom{bn+k}{k}^3}$$

$$=\sum_{r=1}^{k}\left[A_r(k!)^3\sum_{n\geq 1}\frac{H^{(1)}_n}{n^3(bn+r)}+B_r(k!)^3\sum_{n\geq 1}\frac{H^{(1)}_n}{n^3(bn+r)^2}+C_r(k!)^3\sum_{n\geq 1}\frac{H^{(1)}_n}{n^3(bn+r)^3}\right]. \tag{2.31}$$

After much algebraic simplification, the following identity is obtained:

$$\sum_{n\geq 1}\frac{H^{(1)}_n}{n^3(bn+r)^3}$$

$$=\sum_{n\geq 1}H^{(1)}_n\left[\frac{1}{r^3n^3}-\frac{3b}{r^4n^2}-\frac{b^3}{r^3(bn+r)^3}-\frac{3b^3}{r^4(bn+r)^2}+\frac{6b^2}{r^4n(bn+r)}\right]$$

$$=\frac{\zeta(4)}{r^3}-\frac{9b\zeta(3)}{r^4}-\frac{\zeta(3)H^{(1)}_{(r/b)-1}}{r^3}+\frac{6b^2\zeta(2)}{r^5}-\frac{3b\zeta(2)H^{(1)}_{(r/b)-1}}{r^4}-\frac{\zeta(2)H^{(2)}_{(r/b)-1}}{r^3}$$

$$+\frac{3b^2}{r^5}\left\{\left(H^{(1)}_{(r/b)-1}\right)^2+H^{(2)}_{(r/b)-1}\right\}+\frac{3b}{r^4}\left\{H^{(1)}_{(r/b)-1}H^{(2)}_{(r/b)-1}+H^{(3)}_{(r/b)-1}\right\}$$

$$+\frac{1}{2r^3}\left\{\left(H^{(2)}_{(r/b)-1}\right)^2+2H^{(1)}_{(r/b)-1}H^{(3)}_{(r/b)-1}+3H^{(4)}_{(r/b)-1}\right\},$$

$$\sum_{n\geq 1}\frac{H^{(1)}_n}{n^3(bn+r)^2}=\frac{5\zeta(4)}{4r^2}-\frac{5b\zeta(3)}{r^3}+\frac{3b^2\zeta(2)}{r^4}-\frac{b\zeta(2)H^{(1)}_{(r/b)-1}}{r^3}$$

$$+\frac{3b^2}{2r^4}\left\{\left(H^{(1)}_{(r/b)-1}\right)^2+H^{(2)}_{(r/b)-1}\right\}+\frac{b}{r^3}\left\{H^{(1)}_{(r/b)-1}H^{(2)}_{(r/b)-1}+H^{(3)}_{(r/b)-1}\right\},$$

$$\sum_{n\geq 1}\frac{H^{(1)}_n}{n^3(bn+r)}=\frac{5\zeta(4)}{4r}+\frac{b^2\zeta(2)}{r^3}-\frac{2b\zeta(3)}{r^2}+\frac{b^2}{2r^3}\left\{\left(H^{(1)}_{(r/b)-1}\right)^2+H^{(2)}_{(r/b)-1}\right\}. \tag{2.32}$$

Now we can substitute (2.32) into (2.31), collecting zeta functions and using (2.12) and (2.13) for (k) and $YR(k)$, respectively, the identity (2.30) is obtained.

Some specific examples of Corollary 2.7 are as follows.

For k=1 and b=1 the following identity is valid,

$$\sum_{n\geq 1}\frac{H_n^{(1)}}{n^3(n+1)^3}=\zeta(4)-9\zeta(3)+\frac{1}{6}\zeta(2),$$

$$\begin{aligned}\sum_{n\geq 1}\frac{H_n^{(1)}}{n^3\binom{4n+8}{8}^3}&=\frac{12729073000}{27}-\frac{2641017400832}{11025}\ln 2+\frac{440380018688}{3675}(\ln 2)^2\\&-\frac{16604790784}{315}G+11272192G^2+\frac{1576747008}{35}G\ln 2\\&-\frac{1390679293952\pi}{33075}-\frac{172233728\pi^3}{105}-\frac{203992775989}{2450}\zeta(2)\\&-\frac{4069970159}{210}\zeta(3)+28974848\ln 2\zeta(3)-\frac{62246737}{2}\zeta(4),\end{aligned} \tag{2.33}$$

where G is Catalan›s constant, defined by

$$G=\frac{1}{2}\int_0^1 K(s)ds=\sum_{r=1}^{\infty}\frac{(-1)^r}{(2r+1)^2}\approx 0.915965\ldots, \tag{2.34}$$

and $K(s)$ is the complete elliptic integral of the first kind. The degenerate case k=0, gives the well-known result

$$\sum_{n\geq 1}\frac{H_n^{(1)}}{n^3}=\frac{5}{4}\zeta(4). \tag{2.35}$$

Remark 2.8. Corollaries 2.3, 2.5, and 2.7 are important and can be evaluated as demonstrated independently of their integral representations. Similarly the proofs of Corollaries 2.3, 2.5, and 2.7 are not obvious therefore their explicit representations is desired.

Remark 2.9. Theoretically it should be possible to obtain an integral representation for the general sum

$$\sum_{n\geq 1}\frac{t^n\binom{n+p-1}{n-1}\left(\psi(j+1+an)-\psi(j+1)\right)}{n^q\binom{an+j}{j}\binom{bn+k}{k}\binom{cn+l}{l}\binom{dn+m}{m}},\quad \text{for } q=1,2,3,\ldots, \tag{2.36}$$

with its associated corollaries. This work will be investigated in a forthcoming paper.

REFERENCES

1. H. Chen, "Evaluations of some Variant Euler Sums," Journal of Integer Sequences, vol. 9, no. 2, article 06.2.3, p. 9, 2006.
2. K. N. Boyadzhiev, "Harmonic number identities via Euler's transform," Journal of Integer Sequences, vol. 12, no. 6, article 09.6.1, p. 8, 2009.
3. L. Euler, Opera Omnia, Series 1, vol. 15, Teubner, Berlin, Germany, 1917.
4. P. Flajolet and B. Salvy, "Euler sums and contour integral representations," Experimental Mathematics, vol. 7, no. 1, pp. 15–35, 1998.
5. A. Basu, "A new method in the study of Euler sums," Ramanujan Journal, vol. 16, no. 1, pp. 7–24, 2008.
6. J. Sondow and E. W. Weisstein, Harmonic number. From MathWorld-A Wolfram Web Rescources, http://mathworld.wolfram.com/HarmonicNumber.html.
7. H. Alzer, D. Karayannakis, and H. M. Srivastava, "Series representations for some mathematical constants," Journal of Mathematical Analysis and Applications, vol. 320, no. 1, pp. 145–162, 2006.
8. T. Mansour, "Combinatorial identities and inverse binomial coefficients," Advances in Applied Mathematics, vol. 28, no. 2, pp. 196–202, 2002.
9. A. Sofo, "Integral forms of sums associated with harmonic numbers," Applied Mathematics and Computation, vol. 207, no. 2, pp. 365–372, 2009.
10. A. Sofo, Computational techniques for the summation of series, Kluwer Academic Publishers/Plenum Publishers, New York, NY, USA, 2003.
11. A. Sofo, "Sums of derivatives of binomial coefficients," Advances in Applied Mathematics, vol. 42, no. 1, pp. 123–134, 2009.
12. A. Sofo, "Harmonic numbers and double binomial coefficients," Integral Transforms and Special Functions, vol. 20, no. 11-12, pp. 847–857, 2009.
13. A. Sofo, "Harmonic sums and integral representations," Journal of Applied Analysis, vol. 16, no. 2, pp. 265–277, 2010.
14. K. S. Kölbig, "The polygamma function and the derivatives of the cotangent function for rational arguments," CERN-IT-Reports CERN-

CN 96-005, 1996.

15. K. S. Kölbig, "The polygamma function ψ(x) for x=1/4 and x=3/4," Journal of Computational and Applied Mathematics, vol. 75, no. 1, pp. 43–46, 1996.
16. J. Choi and D. Cvijović, "Values of the polygamma functions at rational arguments," Journal of Physics. A, vol. 40, no. 50, pp. 15019–15028, 2007.
17. Wolfram Research Inc., Mathematica, Wolfram Research Inc., Champaign, Ill, USA.
18. Y. A. Brychkov, Handbook of Special Functions, CRC Press, Boca Raton, Fla, USA, 2008.

4

A Generalization of a Combinatorial Identity by Chang and Xu

Ulrich Abel[1], Vijay Gupta[2], and Mircea Ivan[3]

[1] Department MND, Technische Hochschule Mittelhessen, Wilhelm-Leuschner-Straße 13, 61169 Friedberg, Germany

[2] Department of Mathematics, Netaji Subhas Institute of Technology, Sector 3 Dwarka, New Delhi 110078, India

[3] Department of Mathematics, Technical University of Cluj-Napoca, Str. Memorandumului nr. 28, 400114 Cluj-Napoca, Romania

ABSTRACT

Recently, Chang and Xu gave a probabilistic proof of a combinatorial identity which involves binomial coefficients. Duarte and Guedes de Oliveira (J Integer Seq 16, 2013) extended the result. Applying a generalization of the Leibniz rule for higher derivatives of the product of functions yields a new short proof and a generalization of the above mentioned identity.

Citation: Abel, U., Gupta, V. & Ivan, M. "A generalization of a combinatorial identity by Chang and Xu" Bull. Math. Sci. (2015) 5: 511. https://doi.org/10.1007/s13373-015-0072-z

Keyword
Combinatorial identities

INTRODUCTION AND MAIN RESULTS

Throughout the paper $\mathbf{k}=(k_1,\ldots,k_r)\in\mathbb{N}_0^r$ denotes a multi-index, $|\mathbf{k}|=k_1+\cdots+k_r$, and the multinomial coefficient is defined by

$$\binom{n}{\mathbf{k}}=\binom{n}{k_1,\ldots,k_r}:=\frac{n!}{k_1!\ldots k_r!\,(n-|\mathbf{k}|)!}.$$

In 2011, Chang and Xu [4, Theorem 1] gave a probabilistic proof of the combinatorial identity

$$\sum_{|\mathbf{k}|=n}\left[\prod_{i=1}^{r}\binom{2k_i}{k_i}\right]=4^n\binom{n+r/2-1}{n}.$$

for n, $r\in\mathbb{N}, r\geq 2$, which involves central binomial coefficients. It consists of showing that the identity essentially computes the moment of order n of the chi-square random variable with r degrees of freedom. The special instance r=2, i.e.,

$$\sum_{k=0}^{n}\binom{2k}{k}\binom{2(n-k)}{n-k}=4^n$$

plays an important role in combinatorics and probability theory (see the introduction of [4]). In 2013, Duarte and Guedes de Oliveira [2, Theorem 2] showed that, for $\mathbf{a}=(a_1,\ldots,a_r)\in\mathbb{R}^r$ with $|\mathbf{a}|=0$,

$$\sum_{|\mathbf{k}|=n}\left[\prod_{i=1}^{r}\binom{2k_i-a_i}{k_i}\right]=4^n\binom{n+r/2-1}{n}. \tag{1}$$

The special case r=2, reveals that, for all constants $a\in\mathbb{R}$,

$$\sum_{k=0}^{n}\binom{2k-a}{k}\binom{2(n-k)+a}{n-k}=4^n$$

where the sum in the left-hand side is independent of a. Identities for similar sums can be found in [3, Theorem 1].

We study the identity (1) without the restriction |a|=0. In particular, we give a presentation of the sum

$$\sum_{|\mathbf{k}|=n}\left[\prod_{i=1}^{r}\binom{ck_i-a_i}{k_i}\right]$$

as a derivative of an explicitly defined function, for arbitrary $c\in\mathbb{R}$ and a= $(a_1,\ldots,a_r)\in\mathbb{R}^r$. In the special case c=2 and when it happens that |a| is a non-negative integer, this sum can be represented in terms of a sum consisting of |a|+1 summands. While Chang and Xu used probabilistic arguments involving the expected value of a χ^2 random variable, and Duarte and Guedes de Oliveira used standard combinatorial tools like generating functions, our proof is based on methods of complex analysis.

The main result is the following identity.

Theorem 1

For $n, r \in \mathbb{N},\ r \geq 2,\ and\ \mathbf{a} = (a_1, \ldots, a_r) \in \mathbb{R}^r$ such that |a| is a nonnegative integer,

$$\sum_{|\mathbf{k}|=n} \left[\prod_{i=1}^{r} \binom{2k_i - a_i}{k_i} \right] = 4^{n-|a|/2} \sum_{j=0}^{|a|} \binom{|\mathbf{a}|}{j} \binom{n + (r - |\mathbf{a}| + j)/2 - 1}{n}. \tag{2}$$

Remark 1

In the special case |a|=0 Theorem 1 reduces to the identity (1) by Duarte and Guedes de Oliveira.

Corollary 1

Under the assumptions of Theorem 1, the condition |a|=0 implies that

$$\sum_{|\mathbf{k}|=n} \left[\prod_{i=1}^{r} \binom{2k_i - a_i}{k_i} \right] = 4^n \binom{n + r/2 - 1}{n}.$$

Proposition 1

For $n, r \in \mathbb{N},\ r \geq 2,\ c \in \mathbb{R}\ and\ \mathbf{a} = (a_1, \ldots, a_r) \in \mathbb{R}^r$,

$$\sum_{|\mathbf{k}|=n} \left[\prod_{i=1}^{r} \binom{ck_i - a_i}{k_i} \right] = \frac{1}{n!} F_n^{(n)}(0),$$

where

$$F_n(x) \equiv F_n(r, c, \mathbf{a}; x) := \frac{(1+x)^{cn+r-1-|\mathbf{a}|}}{(1+(1-c)x)^{r-1}}. \tag{3}$$

Remark 2

Application of the Leibniz Rule for the derivatives of products of differentiable functions implies as an immediate consequence the formula

$$\sum_{|\mathbf{k}|=n} \left[\prod_{i=1}^{r} \binom{ck_i - a_i}{k_i} \right] = \sum_{k=0}^{n} \binom{cn - |\mathbf{a}| + r - 1}{k} \binom{r - 2 + n - k}{n - k} (c-1)^{n-k},$$

which reduces the number of multiplied binomial coefficients from r to two.

Recently, Michael Z. Spivey gave a combinatorial proof of the alternating sum

$$\sum_{k=0}^{n}(-1)^k\binom{2k}{k}\binom{2(n-k)}{n-k}=\begin{cases}2^n\binom{n}{n/2}, & n\ even,\\ 0, & n\ odd\end{cases}$$

by applying an involution to certain colored permutations.

PROOFS AND AUXILIARY RESULTS

The proofs of the results in the preceding section are essentially based on the following formula.

Lemma 1

Let $n\in\mathbb{N}_0,\ r\in\mathbb{N},\ and\ let\ h,\ f_i\ (i=1,\ldots,r)$ be functions which are *n* times differentiable in $x_0\in\mathbb{R}\ with\ h(x_0)\neq 0$. Then

$$\sum_{|\mathbf{k}|=n}\binom{n}{\mathbf{k}}\prod_{i=1}^{r}\left(h^{k_i}f_i\right)^{(k_i)}(x_0)=\left(\left(\frac{\mathrm{d}}{\mathrm{d}x}\right)^n\frac{h^{n+r-1}(x)\prod_{i=1}^{r}f_i(x)}{(h(x)-h'(x)(x-x_0))^{r-1}}\right)\Bigg|_{x=x_0}. \tag{4}$$

A proof of the intriguing formula (4) can be found in [1]. If *h* is a constant function, Eq. (4) obviously reduces to the well-known Leibniz Rule

$$\left(\prod_{i=1}^{r}f_i\right)^{(n)}=\sum_{|\mathbf{k}|=n}\binom{n}{\mathbf{k}}\prod_{i=1}^{r}f_i^{(k_i)}\qquad(n=0,1,2,\ldots),$$

for several *n* times differentiable functions f_i (i=1,…,r).

Proof of Prop. 1

We put

$$h(x)=x^c,$$

$$f_i(x)=x^{-a_i},\qquad \text{for } 1\leq i\leq r.$$

The left-hand side of Eq. (4) is equal to

$$\text{LHS}=\sum_{|\mathbf{k}|=n}\binom{n}{\mathbf{k}}\prod_{i=1}^{r}\left(h^{k_i}f_i\right)^{(k_i)}(x_0)$$

$$=n!x_0^{(c-1)n-|\mathbf{a}|}\sum_{|\mathbf{k}|=n}\left[\prod_{i=1}^{r}\binom{ck_i-a_i}{k_i}\right].$$

The right-hand side of Eq. (4) is equal to

$$\text{RHS}=\left(\left(\frac{\mathrm{d}}{\mathrm{d}x}\right)^n\frac{x^{c(n+r-1)-|\mathbf{a}|}}{((1-c)x+cx_0)^{r-1}}\right)\Bigg|_{x=x_0},$$

where we used that

$$h(x) - h'(x)(x - x_0) = x^{c-1}((1-c)x + cx_0).$$

Comparison of both sides with $x_0=1$ leads to

$$n! \sum_{|\mathbf{k}|=n} \prod_{i=1}^{r} \binom{ck_i - a_i}{k_i} = \left(\left(\frac{d}{dx} \right)^n \frac{x^{cn+r-1-|\mathbf{a}|}}{((1-c)x+c)^{r-1}} \right) \Bigg|_{x=1}$$

which implies the assertion.

Proof of Theorem 1

Put c=2 in Prop. 1. Taking advantage of the Cauchy integral formula we obtain, for the function *F* as defined in (3),

$$F_n^{(n)}(0) = \frac{n!}{2\pi i} \int_{W_1} \frac{(1+z)^{2n-|\mathbf{a}|}}{z^{n+1}} \left(\frac{1+z}{1-z} \right)^{r-1} dz,$$

where the integration path $W_1 = \partial D_\rho = \{z \mid |z| = \rho\}$ with sufficiently small $\rho>0$ encircles the origin counterclockwise. The change of variables

$$z = \frac{w-1}{w+1}, \qquad dz = 2(w+1)^{-2} dw,$$

i.e., $w = (1+z)/(1-z)$, yields

$$F_n^{(n)}(0) = 2^{2n-|\mathbf{a}|+1} \frac{n!}{2\pi i} \int_{W_2} \frac{w^{2n-|\mathbf{a}|+r-1}(w+1)^{|\mathbf{a}|}}{(w^2-1)^{n+1}} dw,$$

where the integration path W_2 encounters $w_0=1$ such that $\mathrm{Re}(w+1)>0$. A second change of variables $w = \sqrt{\varsigma}$ yields

$$F_n^{(n)}(0) = 2^{2n-|\mathbf{a}|} \frac{n!}{2\pi i} \int_{W_3} \frac{\varsigma^{n+\frac{-|\mathbf{a}|+r}{2}-1}\left(\sqrt{\varsigma}+1\right)^{|\mathbf{a}|}}{(\varsigma-1)^{n+1}} d\varsigma,$$

where the integration path W_3 encounters $\varsigma_0=1$ such that $\mathrm{Re}\,\varsigma>0$. When it happens that $|\mathbf{a}| \in \mathbb{N}_0$ expansion by the binomial formula and application of the Cauchy integral formula completes the proof of Theorem 1.

AN ALTERNATIVE PROOF OF THEOREM 1

In this section we present an alternative derivation of Theorem 1. The referee pointed out that our main result (2) is a consequence of results by Duarte and Guedes de Oliveira [2] and suggested a proof. It is based on the technique of generating functions and will be outlined below. To this end denote, as in [2, Sect. 4], by

$$g(x) := \frac{1}{\sqrt{1-4x}} = \sum_{n \geq 0} \binom{2n}{n} x^n$$

and

$$C(x) := \frac{2}{1+\sqrt{1-4x}} = \sum_{n \geq 0} \frac{1}{n+1} \binom{2n}{n} x^n$$

the generating functions of the central binomial coefficients and of the Catalan numbers, respectively. By [2, Theorem 11], we have, for all real numbers s and t,

$$g(x)\, C(x)^s = \sum_{n \geq 0} \binom{2n+s}{n} x^n$$

and

$$g(x)^t = \sum_{n \geq 0} 4^n \binom{n+t/2-1}{n} x^n.$$

Since

$$1 + \frac{1}{g(x)} = \frac{2}{C(x)},$$

we deduce that

$$\prod_{i=1}^{r} g(x)\, C(x)^{-a_i} = g(x)^r\, C(x)^{-|\mathbf{a}|} = 2^{-|\mathbf{a}|} g(x)^{r-|\mathbf{a}|} (1+g(x))^{|\mathbf{a}|}. \tag{5}$$

Note that the left-hand side of (2) is the general term (i.e., the coefficient of the term in xnxn) of the left-hand side of (5). Moreover, since the general term of $g(x)^{r-|\mathbf{a}|+j}$ is $4^n \binom{n+(r-|\mathbf{a}|+j)/2-1}{n}$, the right-hand side of (2) is the general term of the right-hand side of (5).

Hence, the new proof of Theorem 1 is complete.

ACKNOWLEDGMENTS

We kindly acknowledge the anonymous reviewer for valuable comments and for providing a new proof of Theorem 1.

REFERENCES

1. Abel, U.: A generalization of the Leibniz Rule. Amer. Math. Mon. 120(10), 924–928 (2013)
2. Duarte, R., Guedes de Oliveira, A.: Note on the convolution of binomial coefficients. J. Integer Seq. 16 (2013) (Article 13.7.6)
3. Duarte, R., Guedes de Oliveira, A.: A short proof of a famous combinatorial identity. arXiv:1307.6693. Accessed 30 October 2013 (preprint)
4. Chang, G., Xu, C.: Generalization and probabilistic proof of a combinatorial identity. Amer. Math. Mon. 118, 175–177 (2011)
5. Spivey, M.Z.: Alternating convolution of the central binomial coefficients. Amer. Math. Mon. 121, 537–540 (2014)
6. Spivey, M.Z.: Combinatorial proof that $\sum_{k=0}^{n} \binom{2k}{k}\binom{2(n-k)}{n-k}(-1)^k = 2^n\binom{n}{n/2}$ is even (version: 2012-12-21). http://math.stackexchange.com/a/98327/

SECTION 2:
GRAPH THEORY AND PARTIALLY ORDERED SETS

5

Total Dominator Chromatic Number of Paths, Cycles and Ladder Graphs

A. Vijayalekshmi and J. Virgin Alangara Sheeba

S. T. Hindu College, Nagercoil, India

ABSTRACT

Let G be a graph with minimum degree atleast one. A total dominator coloring of G is a proper coloring of G with the extra property that every vertex in G properly dominates a color class. The total dominator chromatic number of G is denoted by $\chi_{td}(G)$ and is defined by the minimum number of colors needed in a total dominator coloring of G. In this paper, we obtain total dominator chromatic number of paths, cycles and ladder graph.

Keywords: Total dominator chromatic number, ladder graph

Citation: A. Vijayalekshmi, J. Virgin Alangara Sheeba "Total dominator chromatic number of paths, cycles and ladder graphs" International Journal of Contemporary Mathematical Sciences, Vol. 13, 2018, no. 5, 199-204 https://doi.org/10.12988/ijcms.2018.8619

INTRODUCTION

All graphs considered in this paper are finite , undirected graphs and we follow standard definition of graph theory as found in F.Harrary [1] Let G = (V, E) be a graph of order n with minimum degree at least one. The open neighborhood N(v) of a vertex v ∈ V (G) consisits of the set of all vertices adjacent to v. The closed neighbourhood of v is N[v] = N(v) ∪ {v}. The path and cycle of order n are denoted by P_n and C_n respectively. An isomorphism of graphs G and H is a bijection f:V(G)−→V(H) such that any two vertices u and v are adjacent in G iff f(u) and f(v) are adjacent in H.A decomposition of a graph G is a set of subgraphs H_1, H_2, ...H_k that partition of edges of G. That is $\forall i \neq j \cup_{1\leq i\leq k} H_i = G$ and $E(H_i) \cap E(H_j) = \varphi$. For any two graphs G and H, we define the cartesian product, denoted by G× H , to be the graph with vertex set V(G) × V(H) and edges between two vertices $(u_1, v_1)$and (u_2, v_2) iff either $u_1 = u_2$ and $v_1v_2 \in E(H)$ or $u_1u_2 \in E(G)$ and $v_1 = v_2$ A ladder graph can be defined as $P_2 \times P_n$, where n ≥ 2 and is denoted by L_n.

A proper coloring of G is an assignment of colors to the vertices of G,such that adjacent vertices have different colors.The smallest number of colors for which there exists a proper coloring of G is called chromatic number of G and is denoted by χ(G). A total dominator coloring (td-coloring) of G is a proper coloring of G with extra property that every vertex in G properly dominates a color class.The total dominator chromatic number is denoted by $\chi_{td}(G)$ and is defined by the minimum number of colors needed in a total dominator coloring of G.This concept was introduced by Dr.A.Vijayalekshmi in [2]. This notion is also referred as a smarandachely k-dominator coloring of G, (k ≥ 1) and was introduced by Dr.A.Vijayalekshmi in [7].For an integer k ≥ 1 , a smarandachely k-dominator coloring of G is a proper coloring of G, such that every vertex in a graph G properly dominates a k color class. The smallest number of colors for which there exists a smarandachely k-dominator coloring of G is called the smarandachely k-dominator chromatic number of G and is denoted by $\chi_{td}^{s}(G)$.

In a proper coloring C of a graph G, a color class of C is a set consisting of all those vertices assigned the same color.Let C' be a minimal td-coloring of G.We say that a color class $c_i \in C'$ is called a non-dominated color class (n-d color class) if it is not dominated by any vertex of G.These color classes are also called repeated color classes.

The total dominator chromatic number of Paths,Cycles were found in[3]. The total dominator chromatic number of Ladder graph was found in [4].

We have the following observations from [3,4]

Theorem A [3]

Let P_n be a path of order n then

$$\chi_{td}(P_n) = \begin{cases} 2\lceil \frac{n}{3} \rceil - 1 & if n = 1(mod3) \\ 2\lceil \frac{n}{3} \rceil & otherwise \end{cases}$$

Theorem B[3]

Let C_n be the cycle of order n≥3.Then

$$\chi_{td}(C_n) = \begin{cases} 2 & if n = 4 \\ 4\lfloor \frac{n}{6} \rfloor + r & if n \neq 4 \; and \; for \; r = 0,1,2,4, n \equiv r(mod6) \\ 4\lfloor \frac{n}{6} \rfloor + r - 1 & if \; n \equiv r(mod6), \; r = 3,5 \end{cases}$$

Theorem C[4]

For any n≥2

$$\chi_{td}(L_n) = \begin{cases} n+1 & if \; n \; is \; odd \\ n & otherwise \end{cases}$$

In this paper,we obtain the smallest value for total dominator chromatic number for Paths,Cycles and Ladder graphs.

MAIN RESULTS

Theorem 1

Let G be P_n or C_n. Then

$$\chi_{td}(P_n) = \chi_{td}(C_n) = \begin{cases} 2\lfloor \frac{n}{4} \rfloor + 2 & if n \equiv 0(mod4) \\ 2\lfloor \frac{n}{4} \rfloor + 3 & if n \equiv 1(mod4) \\ 2\lfloor \frac{n+2}{4} \rfloor + 2 & otherwise \end{cases}$$

Proof:

Let $V(C_n) = \{v_1, v_2, ..., v_n\}$; n≥11

We take $N(v_i) = \{v_{i-1}, v_{i+1}\}$ ∀ i=2,3,.....(n-1) and $N(v_1) = \{v_2, v_n\}$,
$N(v_n)=\{v_1, v_{(n-1)}\}$

We consider 3 cases

Case(i) n ≡ 0(mod 4)

Assign 2 repeated colors say 1,2 to the vertices v_i , i ≡ 1 (mod 4) and v_j , j

≡ 0 (mod 4) respectively.Also assign one new color say 3,4,...... $(2\lfloor\frac{n}{4}\rfloor+2)$ to each of the remaining vertices and so we obtain $\chi_{td}(C_n) = 2\lfloor\frac{n}{4}\rfloor+2$.

Case (ii) n ≡ 1(mod 4)

In this case, we assign 2 repeated colors say 1,2 to the vertices v_i , i ≡ 0(mod 4),i≠ 1 and v_j , j ≡ 0(mod 4)respectively. Also assign one new color $3,4,\ldots(2\lfloor\frac{n}{4}\rfloor+3)$ to each of the remaining vertices.

Thus $\chi_{td}(C_n) = 2\lfloor\frac{n}{4}\rfloor+3$.

Case (iii) n ≡ 2,3(mod 4)

By case (ii) we assign the two repeated colors to the vertices v_i , v_j i ≡ 1(mod 4),i ≠1 and j ≡ 0(mod 4) respectively.Assign one new color $3,4,\ldots(2\lfloor\frac{n+2}{4}\rfloor+2)$ to each of the remaining vertices.

Hence $\chi_{td}(C_n) = 2\lfloor\frac{n+2}{4}\rfloor+2$.

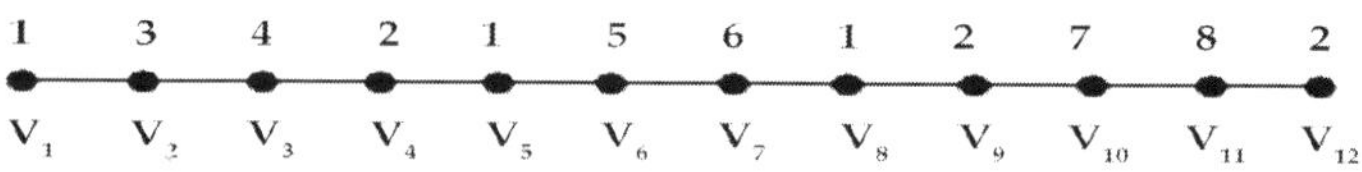

Figure 1: P_{12}.

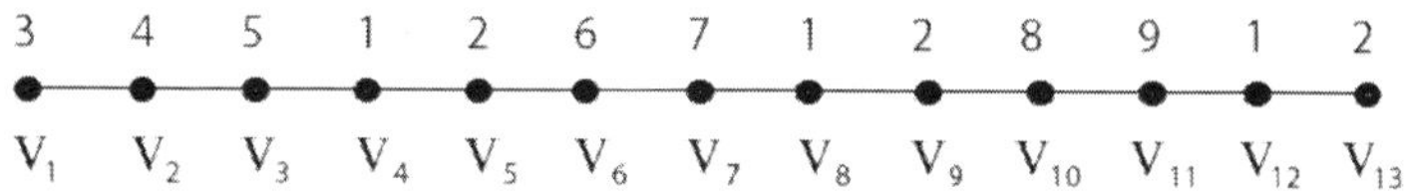

Figure 2: P_{13}.

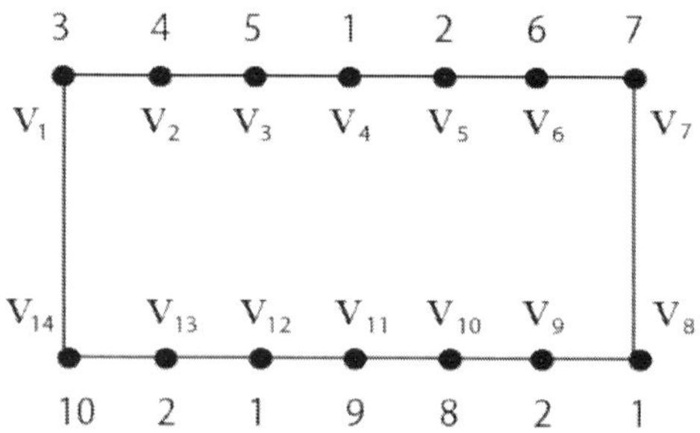

Figure 3: C_{14}.

Theorem 2

For every $n \geq 2$, the total dominator chromatic number of a ladder graph L_n is

$$\chi_{td}(L_n) = \begin{cases} 2\lfloor \frac{p}{6} \rfloor + 2 & if p \equiv 0 (mod 6) \\ \begin{cases} 2\lfloor \frac{p-2}{6} \rfloor + 4 \\ 2\lfloor \frac{p-4}{6} \rfloor + 4 \end{cases} & otherwise \end{cases}$$

Proof:

Let L_n be a ladder graph of order p = 2n and let $V(L_n) = \{v_1, v_2, ...v_n, v_{n+1}, ...v_p\}$.

$deg(v_i) = 3 \ \forall i = 3,4,......(p-3),(p-2)$ and $deg(v_j) = 2 \ \forall j = 1,2,(p-1),p$.

Let $N(v_i) = \{v_{i-3}, v_{i-1}, v_{i+1}\}$ for i = 4,6,8,....(p-2) and $N(v_j) = \{v_{i-1}, v_{i+1}, v_{i+3}\}$ for i = 3,5,7,....(p-3).

Therefore $| N[v_i] \cap N[v_j] | = 2$

We consider 2 cases

Case (i) $p \equiv 0 (mod 6)$

Since $L_n = P_2 \times P_n$, decompose Ln into p/6 copies of $P_2 \times P_3$ and $\chi_{td}(P_2 \times P_3) = 4$. Among these 4 colors, 2 colors are repeated and two of them are non-repeated colors. So $\chi_{td}(L_n) = 2\lfloor \frac{p}{6} \rfloor + 2$.

Case(ii) $p \not\equiv 0 (mod 6)$

Let L_m, $m \equiv 0 (mod 6)$ be a largest ladder subgraph of order either 2(n-1) when $p \equiv 2 (mod 6)$ or 2(n-2) when $p \equiv 4 (mod 6)$ and H be a sub graph of L_n with either H ~= P_2 or H ~= C_4

By case(i) $\chi_{td}(L_m)$ is either $\chi_{td}(Ln-1)$ or $\chi_{td}(L_{n-2})$

Therefore $\chi_{td}(L_m)$ is either $2\lfloor \frac{p-2}{6} \rfloor + 2$ or $2\lfloor \frac{p-4}{6} \rfloor + 2$. If $H \cong P_2$

Then $\chi_{td}(L_n) = 2\lfloor \frac{p-2}{6} \rfloor + 4$. Also if $H \cong C_4$ then $\chi_{td}(L_n) = 2\lfloor \frac{p-4}{6} \rfloor + 4$

REFERENCES

1. F. Harrary, Graph Theory, Addition-Wesley Reading Mass, 1969. https://doi.org/10.21236/ad0705364
2. M. I. Jinnah and A. Vijayalekshmi, Total Dominator Colorings in Graphs, Diss, University of Kerala, 2010.
3. A.P. Kezemi, Total dominator chromatic number of a graph, Trans. Combin., 4 (2015), no. 2, 57-68.
4. Saeid Alikhani and Nima Ghanbari, Total dominator chromatic number of soecific graphs, arXiv:1511.01652v1[math.CO].
5. Terasa W. Haynes, Stephen T. Hedetniemi, Peter J. Slater, Domination in Graphs, Marcel Dekker, New York, 1998.
6. Terasa W. Haynes, Stephen T. Hedetniemi, Peter J. Slater, Domination in Graphs - Advanced Topics, Marceel Dekker, NewYork, 1998.
7. A. Vijayalekshmi, Total dominator colorings in paths, International Journal of Mathematical Combinatorics, 2 (2012), 89-95
8. A. Vijayalekshmi, Total dominator colorings in cycles, International Journal of Mathematical Combinatorics, 4 (2012), 92-96

6

Modular Leech Trees of Order at Most 8

David Leach

Department of Mathematics, University of West Georgia, 1601 Maple Street, Carrollton, GA 30118, USA

ABSTRACT

In 1975, John Leech asked when can the edges of a tree on n vertices be labeled with positive integers such that the sums along the paths are exactly the integers $1, 2, \ldots, \binom{n}{2}$. He found five such trees, and no additional trees have been discovered since. In 2011 Leach and Walsh introduced the idea of labeling trees with elements of the group $\mathbb{Z}_k$ where $k = \binom{n}{2} + 1$ and examined the cases for $n \leq 6$. In this paper we show that no modular Leech trees of order 7 exist, and we find all modular Leech trees of order 8.

Citation: David Leach, "Modular Leech Trees of Order at Most 8," International Journal of Combinatorics, vol. 2014, Article ID 218086, 2 pages, 2014. https://doi.org/10.1155/2014/218086

INTRODUCTION

A tree on n vertices is said to be a Leech tree if its edges can be weighted with positive integers in such a way that each of the $\binom{n}{2}$ paths has a distinct weight from the set$\{1, 2, \ldots, \binom{n}{2}\}$.The weight of a path is found by summing all of its edge weights. Leech [1] found the five examples shown in Figure 1, which are to date the only ones known. In 1977, Taylor [2] proved that, in order for a Leech tree of order n to exist, it must be that $n=k^2$ or $n=k^2+2$ for some integer k. Since then it has been shown by several authors [3–5] that no Leech trees of order 9, 11, or 16 exist, leaving $n = 18$ as the smallest open case. Szekely et al. [´ 4] have conjectured that no additional Leech trees exist. Since Leech trees are so difficult to come by, we consider the generalization to modular Leech trees.

Let T be a tree on n vertices and let $k = \binom{n}{2} + 1$. We say that T is a modular Leech tree if there exists an edge weighting function $: E(T) \rightarrow \mathbb{Z}_k$ such that each of the $\binom{n}{2}$ paths within T has a distinct weight from $1, 2, \ldots, \binom{n}{2}$ with the sums taken modulo $\binom{n}{2} + 1$. We call such an edge weighting function a $\mathbb{Z}_k$-Leech labeling. Since the pathweights are all distinct, the function w induces a bijection between the paths of T and the elements of the group $\mathbb{Z}_k$. We use w to refer to this bijection as well.

Note that a "normal" Leech tree of order n is also a modular Leech tree over $\mathbb{Z}_{\binom{n}{2}+1}$ in which none of the path sums, before applying the mod operation, have a weight greater than $\binom{n}{2}$. Thus Leech's original five examples provide us with five modular Leech trees. The only other example previously known is the tree of order 6 found in [6] shown in Figure 2. It was also shown in [6] that no modular Leech tree of order 5 exists.

In Section 2 we will see how Taylor's condition applies to modular Leech trees, and in Section 3 we will enumerate all Leech trees of order at most 8.

TAYLOR'S CONDITION FOR MODULAR LEECH TREES

For the normal Leech trees, Taylor's condition restricts the possible orders severely. However, with modular Leech trees over $\mathbb{Z}_k$, Taylor's condition only applies when k is even. The proof is very similar to half of Taylor's proof.

Theorem 1. Suppose that T is a modular Leech tree of order n and $n \equiv 2$ or 3 (mod 4); then $n = m^2 + 2$ for some integer m.

Proof. Assume that T is a modular Leech tree of order n and that $n \equiv 2$ or 3 (mod 4). Since $n \equiv 2$ or 3 (mod 4), we have that $\binom{n}{2}$ is odd and thus the modulus $\binom{n}{2} + 1$ is even.

We color each vertex of the tree black or white as follows: start at any vertex V and color it black. From V we color all other vertices by traversing the edges of the graph. We keep the same color across edges with even weight and change colors across edges with odd weight. When all vertices are colored, an edge connects different colored vertices if and only if its weight is odd. Furthermore, since the modulus is even, for any vertices u, V $\in (T)$, the path from u to V has odd weight if and only if u and V are colored with opposite colors.

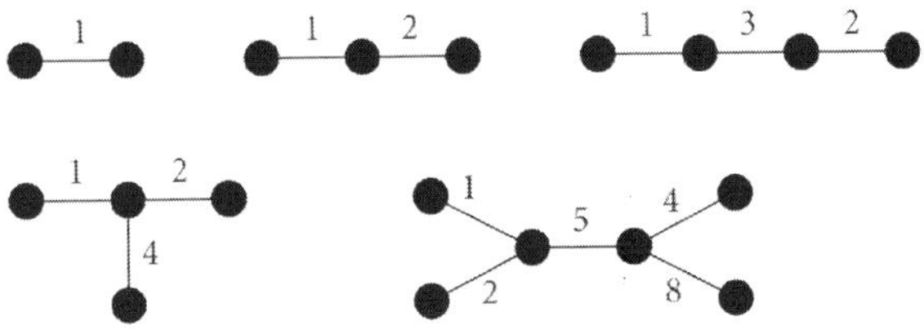

Figure 1: The five known Leech trees.

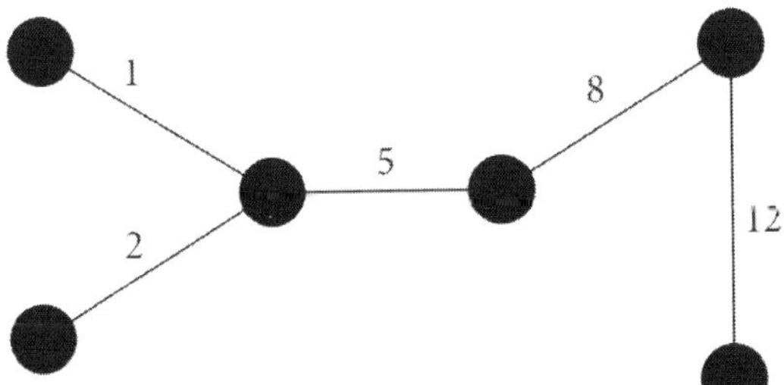

Figure 2: Leech tree of order 6 over $\mathbb{Z}_{16}$.

Now we count the number of odd paths in two ways: let b and w be the number of black and white vertices, respectively. Thus the number of odd paths is bw. Also, since there are $\binom{n}{2}$ paths and $\binom{n}{2}$ is odd, the number of odd paths is $(1/2)((n(n-1)/2)+1)$. Putting these together gives $n-2 = n^2 - 4bw$. Substituting $b+w$ for n on the right side leads to $n = (b-w)^2 + 2$ and the theorem is proved.

The proof of Theorem 1 makes use of the fact that, under the assumptions of the theorem, a path has odd weight if and only if it contains an odd number of odd-weight edges. This fact does not hold when the modulus is odd. (For example, consider a path on four vertices with edges weighted 3, 5, and 1 and examine the path-weights mod 7.) Recall that a normal Leech tree on n vertices is also a modular Leech tree over $\mathbb{Z}_k$ where $k = \binom{n}{2} + 1$ and by Taylor's condition $n=m^2$ or $m^2 + 2$ for some integer m. We can conclude the following.

Theorem 2. If there exists a modular Leech tree of order n, and n is not $m^2 + 2$ for some integer m, then $n \equiv 0$ or 1 (mod 4).

COMPUTATIONAL RESULTS

We have already seen modular Leech trees of orders 2, 3, 4, and 6. By Leach and Walsh [6] and Theorem 2 we know that none exist for 5 or 7. To examine larger values of n, we use computer search.The following theorem and corollary reduce the space that must be searched.

Theorem 3. Let n be an integer and $k = \binom{n}{2} + 1$. Suppose that $: E(T) \rightarrow \mathbb{Z}_k$ is a $\mathbb{Z}_k$-Leech labeling of a tree T. Then for any $b \in \mathbb{Z}_k$ satisfying $gc(b, k) = 1$, the function $g : E(T) \rightarrow \mathbb{Z}_k$ defined by $g(e) = b \cdot f(e)$ is also a $\mathbb{Z}_k$-Leech labeling of T.

Proof. Let (T) be the set of all paths in T. We can consider f as a bijection between (T) and $\mathbb{Z}_k$, so for every nonzero $b \in \mathbb{Z}_k$, there exists a path with weight b. Now define $: P(T) \rightarrow \mathbb{Z}_k$ by $g(e) = b \cdot f(e)$.

We now show that $: P(T) \rightarrow \mathbb{Z}_k$ is bijective: let P_1 and P_2 be paths in T and suppose that $g(P_1) = g(P_2)$. Then $b \cdot (P_1) = b \cdot f(P_1)$. Since $gc(b, k) = 1$, b^{-1} exists and thus $f(P_1) = f(P_2)$. Since f is a bijective, $P_1 = P_2$ and thus g is also bijective. Since $: P(T) \rightarrow \mathbb{Z}_k$ is bijective, g is a Leech labeling of T.

Corollary 4. If there exists a $\mathbb{Z}_k$-Leech labeling of a tree T and $e \in (T)$, then there exists a $\mathbb{Z}_k$-Leech labeling of T in which edge e has weight 1.

For n=8, there are 23 distinct unlabeled trees. By computer search, we find that there is one modular Leech tree for n=8 and the edge-weighting function is unique, up to group and graph isomorphism. It is shown in Figure 3.

REFERENCES

1. J. Leech, "Research problems: another tree labelling problem," The American Mathematical Monthly, vol. 82, no. 9, pp. 923–925, 1975.
2. H. Taylor, "Odd path sums in an edge-labeled tree," Mathematics Magazine, vol. 50, no. 5, pp. 258–259, 1977.
3. H. Taylor, "A distinct distance set of 9 nodes in a tree of diameter 36," Discrete Mathematics, vol. 93, no. 2-3, pp. 167–168, 1991.
4. L. A. Székely, H. Wang, and Y. Zhang, "Some non-existence results on Leech trees," Bulletin of the Institute of Combinatorics and its Applications, vol. 44, pp. 37–45, 2005.
5. B. Calhoun, K. Ferland, L. Lister, and J. Polhill, "Minimal distinct distance trees," Journal of Combinatorial Mathematics and Combinatorial Computing, vol. 61, pp. 33–57, 2007.
6. D. Leach and M. Walsh, "Generalized Leech trees," Journal of Combinatorial Mathematics and Combinatorial Computing, vol. 78, pp. 15–22, 2011.

7

Recursive Algorithms for Phylogenetic Tree Counting

Alexandra Gavryushkina[1], David Welch[1] and Alexei J Drummond[1,2]

[1]Department of Computer Science, The University of Auckland, Auckland, New Zealand

[2]Allan Wilson Centre for Molecular Ecology and Evolution, University of Auckland, Auckland, New Zealand

ABSTRACT

Background

In Bayesian phylogenetic inference we are interested in distributions over a space of trees. The number of trees in a tree space is an important characteristic of the space and is useful for specifying prior distributions. When all samples come from the same time point and no prior information available on divergence times, the tree counting problem is easy. However,

Citation: Alexandra Gavryushkina, David Welch and Alexei J Drummond "Recursive algorithms for phylogenetic tree counting" Algorithms for Molecular Biology 2013 8:26. https://doi.org/10.1186/1748-7188-8-26

when fossil evidence is used in the inference to constrain the tree or data are sampled serially, new tree spaces arise and counting the number of trees is more difficult.

Results

We describe an algorithm that is polynomial in the number of sampled individuals for counting of resolutions of a constraint tree assuming that the number of constraints is fixed. We generalise this algorithm to counting resolutions of a fully ranked constraint tree. We describe a quadratic algorithm for counting the number of possible fully ranked trees on n sampled individuals. We introduce a new type of tree, called a fully ranked tree with sampled ancestors, and describe a cubic time algorithm for counting the number of such trees on n sampled individuals.

Conclusions

These algorithms should be employed for Bayesian Markov chain Monte Carlo inference when fossil data are included or data are serially sampled.

Keywords: Ranked tree, Constraint tree, Resolution, Counting trees, Dynamic algorithms, Bayesian tree prior, Phylogenetics

BACKGROUND

A phylogenetic tree is the common object of interest in many areas of biological science. The tree represents the ancestral relationships between a group of individuals. Given molecular sequence data sampled from a group of organisms it is possible to infer the historical relationships between these organisms using a statistical model of molecular evolution. At present, Bayesian Markov chain Monte Carlo (MCMC) methods are the dominant inferential tool for inferring molecular phylogenies [1].

It is a recent trend to include fossil evidence into the inference to obtain absolute estimates of divergence times [2, 3]. Fossils may restrict the age of the most resent common ancestor of a subgroup of individuals. This imposes a constraint on the tree topology (the discrete component of a genealogy) and therefore reduces the space of allowable genealogies.

Another trend in phylogenetic analyses is serial (or heterochronous) sampling in which molecular data is obtained from significantly different time points and analysed together. This type of data arises most frequently with ancient DNA and rapidly evolving pathogens [4]-[6]. In this case tip dates become a part of the genealogy.

Including serially sampled or fossil data modifies or restricts the shape of a phylogenetic tree. Little has been done to describe and classify these modified trees. In this paper, we aim to explore the new spaces formed by these trees.

A genealogy consists of discrete and continuous components — the tree topology and the divergence times. The tree topologies form a finite tree space when the number of tips is bounded. An important characteristic of this space is the number of trees in it and we aim to find an efficient way to calculate this number.

In the case that fossil data restricts the tree topology, counting the number of trees that satisfy the imposed constraints reveals how much the constraints reduce the tree space.

The number of trees arises as a constant in tree prior distributions. Typically we model the distribution of tree topologies as independent of the distribution of divergence times. The density function of the distribution of genealogies is then a product of the density function for the divergence times and the distribution function for tree topologies. A common prior on tree topologies is uniform over all allowable topologies so the distribution function is a constant that is equal to one over the number of tree topologies. When inferring tree topologies using Bayesian MCMC methods, we do not usually need to know this constant but in some cases, as described below, the absolute value of the prior distribution is of interest and the constant has to be calculated.

When fossils are used to restrict the age of internal nodes, the tree prior should accurately account for this fact. Heled and Drummond [3] introduced a natural approach for tree prior specification when fossil evidence is employed in the inference. Their method requires counting of ranked phylogenetic trees that obey a number of constraints that arise from including the fossil evidence. The construction requires calculation of the marginal density for the time of the calibration node, the node representing the most recent common ancestor of a clade which may or may not be monophyletic. For a particular location of the calibration node, or particular constraints on the tree topology, the marginal density function is the marginal density function for the divergence times weighted by the number of trees satisfying the constraints. In this case, the weight constants do not cancel in the MCMC scheme and therefore have to be calculated.

Tree counting has a long history. For phylogenetic trees, the counting problem is to find the number of all possible trees on n leaves. For some types of phylogenetic tree, there are known closed form solutions to this problem. For other types, only recursive equations have been derived. In this paper, we consider only rooted trees.

A survey of results on counting different types of rooted trees is presented in [7] where trees with different combinations of the following properties are considered: trees are either labeled (only leaves are labeled) or unlabeled, ranked or non-ranked, and bifurcating or multifurcating. The results presented in the survey can also be found in [6, 8, 9].

In [10], Griffiths considered unlabeled, non-ranked rooted trees such that interior nodes can have one child or more and the root has at least two children. Using generating functions, he derived recursive equations for counting the number of all possible such trees on n leaves with s interior nodes. In [11], Felsenstein considered partially labeled trees, i.e., a tree in which all the leaves are labeled and some interior nodes also may be labeled. He derived the recursive equations for counting the number of rooted, non-ranked, partially labeled trees with n labeled nodes.

In this paper, we consider a number of counting problems for different classes of phylogenetic trees. First, we describe an effective way of counting the number of all possible fully ranked trees on n leaves, that is, trees on n leaves in which all internal and leaf nodes are ranked.

Second, we find the number of bifurcating trees that resolve a given multifurcating tree with n leaves. We give a solution to this problem for rooted, ranked, labeled trees and generalise the algorithm to count resolutions to fully ranked trees.

Finally, we introduce and formally describe a new type of phylogenetic tree and describe an algorithm for counting the number of all such trees on n leaves. This type of tree is important when we have a serial sample and sampled individuals can be direct ancestors of later sampled individuals. When the population size is small or the fraction of individuals sampled from the population is large, this type of tree should be included in the inference [12, 13].

SERIAL SAMPLING

We mainly follow the terminology from [9] for the definitions of phylogenetic trees. A tree is a finite connected undirected graph with no cycles. A rooted tree is a tree with a single node ρ designated as a root. Every rooted tree T=(V,E,ρ) imposes a partial order on V that is defined as follows: $v_1 \leq_T v_2$ if a unique simple path from the root to v_2 passes through v_1. So the root is the smallest element. If $v_1 \leq_T v_2$ then we say that v_1 is an ancestor of v_2 and v_2 is a descendant of v_1. A node in a rooted tree is called interior if it has descendants and a leaf if it has no descendants. The root is considered interior. Denote $\mathring{V}$ the set of interior nodes of T. A node u is a parent of a node v and v is a child of u if $v <_T u$ and there is no w∈V such that $v <_T w <_T u$. A rooted tree is called binary if every interior node has exactly two children. It is called weakly binary if every interior node has at most two children. We have chosen this terminology to fit with the usage of "binary" in the phylogenetics literature which may not agree with that in other literatures.

Let X be a finite non-empty set of labels. A phylogenetic X-tree is a pair T=(T,ϕ), where T is a tree and ϕ is a bijection from X onto the set of leaves of T (we may omit X and say "tree" instead of "X-tree" if the set of labels is not specified). The tree T is called an underlying tree or a shape of the phylogenetic tree $\mathcal{T}$ and ϕ is a labeling function. If the underlying tree of $\mathcal{T}$ is rooted then $\mathcal{T}$ is called a rooted phylogenetic tree. In what follows, we consider only rooted trees unless explicitly stated otherwise. A phylogenetic tree is binary (weak binary) if its underlying tree is binary (weak binary). A ranked phylogenetic tree is a pair (T,h), where $\mathcal{T}$ is a rooted phylogenetic tree and h is an injective function (ranking function) from the set $\mathring{V}$ into the set $\{1,\ldots,|\mathring{V}|\}$ such that $v_1 \leq_T v_2$ implies $h(v_1) \leq h(v_2)$ for every $v_1, v_2 \in \mathring{V}$. In other words, there is a linear order on the interior nodes of T that is consistent with the partial order of T.

Definition 1

A ranked X-tree is a binary ranked phylogenetic X-tree.

An example of a ranked tree is given in Figure 1.

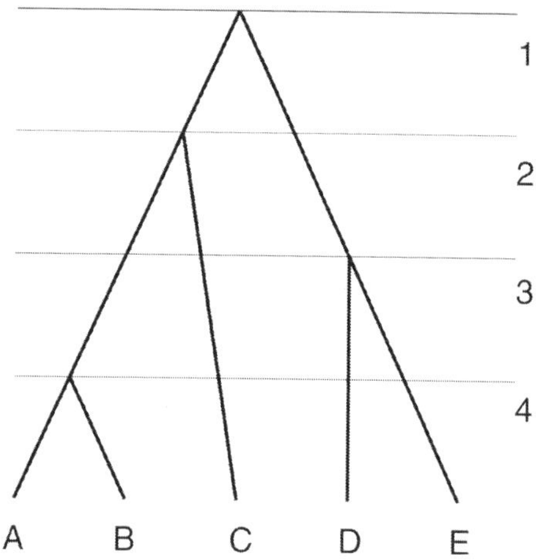

Figure 1: Ranked tree. Ranked X-tree, X={A,B,C,D,E}. The numbers on the right are values of the ranking function.

In biology, a phylogenetic tree represents the evolutionary history of a collection of sampled individuals. The collection of individuals is represented by the set X. The root of the tree is the most recent common ancestor of X and interior nodes are bifurcation events. The ranking function represents the time order of the bifurcation events. A general problem in evolutionary biology is how to reconstruct the phylogenetic tree from sequence data obtained from sampled individuals. Tackling this problem in a Bayesian framework may require counting the number of all possible histories on a sample of individuals.

When all individuals are sampled at the same time (as in Figure 1) counting tree problem has a simple solution.

Let X be a fixed label set such that |X|=n. The number of all ranked X-trees up to isomorphism is

$$R(n) = \frac{n!\,(n-1)!}{2^{n-1}}$$

This formula has been derived by many authors. Proofs can be found in [6, 7], or [9]. The letter R in the equation comes from the word “ranked”.

The situation is different when individuals are sampled at different times (serially sampled). In this case, we need to define another kind of phylogenetic tree in which leaves are also ranked.

Definition 2

A fully ranked (FR) X-tree is a pair(T,h), where $\mathcal{T}$ is a binary rooted phylogenetic X-tree and h:V→{1,…,l} with $|\mathring{V}| < l \leq |V|$ is a surjective function such that

- $v_1 \leq_T v_2$ implies $h(v_1) \leq h(v_2)$ and
- $h(v_1)=h(v_2)$ implies $v_1=v_2$ or $v_1, v_2 \in V \setminus \mathring{V}$.

An example of a fully ranked X-tree is given in Figure 2.

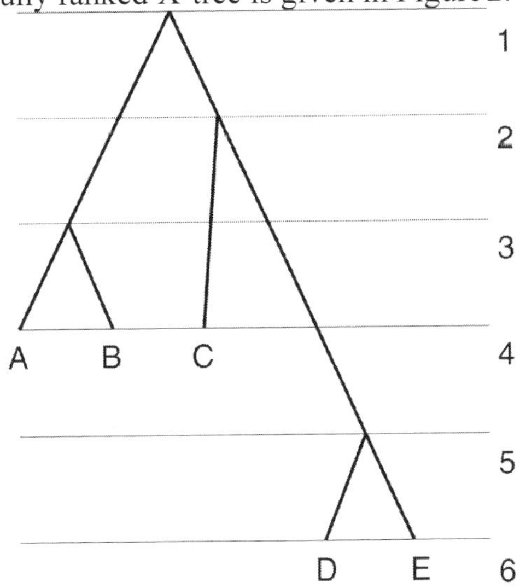

Figure 2: Fully ranked tree. Fully ranked X-tree. X={A,B,C,D,E}. The numbers on the right are values of the ranking function.

Before the tree is reconstructed we observe only leaves (sampled individuals) of the tree that are grouped (pre-ranked) according to the times they were sampled. For the tree shown in Figure 2, we have two sampling times and hence two groups: A, B, and C form the first group, D and E form the second group.

Let $\mathfrak{T} = (\mathcal{T}, h)$ be a fully ranked X-tree with h:V→{1,…,l}. Let m=|h(ϕ(X))|, that is, the number of sampling times. Define a pre-ranking function ĥ from X onto {1,…,m} for tree $\mathfrak{T}$ such that for all $x_1,x_2 \in X$

- $h(\phi(x_1)) \leq h(\phi(x_2))$ implies $\hat{h}(x_1) \leq \hat{h}(x_2)$ and
- $h(\phi(x_1)) = h(\phi(x_2))$ iff $\hat{h}(x_1)=\hat{h}(x_2)$.

For the tree given in Figure 1, ĥ(A)=ĥ(B)=ĥ(C)=1 and ĥ(D)=ĥ(E)=2.

Let X and ĥ:X→{1,…,m} be fixed. We are interested in the number of all fully ranked X-trees that have ĥ as a pre-ranking function. Note that this number depends only on the numbers $n_i=|\{x|\hat{h}(x)=i\}|$, the number of individuals sampled at the ith time point, not on X and $\hat{h}$ directly. We denote this quantity by $F(n_1,\ldots,n_m)$, where F stands for "fully ranked". Then

$$F(n_1,\ldots,n_m)=\sum_{i=1}^{n_m}\frac{R(n_m)}{R(i)}F(n_1,\ldots,n_{m-1}+i) \tag{1}$$

and F(n)=R(n).

Proof

Consider a continuous process of bifurcation in which lineages may bifurcate in time or be cut and labeled (sampled). The process finishes when all lineages are cut producing a tree. The discrete structure of the tree produced by this process is a fully ranked X-tree. It is easy to see that every fully ranked X-tree can be obtained as a result of this process. To count the required number we can count the number of different trees which can be produced by the process if we know that after it finishes there are n_i sampled individuals (i.e., cut and labeled lineages) at the ith time point, i.e., we have the sequence $(n_1,\ldots,n_m)$.

Suppose that at the (m−1)th time point there are i lineages that are ancestral to n_m individuals sampled at time m. When we look at this process backwards in time the bifurcation events become coalescence events. The number of different ways these n_m lineages coalesce to i lineages is $\frac{R(n_m)}{R(i)}$. This is the number of all possible ranked X-trees on n_m individuals but since we are not interested in the structure of the coalescent after we reach i lineages, it is divided by the number of ways in which the remaining i lineages can coalesce. Note that if coalescence patterns are different between the (m−1) th and mth time points then the trees are also different.

Further, for each of these coalescence patterns, we need to count the number of different ways these i lineages and other $n_1,\ldots,n_{m-1}$ lineages can coalesce. This is where we can apply the recursion. We can consider that we also cut these i lineages at time m−1 and label them with the ranked subtrees descendant from these lineages. Then, at time m−1, we have n_{m-1} sampled individuals and another i sampled individuals and it remains to count the number of trees on the sequence $(n_1,\ldots,n_{m-1}+i)$.

Note that two trees are different if they have different numbers of lineages at time m−1. The number of additional i lineages can be between 1 and n_m and we need to sum over all possible i to complete the recursion.

We introduce a third type of tree in which sampled individuals may be direct ancestors of later sampled individuals. We call it a tree with sampled ancestors. This type of tree is not usually considered in phylogenetics since the probability of sampling a direct ancestor is often negligible. In small populations or when a large portion of the population is sampled, however, this can not be ignored.

Let T=(V,E,ρ) be a weak binary tree. Define a set $\hat{V}$ as follows:

$$\hat{V} = \{v \in V | deg(v) = 1 \text{ or } [deg(v) = 2 \text{ and } v \text{ is not the root}]\}$$

A rooted S-phylogenetic X-tree is a pair $\mathcal{T}$=(T,ϕ), where T is a weak binary tree and ϕ:X→$\hat{V}$ is a bijection.

Definition 3

A fully ranked X-tree with sampled ancestors (FRS X-tree) is a pair (T,h), where $\mathcal{T}$ is a rooted S-phylogenetic X-tree and h:V→{1,…,l} is a surjective function such that

- $v_1 <_T v_2$ implies $h(v_1) < h(v_2)$ and
- $h(v_1) = h(v_2)$ implies $v_1 = v_2$ or $v_1, v_2 \in \hat{V}$,

(see Figure 3).

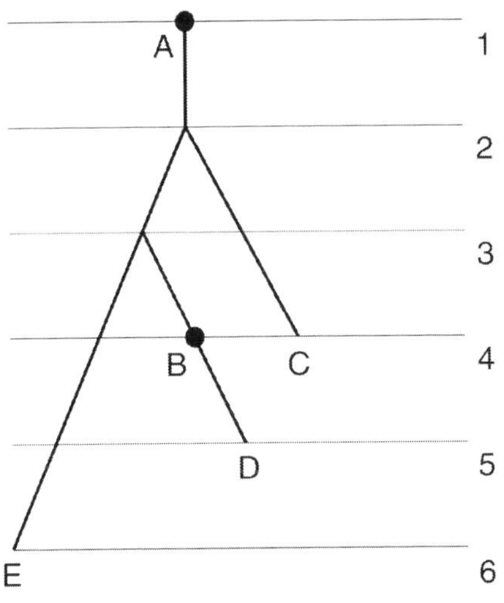

Figure 3: FRS tree. FRS X-tree with the labeled 1-degree root. X={A,B,C,D,E}. The numbers on the right are values of the ranking function.

The definition of a pre-ranking function remains the same for FRS trees. Let $S(n_1,\ldots,n_m)$ (with S standing for 'sampled ancestors') denote the number of all FRS X-trees that have the same pre-ranking function $\hat{h}$, where $n_i=|\{x|\hat{h}(x)=i\}|$. Then

$$S(n_1,\ldots,n_m)=\sum_{i=1}^{n_m}\sum_{j=0}^{min\{i,n_{m-1}\}}\binom{i}{j}\binom{n_{m-1}}{j}\times\frac{R(n_m)}{R(i)}S(n_1,\ldots,n_{m-1}+i-j) \tag{2}$$

and S(n)=R(n).

Proof

Consider the same process as before with only change of sampling events. Now, at some points of time, some lineages are cut and labeled and others are only labeled but not cut.

Then this equation can be obtained as follows. We have n_m individuals that are sampled at the mth time point. At time m−1, there are between 1 and n_m ancestral lineages of those n_m individuals, depending on the number of coalescence between times m and m−1. Let i denote the number of these lineages. Then there are $\frac{R(n_m)}{R(i)}$ different possible coalescent patterns that can lead to this situation. Some of these i ancestral lineages may be among the individuals sampled at time m−1, i.e. lineages that are labeled but not cut at time m−1. Let j be the number of those ancestral lineages that are sampled at time m−1. There are $\binom{i}{j}$ ways to chose these j lineages out of i and there are $\binom{n_{m-1}}{j}$ possible ways to chose j sampled at time m−1 individuals that are not cut at time m−1.

Further, at time m−1, there are n_{m-1} sampled lineages and i−j ancestral lineages that are not sampled and it remains to count the number of FRS trees on the sequence $(n_1,\ldots,n_{m-1}+i-j)$.

Finally, we sum over all possible i and j to complete the recursion.

Dynamic counting

Calculating the recursions (1) and (2) directly is inefficient and impractical. Here we describe a more efficient algorithm for counting fully ranked trees using these recursions. Rewrite equation (1) as

$$F(n_1,\ldots,n_m)=R(n_m)\sum_{i=1}^{n_m}\frac{F(n_1,\ldots,n_{m-1}+i)}{R(i)}$$

Then instead of calculating $F(n_1,\ldots,n_j+\alpha)$ for $j\in\{1,\ldots,m-1\}$ and $\alpha\in\{0,\ldots,n_{i+1}+\ldots+n_m\}$ we can calculate

$$A^j(\alpha) = \frac{F(n_1,\ldots,n_j+\alpha)}{R(\alpha)} \quad (3)$$

using recurrence equations:

$$A^j(0) = R(n_j)\sum_{i=1}^{n_j} A^{j-1}(i) \text{ and} \quad (4)$$

$$A^j(\alpha+1) = \frac{(n_j+\alpha)(n_j+\alpha+1)}{\alpha(\alpha+1)}A^j(\alpha) + \frac{R(n_j+\alpha+1)}{R(\alpha+1)}A^{j-1}(n_j+\alpha+1). \quad (5)$$

This leads to Algorithm 1 to calculate $F(n_1,\ldots,n_m)$. Let n be the number of sampled individuals, i.e., $n = \sum_{i=1}^{m} n_i$. Calculation of all the R(i) takes O(n) steps and calculation of all the A^j takes at most O(n) steps. In total, the algorithm takes O(m n) steps.

Algorithm 1: Calculating the number of fully ranked trees

for $i = 1 \to n$ **do**
  calculate $R(i)$ using $R(i) = R(i-1)\frac{i(i-1)}{2}$
end for
for $i = 1 \to n_2 + \ldots + n_m$ **do**
  calculate $A^1(i)$ using equation (3) and equality $F(x) = R(x)$
end for
for $j = 2 \to m-1$ **do**
  for $\alpha = 0 \to n_{j+1} + \ldots + n_m$ **do**
    calculate $A^j(\alpha)$ using equations (4) and (5)
  end for
end for
compute $F(n_1,\ldots,n_m) = A^m(0)$

A similar approach leads to an O(m n^2)-time algorithm for counting FRS trees. The description of this algorithm is in Appendix Appendix 1: Algorithm for counting FRS trees.

CONSTRAINTS

In phylogenetic analysis, it is common to have some limited information

When k=1, i.e., there is only one interior node, f=∅. See Figure 4 for an example of introduced notation. Note that a tuple $(n_1,\ldots,n_k,f)$ completely defines a pair (T,h).

Let $R^r(n_1,\ldots,n_k,f)$ be the number of ranked trees resolving a constraint tree defined by the tuple $(n_1,\ldots,n_k,f)$. The superscript r stands for "resolution". Then the following equations hold.

$$R^r(2,\emptyset) = 1, \tag{6}$$

$$R^r(n_1,\ldots,n_{k-1},2,f) = \sum_{i\in C}\binom{n_i}{2}R^r(n_1,\ldots,n_i-1,\ldots,n_{k-1},2,f)$$
$$+ R^r(n_1,\ldots,n_{f(k)}+1,\ldots,n_{k-1},f|_{\{2,\ldots,k-1\}}) \text{ and} \tag{7}$$

$$R^r(n_1,\ldots,n_k,f) = \sum_{i\in C}\binom{n_i}{2}R^r(n_2,\ldots,n_i-1,\ldots,n_k,f),$$
$$\text{if } n_k > 2, \tag{8}$$

where C is the collection of nodes that have more than 2 children and at least 2 of them are leaves, i.e., $C=\{i<k \mid n_i \geq 2 \text{ and } n_i+\alpha_i>2\}$ and $\alpha_i=|\{a| f(a)=i\}|$. Note that $n_i+\alpha_i$ is the number of children of node i.

Proof

When a constraint tree has 2 leaves, it is unique and is resolution of itself. So Equation (6) is trivial. To explain the main sum in Equation (7) and (8) we consider the constraint tree which is defined by (3,3,{(2,1)}) and shown in Figure 5, left. The last interior node of a resolving tree (that is, the interior node with the highest rank or the furthest node from the root) is either a parent to leaves in subtree 1 or leaves in subtree 2. Suppose it is the first case (see Figure 5, centre). Since leaves have distinct labels from X, there are $\binom{3}{2}$ ways to chose two leaves that are children of that last node. We can partition all the resolving trees for which the last node is in subtree 1 in $\binom{3}{2}$ groups. The number of trees in each group is the number of trees that resolve a constraint tree defined by (2,3,{(2,1)}) and shown on the right of Figure 5. A similar argument holds if the last node in a resolving tree is a parent to nodes from subtree 2.

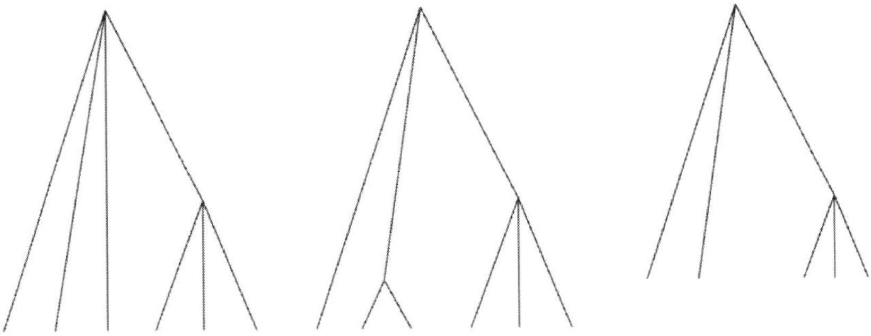

Figurc 5: Recursive approach. The last interior node in a resolving tree locates in subtree 2.

So in the general case, there are k subtrees and if the last node is in subtree iwe have $\binom{n_i}{2}$ ways to choose two lineages that coalesce and then we should count the number of trees resolving the tree defined by $(n_1,\ldots,n_i-1,\ldots,n_k,f)$. Note that in the example above, we consider the tree with more than 2 leaves in each subtree. However, the last interior node of a resolving tree can not be in subtree i for i<k if there is not enough leaves in this subtree. This can happen either if there are less than 2 leaves in subtree i or if there are 2 leaves in subtree i and node i has only these two leaves as its children. Both cases imply that any parent to leaves of subtree i in a resolving tree has a lesser rank than the rank of node k. This explains why we sum only over the elements of the set C.

Finally, we should consider one more case which explains why there is one more summand in equation (7). If the last node in a constraint tree has only two children, i.e., there are 2 leaves in subtree k, then there is one more group of resolving trees, the group that consists of resolving trees that have this node as the last node.

Dynamic counting

We will calculate $R^r(n_1,\ldots,n_k,f)$ for the corresponding constraint tree. In order to find $R^r(n_1,\ldots,n_k,f)$, at each step s, we will calculate numbers $R^r(x_1,\ldots,x_t,f|_{t-1})$ with $\sum_{i\le t} x_i = s, \mathbf{t}_{-1} = \{2,\ldots,t\}$, and $t\le k$. Note that we do not have to calculate all such numbers. To determine which numbers are required we define two upper triangular matrices m and M of size k×k.

Suppose we draw a horizontal line which is strictly below the line that passes through node j and strictly above the line that passes through node j+1 (or all the leaves if j=k). Then $m_{i,j}$ is the minimal possible number of intersections of this line with branches of subtree i in a resolving tree and $M_{i,j}$ is the maximal possible number. An example is given in Figure 6.

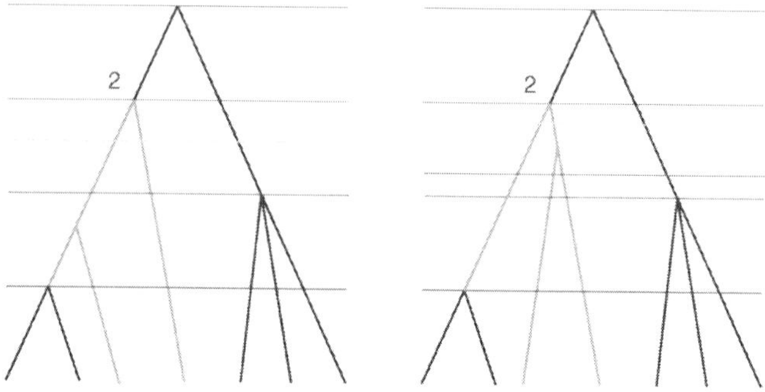

Figure 6: Defining matrices m and M . Two trees that resolve a constraint tree from Figure 4 (only resolutions of subtree 2 are shown) and three ways to draw a horizontal line. The yellow lines correspond to the minimal number of intersections and the red line, to maximal. Thus, $m_{2,2}=2$ and $M_{2,2}=3$. We can also note that $m_{1,2}=M_{1,2}=1$.

Let $a_{i,j}=|\{x\leq j \mid f(x)=i\}|$ for $i\leq j$. So $a_{i,j}$ is the number of children of node i with ranks at most j. Then

$$M_{i,j} = n_i + \alpha_i - a_{i,j} \text{ and}$$

$$m_{i,j} = \begin{cases} 2 \text{ if } a_{i,j} = 0, \\ 1 \text{ if } a_{i,j} > 0 \text{ and } M_{i,j} > 0, \\ 0 \text{ otherwise.} \end{cases}$$

Let $t\leq k$ and $x_1,\ldots,x_t \in \mathbb{N}$. We call a tuple $(x_1,\ldots,x_t,f|_{t-1})$ eligible if $m_{i,t}\leq x_i\leq M_{i,t}$ for $1\leq i\leq t$.

We now turn to Algorithm 2 to count resolutions. At each step $s\leq n$, we construct a set S_s. A unique element of S_n is $R^r(n_1,\ldots,n_k,f)$ and calculating elements of S_s only requires elements of S_{s-1}.

Algorithm 2: Calculating the number of resolutions of a constraint tree

$S_2 = \{R^r(2, \emptyset)\}$
for $s = 3 \rightarrow n-1$ **do**
  while there is a new element $R^r(x_1, \ldots, x_t, f|_{t-1})$ in the set S_{s-1} **do**
    if $t < k$ and *eligible* $(x_1, \ldots, x_{f(t+1)} - 1, \ldots, x_t, 2, f|_{t-1})$ **then**
      calculate $R^r(x_1, \ldots, x_{f(t+1)} - 1, \ldots, x_t, 2, f|_{t-1})$ and add it to S_s
    end if
    for $i = 1 \rightarrow t$ **do**
      if *eligible*$(x_1, \ldots, x_i + 1, \ldots, x_t, f|_{t-1})$ **then**
        $R^r(x_1, \ldots, x_i + 1, \ldots x_t, f|_{t-1})$ and add it to S_s
      end if
    end for
  end while
end for

Proposition 1

When k is fixed Algorithm 2 does at most $O(n^k)$ steps.

Proof

The algorithm does O(k) steps for each eligible tuple and, since we assume k is a constant, it is O(1). For given j there are

$$\prod_{i=1}^{j}(M_{i,j} - m_{i,j} + 1)$$

eligible tuples of size j. Since $M_{i,j} - m_{i,j} + 1 \leq n_i + \alpha_i - 1$, for the total number of eligible tuples we have

$$\sum_{j=1}^{k}\prod_{i=1}^{j}(M_{i,j} - m_{i,j} + 1) \leq \sum_{i=1}^{k}(n_1 + \alpha_1 - 1) \cdot \ldots \cdot (n_i + \alpha_i - 1) <$$
$$k(n_1 + k - 1) \cdot \ldots \cdot (n_k + k - 1) \leq k\left(\frac{n}{k} + k - 1\right)^k = O(n^k)$$

The Number of Resolutions of a Fully Ranked Constraint Tree

We can generalise the results of the previous section to fully ranked trees. Now we replace the word "binary" with "multifurcated" in the definition of a fully ranked tree to get

Definition 5

A fully ranked constraint X-tree is a pair $(\mathcal{T},h)$, where $\mathcal{T}$ is a multifurcated rooted phylogenetic X-tree and h is a function such that $h:V\rightarrow\{1,\ldots,l\}$, where $|\mathring{V}| < l \leq |V|$, and

- $v_1 \leq_T v_2$ implies $h(v_1)\leq h(v_2)$,
- $h(v_1)=h(v_2)$ implies $v_1=v_2$ or $v_1, v_2 \in V \setminus \mathring{V}$.

We say that a fully ranked X-tree $\mathfrak{T}_1 = (\mathcal{T}_1, h_1)$ resolves a fully ranked constraint X-tree $\mathfrak{T}_2 = (\mathcal{T}_2, h_2)$ if there is an isomorphic embedding of $\mathfrak{T}_2$ into $\mathfrak{T}_1$, i.e., there is an injective mapping $f:V_2\rightarrow V_1$ such that

- $\phi_1(x)=f(\phi_2(x))$ for each $x\in X$;
- $v\leq T_2 u$ iff $f(v)\leq T_1 f(u)$ for each $u,v\in V_2$,
- $h_2(v)\leq h_2(u)$ implies $h_1(f(v))\leq h_1(f(u))$ for each $u,v\in V_2$, and
- $h_2(v)=h_2(u)$ iff $h_1(f(v))=h_1(f(u))$ for each $u,v\in V_2$.

The problem is to count the number of fully ranked resolutions of a fully ranked constraint tree.

Let $\mathfrak{T} = (\mathcal{T}, h)$ be a fully ranked constraint tree with $h:V\rightarrow\{1,\ldots,l\}$ and $|\mathring{V}| = k$. Again we label interior nodes with numbers from $\{1,\ldots,k\}$. However, labels and ranks of interior nodes do not necessary coincide now. Let $r : \{1, \ldots, k\} \rightarrow h(\mathring{V})$ be an injective increasing function which maps the kth interior node to its rank. Note that r(1) is always equal to 1 because the root node always has rank 1 and that $r(k)<l$ because only leaves can have rank l. Now, node i is the node that has rank r(i) and subtree i is the subtree which is induced by node i and its children. Since leaves in subtrees are now ranked, we need more parameters to encode the tree. Let $\mathbf{n}= [n_{i,j}]$ be a matrix of size $k\times l$, where $n_{i,j}$ is the number of leaves of rank j in subtree i or the number of children of node i with rank j. Note that if node i is a parent of node j then $n_{i,r(j)}=1$ and $n_{i,x}=0$ for $x \neq r(j)$ and this means that the parent function is uniquely defined by the matrix **n** and function r and the number of resolutions of T depends only on (**n**,r).

If a constraint tree has only 2 leaves then there is only 1 tree resolving it:

$$F^r((2), r_0) = F^r((1, 1), r_0) = 1$$

with r_0 mapping 1 to 1.

For the main recursion we need to determine the location of the last interior node in a resolving tree. As before, this node can be a parent to leaves in different subtrees. Also, since leaves now may have different ranks, the last interior node in a resolving tree may be ranked in different ways with respect to ranks of the leaves. Let $N_{c,x}$ denote the number of children of node c with ranks at least x, that is, $N_{c,x} = \sum_{j=x}^{l} n_{c,j}$. For each $p \in \{r(k)+1, \ldots, l\}$, we define a set C^p of candidate subtrees in which the last node can be placed at level p, i.e. between the (p−1)th and pth time points.

$$C^p = \{c \mid N_{c,p} \geq 2 \text{ and } N_{c,r(c)+1} > 2\}$$

See Figure 7 for an example of subtree where the last node can be placed between the (p−1)th and pth time points.

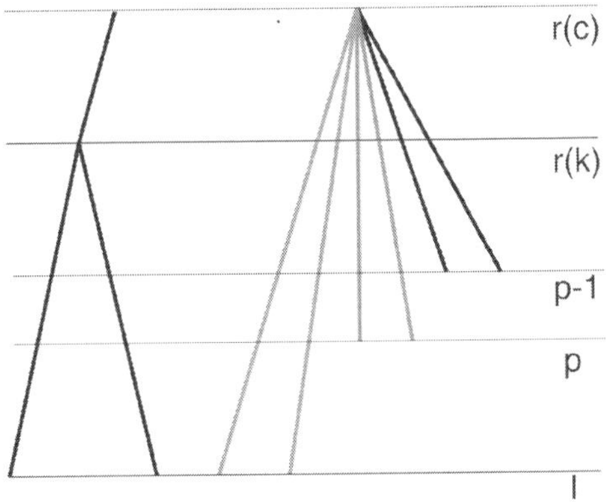

Figure 7: Location of a new node between the (p−1)-th and p-th time points. A new node can be placed in subtree c between the (p−1)-th and p-th time points if the number of green branches is greater or equal to 2 and the number of all branches in subtree c is greater than 2.

For each p andc such that $c \in C^p$, there is a distinct group of resolving trees. The quantity of trees in this group is equal to the number of resolutions of a constraint tree which is defined by matrix $\mathbf{n}^{c,p} = [n_{i,j}^{c,p}]$ of sizek×p such that

- $n_{i,j}^{c,p} = n_{i,j}$ for 1≤i≤k andj<p,
- $n_{c,p}^{c,p} = N_{c,p} - 1$,
- $n_{i,p}^{c,p} = N_{i,p}$ for i≤k and i≠c.

If node k has only 2 children then, as before, there is one more group of resolving trees. This group consists of the resolving trees in which the last interior node coincides with node k. Let $\mathbf{n}' = [n'_{i,j}]$ be a matrix of size (k−1)×r(k) such that

- $n'_{i,j} = n_{i,j}$ for 1≤i≤k and j<r(k),
- $n'_{i,r(k)} = N_{i,r(k)}$ for 1≤i≤k−1.

This matrix defines a constraint tree for which the number of resolutions is equal to the number of trees in the last group.

Then the main recursion is as follows:

$$F^r(\mathbf{n}, r) = \sum_{p=r(k)+1}^{l} \sum_{c \in C^p} \binom{N_{c,p}}{2} F^r(\mathbf{n}^{c,p}, r) + F^r(\mathbf{n}', r|_{\{1,\ldots,k-1\}}) \text{ if } N_{k,r(k)+1} = 2; \tag{9}$$

$$F^r(\mathbf{n}, r) = \sum_{p=r(k)+1}^{l} \sum_{c \in C^p} \binom{N_{c,p}}{2} F^r(\mathbf{n}^{c,p}, r) \quad \text{otherwise;} \tag{10}$$

Using these equations, we can calculate $F^r(\mathbf{n},r)$ for $O(m^2 n^k)$ steps, where m is the number of sampling points, i.e.,m=l−k. The calculation is described in Appendix Appendix 2: Algorithm for counting fully ranked resolutions of a fully ranked constraint tree.

CONCLUSIONS

In Table 1, we summarised the complexity (upper bounds) of the counting algorithms. We can see that counting fully ranked and FRS trees is reasonably fast, particularly when the number of sampling points is small. Counting of resolutions of a constraint tree is an expensive procedure but for small k it remains possible. In practice, k is typically small so this algorithm will be of practical use. Counting resolutions of an FRS constraint tree, using the same methods, has a greater cost and is less elegant, but its practicality and bounds on its complexity remain to be assessed.

Table 1: The complexity of algorithms

	No constraints (k=1)	**k constraints**
Contemporaneous		
sampling	O(n)	$O(n^k)$
(m=1)		
Serial sampling		

with no sampled	O(m n)	O(m ^{2} n ^{k})
ancestors		
Serial sampling		
with sampled	O(m n ^{2})	-
ancestors		

The table summaries the complexity of the counting algorithms, where n is the sample size, k is the number of constraints, and m is the number of sampling time-points. We assume that k is fixed.

These algorithms can be implemented in software for phylogenetic analysis that involves serial sampling scheme or limited prior knowledge about ancestors of particular clades for calculating tree prior distributions.

APPENDIX 1: ALGORITHM FOR COUNTING FRS TREES

Rewrite equation (2) as follows:

$$S(n_1,\ldots,n_m)=\sum_{i=0}^{n_m}A(i,n_{m-1},n_m)S(n_1,\ldots,n_{m-1}+i)$$

where $A(i,a,b)=\sum_{x=0}^{min\{a,b-i\}}\frac{R(b)}{R(i+x)}\binom{i+x}{x}\binom{a}{x}$. Then we can calculate S(n $_1$,…,n $_m$) recursively using Algorithm 3. We use the notation $N_j=\sum_{i=j}^{m}n_i$.

Algorithm 3: Calculating the number of FRS trees

for $i=1\rightarrow N_1$ **do**
  calculate $R(i)$ using $R(i)=R(i-1)\frac{i(i-1)}{2}$
end for
for $j=2\rightarrow m$ **do**
  calculate $A(0,n_{j-1},n_j),\ldots,A(n_j,n_{j-1},n_j)$
  calculate $S(n_1,\ldots,n_j)$
  for $\alpha=1\rightarrow N_{j+1}$ **do**
    calculate $A(0,n_{j-1},n_j+\alpha),\ldots,A(n_j+\alpha,n_{j-1},n_j+\alpha)$
    calculate $S(n_1,\ldots,n_j+\alpha)$
  end for
end for

At each step j>1 of the algorithm, we calculate $S(n_1,\ldots,n_j+\alpha)$ for $0\leq\alpha\leq N_{j+1}$. We skip step j=1 and do not calculate $S(n_1+\alpha)$ for $0\leq\alpha\leq N_2$ because $S(x)=R(x)$ and we have already calculated all the necessary R(x). Further, for calculating each $S(n_1,\ldots,n_j+\alpha)$, we need to calculate the coefficients A and this is the most expensive part of the algorithm. So, at each step $j>1$, we need to calculate $A(i,n_{j-1},n_j+\alpha)$ for $0\leq i\leq n_j+\alpha$ and $0\leq\alpha\leq N_{j+1}$. First, we calculate these values for $\alpha=0$. Denote

$$B(i,a,b,x)=\frac{R(b)}{R(i+x)}\binom{i+x}{x}\binom{a}{x}, \text{for } x\leq a \text{ and } i+x\leq b$$

then

$$A(i,a,b)=\sum_{x=0}^{\min\{a,b-i\}} B(i,a,b,x)$$

where $0\leq i\leq b, a=n_{j-1}$, and $b=n_j$. When i is fixed, calculation of $A(i,n_{j-1},n_j)$ requires values of $B(i,n_{j-1},n_j,x)$ for $0\leq x\leq \min\{n_{j-1},n_j-i\}$. Rewrite this as $B(n_j-\beta,n_{j-1},n_j,x)$ for $0\leq x\leq \min\{n_{j-1},\beta\}$ with $\beta=n_j-i$. Now we can calculate these values for $0\leq\beta\leq n_j$ and, hence, for $0\leq i\leq n_j$ using recursive equations:

$$B(n_j,n_{j-1},n_j,0)=1$$

$$B(n_j-(\beta+1),n_{j-1},n_j,x) = \frac{(n_j-(\beta+1)+x)(n_j-\beta)B(n_j-\beta,n_{j-1},n_j,x)}{2} \tag{11}$$

$$B(n_j-(\beta+1),n_{j-1},n_j,(\beta+1)) = \frac{(n_j-\beta)(n_{j-1}-\beta)B(n_j-\beta,n_{j-1},n_j,\beta)}{(\beta+1)^2} \tag{12}$$

We use equation (11) for $0\leq x\leq \min\{n_{j-1},\beta\}$. If $\min\{n_{j-1},\beta+1\}\neq\min\{n_{j-1},\beta\}$ and therefore $\min\{n_{j-1},\beta+1\}=\min\{n_{j-1},\beta\}+1=\beta+1$ then we also use equation (12). The cost of this calculation is dominated by the number of the summands $B(n_j-\beta,n_{j-1},n_j,x)$ for $0\leq x\leq \min\{n_{j-1},\beta\}$ and $0\leq\beta\leq n_j$. This number is $O(n_j^2)$.

Now we need to calculate $A(0,n_{j-1},n_j+\alpha),\ldots,A(n_j+\alpha,n_{j-1},n_j+\alpha)$ for $1\leq\alpha\leq N_{j+1}$. Having $A(0,n_{j-1},n_j+\alpha),\ldots,A(n_j+\alpha,n_{j-1},n_j+\alpha)$ calculated (note that we have already calculated these values for $\alpha=0$), we can calculate $A(0,n_{j-1},n_j+(\alpha+1)),\ldots,A(n_j+(\alpha+1),n_{j-1},n_j+(\alpha+1))$ using equations

$$A(i,a,b+1)=\begin{cases}\frac{(b+1)b}{2}A(i,a,b)+\binom{b+1}{i}\binom{a}{b+1-i} & \text{if } b-i<a,\\ \frac{(b+1)b}{2}A(i,a,b) & \text{if } a\leq b-i \text{ and}\end{cases}$$
$$A(b+1,a,b+1)=1 \tag{13}$$

where $0\leq i\leq b, a=n_{j-1}$, and $b=n_j+\alpha$.
Moreover, for each α, we can optimize calculation of the second summands in the first case of equation (13) using recursion

$$\binom{n_j+\alpha}{i+1}\binom{n_{j-1}}{n_j+\alpha-(i+1)}$$
$$=\frac{(n_j+\alpha-i)^2}{(i+1)(n_{j-1}-n_j-\alpha+i+1)}\binom{n_j+\alpha}{i}\binom{n_{j-1}}{n_j+\alpha-i} \tag{14}$$

To apply this recursion we need an initial value. This value depends on n_{j-1}, n_j, and α and, since only the first case of equation (13) contains the second summand, it is not necessary the value of this summand for i=0. There are 3 cases for calculation of all the initial values that are necessary at step j.

Case 1: $N_j<n_{j-1}$. That means that we use the first case of equation (13) for all α. In this case, for all α, the initial value for recursion (14) is the value of the second summand in (13) when i=0. So we need to calculate $\binom{n_j+\alpha}{0}\binom{n_{j-1}}{n_j+\alpha+0}$ for $1\leq\alpha\leq N_{j+1}$ and those are $\binom{n_{j-1}}{n_j+1},\ldots,\binom{n_{j-1}}{n_j+N_{j+1}}$. This takes $O(N_j)$ steps.

Case 2: $n_j+\alpha_0=n_{j-1}$ for some $\alpha_0\in\{0,\ldots,N_{j+1}\}$. If $\alpha_0>0$ then, for all $\alpha\leq\alpha_0$, we use the first case of (13) and therefore we need to calculate $\binom{n_{j-1}}{n_j+1},\ldots,\binom{n_{j-1}}{n_j+\alpha_0}$, this takes $O(\alpha_0)$ steps. For all $\alpha>\alpha_0$, we use the second case first and, starting from $i=\alpha-\alpha_0$, we use only the first case. So we need to calculate $\binom{n_j+\alpha}{\alpha-\alpha_0}\binom{n_{j-1}}{n_j+\alpha-(\alpha-\alpha_0)}$ for $\alpha\in\{\alpha_0+1,\ldots,N_{j+1}\}$, which are $\binom{n_{j-1}+1}{1},\ldots,\binom{n_{j-1}+(N_{j+1}-\alpha_0)}{(N_{j+1}-\alpha_0)}$. This takes $O(N_{j+1}-\alpha_0)$ steps. In total, we have $O(N_{j+1})$ steps.

Case 3: $n_{j-1}<n_j$. That means that, for all α, we use the second case first and, starting from $i=n_j+\alpha-n_{j-1}$, we use only the first case. Calculate $\binom{n_j+1}{n_{j-1}},\ldots,\binom{n_j+N_{j+1}}{n_{j-1}}$ for $O(n_{j-1}+N_{j+1})$ steps.

Each case costs at most $O(N_j)$. Provided that the initial values are calculated, the cost of the rest calculation at step j is dominated by the number of the coefficients A ($i,n_{j-1},n_j + a$) for $0 \le i \le n_j + a$ and $1 \le a \le N_{j+1}$ and it is $O(N_j^2)$. Summing up $O(n_j^2)$ X, O (N_{j-1}), and $O(N_j^2)$ gives us the cost of each step j, which is $O(N_j^2)$. Since $1 < j \le m$ and calculation of R (i) for $0 \le i \le N_1 = n$ takes $O(n)$, the algorithm does $O(mn^2)$ steps in total.

APPENDIX 2: ALGORITHM FOR COUNTING FULLY RANKED RESOLUTIONS OF A FULLY RANKED CONSTRAINT TREE

The algorithm for counting resolutions of a constraint tree requires a few changes to count resolutions of a fully ranked constraint tree. Recall that, at each step s, we calculated the set S_s which consists of the numbers of resolutions of intermediate trees with s leaves. To construct S_s, for each element of S_{s-1}, we proposed a collection of tuples that may define intermediate trees with s leaves. To accept eligible tuples we defined two matrices m and M. The general scheme for counting resolutions of a fully ranked constraint tree is the same. However, we need to define matrices m and M for a fully ranked constraint tree and describe the procedure of proposing new tuples. The procedure becomes more technical because of additional ranking.

We will calculate S^r (**n**,r) for the corresponding fully ranked constraint tree T. Let

$$a_{i,j} = |\{x \mid \text{ node } i \text{ is a parent of node } x \text{ and } r(x) \le j\}|$$

for $1 \le i \le k$ and $r(i) \le j \le l-1$, i.e., $a_{i,j}$ is the number of interior nodes that are children of node i and have rank at most j. Define two matrices m and M of size $k \times (l-1)$ such that

$$\mathrm{M}_{i,j} = N_{i,j+1} \quad \text{and} \quad \mathrm{m}_{i,j} = \begin{cases} 2 \text{ if } a_{i,j} = 0; \\ 1 \text{ if } a_{i,j} > 0 \text{ and } \mathrm{M}_{i,j} > 0; \\ 0 \text{ otherwise.} \end{cases}$$

As before, if we consider a horizontal line that is strictly between the horizontal line that passes through the nodes of rank j and the horizontal line that passes through the nodes of rank (j+1) then these matrices determine the minimal and maximal possible numbers of intersections of this line with branches of subtree i.

Let $\mathbf{x}= [x_{i,j}]$ be a matrix of size $t \times q$, where $t \leq k$ and $q \leq l$, and $\mathbf{t}=\{1,\ldots,t\}$. A tuple $(\mathbf{x},r\,|_{\mathbf{t}})$ is eligible if

- $x_{i,j}$x=$n_{i,j}$ for $1 \leq i \leq t$ and $1 \leq j < q$, and
- $\mathfrak{m}_{i,q-1} \leq x_q^i \leq \mathrm{M}_{i,q-1}$ for $1 \leq i < t$.

Having an intermediate tree with$s-1$ leaves, we need to consider all the possible ways to transform this tree to a tree with s leaves. First we need to add a new leaf to some subtree. We give the highest rank to this leaf. After that we may add new time points below the last time point. If we add new timc points, we should rerank the leaves with the highest rank such that the new tree has enough leaves at each time point. Let $\mathbf{x}= [x_{i,j}]$ be a matrix of size $t \times q$ for $t \leq k$ and $q \leq l$. This matrix represents an intermediate tree with$s-1$ leaves. Suppose we add a new leaf to subtree i_0, where $1 \leq i_0 \leq t$. Let q_0 be the number of time points in a new tree and, therefore, $q \leq q_0 \leq r(t+1)$ (or $q \leq q_0 \leq l$ if $t=k$). Define a function a***ddone***$((\mathbf{x},r|_{\mathbf{t}}),i_0, q_0)$ which returns a tuple $(x',r|_t)$, where $\mathbf{x}' = [x'_{i,j}]$ is a matrix of size $t \times q_0$ such that

- $x'_{i_0,q_0} = x_{i_0,q} - \sum_{j=q}^{q_0-1} n_{i_0,j} + 1,$
- $x'_{i,q_0} = x_{i,q} - \sum_{j=q}^{q_0-1} n_{i,j}$ for $i \neq i_0$,
- $x'_{i,j} = x_{i,j}$ for $1 \leq i \leq t$ and $j < q$, and
- $x'_{i,j} = n_{i,j}$ for $1 \leq i \leq t$ and $q \leq j < q_0$.

If this tuple is eligible then it represents a new intermediate tree with s leaves.

At each steps of the algorithm, for each tuple $(\mathbf{x},r\,|_{\mathbf{t}})$ such that $F^r(\mathbf{x},r\,|_{\mathbf{t}}) \in S_{s-1}$, we need to propose a collection of new tuples. The first part of this procedure is Algorithm 4. To stop the procedure, when we recognise that a new tree can not have x time points, we use a predicatee x t(x) which is true if we can extend the number of time points in a new tree to x. The number of time points in a new tree can not always be extended to any arbitrary number because there may not be enough leaves of the highest rank to rerank them in such a way that there will be $n_{i,j}$ leaves of each new rank j. So e x t(q) is always true because the tree to which we add a new leaf already has q time points.

Algorithm 4: Working with a tuple $(x,r|_t)$ such that $F^r(x,r|_t) \in S_{s-1}$ (part 1)

$q_0 = q$
while $q_0 \le r(t+1)$ (or $q_0 \le l$ if $t = k$) and $ext(q_0)$ **do**
  for $i = 1 \to t$ **do**
    $(\mathbf{x}', r|_\mathbf{t}) = addone((\mathbf{x}, r|_\mathbf{t}), i, q_0)$
    if $eligible((\mathbf{x}', r|_\mathbf{t}))$ **then**
      calculate $F^r(\mathbf{x}', r|_\mathbf{t})$ and add it to S_s
    end if
    if all the proposed tuples were not eligible **then**
      $ext(q_0 + 1) = false$
    end if
    $q_0 = q_0 + 1$
  end for
end while

If t < k then the proposing procedure contains the second part. In this case, we also can increase the number of leaves in an intermediate tree by adding a new subtree. That means that one of the leaves of the tree becomes a parent to two new leaves. Again we may add new time points. Let q_0 be the number of time points in a new tree and, therefore, $r(t+1)<q_0\le r(t+2)$ (or $r(t+1)<q_{x0}\le l$ when $t+1=k$). We define another function ***add constraint***$((x,r|_t),q_0)$ which returns a tuple $(x',r|t+1)$, where $x' = [x'_{i,j}]$ is a matrix of size $(t+1)\times q_0$ such that

- $x'_{t+1,q_0} = 2 - \sum_{j=r(t+1)+1}^{q_0-1} n_{t+1,j}$,
- $x'_{t+1,j} = n'_{t+1,j}$ for $1 \le j < q_0$,
- $x'_{i,q_0} = x_{i,q} - \sum_{j=q}^{q_0-1} n_{i,j}$ for $1 \le i \le t$,
- $x'_{i,j} = x'_{i,j}$ for $1 \le i \le t$ and $j < q$, and
- $x'_{i,j} = n'_{i,j}$ for $1 \le i \le t$ and $q \le j < q_0$.

This tuple defines a new tree if it is eligible. The second part of the procedure is Algorithm 5. The values of q_0 and $ext(q_0)$ are as after running Algorithm 4.

Algorithm 5: Working with a tuple $(x,r|_t)$ such that $F^r(x,r|_t) \in S_{s-1}$ (part 2)

```
while q0 ≤ r(t + 2) (or q0 ≤ l when t + 1 = k) and
ext(q0) do
   (x', r|t+1) = addconstraint((x, r|t), q0)
   if eligible(x', r|t+1) then
      calculate F^r(x', r|t+1) and add it to S_s
   else
      ext(q0 + 1) = false
   end if
   q0 = q0 + 1
end while
```

Finally, as before, the main algorithm calculates sets S_s recursively for $0 \leq s \leq n$, where n is the number of leaves in a constraint tree T, and a unique element of S_n is$S^r(\mathbf{n},r)$.

Let $\boldsymbol{m}$ be a number of sampling points, i.e., $\boldsymbol{m}=l-k$.

Proposition 2

If k is fixed then the algorithm does at most O(m) steps for each eligible tuple.

Proof

Let $(\mathbf{x}\ \boldsymbol{r}\,|_t)$ be an eligible tuple. First, having all the necessary summands for equations (9) and (10), we need to calculate $\boldsymbol{F}^r(\mathbf{x},\boldsymbol{r}\,|_t)$. This takes at most $\boldsymbol{O}(t\times[\,r(t+1)-r(t)])$ steps. Since we assume $\boldsymbol{k}$ is a constant and since $t\leq k$ and $r(t+1)-r(t)\leq m$, it is $\boldsymbol{O}(\boldsymbol{m})$. Second, we need to perform the procedure of proposing new tuples for this tuple. The most expensive part here is calculation of the last columns of matrices x′ because it involves calculation of the sums ∑q0−1j=qni,j. Note that, storing the values of $\sum_{j=q}^{q_0-1} n_{i,j}$ for $1\leq i \leq t$ at each step q_0, we can optimise these calculations. Then it takes $O(t\times(q-r(t+2))$ steps and it is $\boldsymbol{O}(\boldsymbol{m})$ again.

Proposition 3

If k is fixed then the algorithm has $O({}^m n^k)$ time complexity.

Proof

Leti_x be such a number that$r(i_x)\leq x < r(i_x+1)$. Then the number of eligible tuples is

$$\sum_{j=1}^{l-1}\prod_{i=1}^{i_j}(\mathrm{M}_{i,j}-\mathrm{m}_{i,j}+1)$$

Note that

$$\mathrm{M}_{i,j}-\mathrm{m}_{i,j}+1\leq N_{i,j}+1\leq N_{i,r(i)+1}+1=b_i+1$$

2where$\boldsymbol{b}_1+\ldots+\boldsymbol{b}_k=\boldsymbol{n}+\boldsymbol{k}-1$. Then

$$\sum_{j=1}^{l-1}\prod_{i=1}^{i_j}(\mathrm{M}_{i,j}-\mathrm{m}_{i,j}+1)<(l-1)(b_1+1)\ldots(b_k+1)$$

$$<(l-1)(\frac{n}{k})^k=O(mn^k)$$

From this and Proposition 2, it follows that the algorithm does at mostO($\mathrm{m}^2\boldsymbol{n}^k$) steps.

ACKNOWLEDGEMENTS

AJD was partially funded by a Rutherford Discovery Fellowship from the Royal Society of New Zealand.

AUTHORS' CONTRIBUTIONS

The authors equally contributed to conceive the work. AG drafted the manuscript and DW and AJD revise it. All authors read and approved the final manuscript.

REFERENCES

1. Yang Z, Rannala B: Bayesian phylogenetic inference using DNA sequences: a Markov chain Monte Carlo method. Mol Biol Evol. 1997, 14 (7): 717-724. 10.1093/oxfordjournals.molbev.a025811.
2. Yang Z, Rannala B: Bayesian estimation of species divergence times under a molecular clock using multiple fossil calibrations with soft bounds. Mol Biol Evol. 2005, 23: 212-226. 10.1093/molbev/msj024.
3. Heled J, Drummond A: Tree priors for relaxed phylogenetics and divergence time estimation. Syst Biol. 2012, 61: 138-149. 10.1093/sysbio/syr087.
4. Rodrigo AG, Felsenstein J: The Evolution of HIV. 1999, Baltimore: Johns Hopkins Univ Press
5. Drummond A, Pybus O, Rambaut A, Forsberg R, Rodrigo A: Measurably evolving populations. Trends Ecol Evol. 2003, 18: 481-488. 10.1016/S0169-5347(03)00216-7.
6. Felsenstein J: Inferring Phylogenies. 2004, Sunderland: Sinauer Associates
7. Murtagh F: Counting dendograms: a survey. Discrete Appl Math. 1984, 7: 191-199. 10.1016/0166-218X(84)90066-0.
8. Gordon AD: A review of hierarchical classification. J R Stat Soc A. 1987, 150 (2): 119-137. 10.2307/2981629.
9. Semple C, Steel M: Phylogenetics. 2003, New York: Oxford University Press
10. Griffiths RC: Counting genealogical trees. J Math Biol. 1987, 25: 423-431. 10.1007/BF00277166.
11. Felsenstein J: The number of evolutionary trees. Syst Zool. 1978, 23: 27-33.
12. Stadler T: Sampling-through-time in birth-death trees. J Theor Biol. 2010, 267 (3): 396-404. 10.1016/j.jtbi.2010.09.010.
13. Stadler T, Kouyos RD, von Wyl, Yerly S, Böni J, Bürgisser P, Klimkait T, Joos B, Rieder P, Xie D, Günthard HF, Drummond A, Bonhoeffer S, the Swiss HIV Cohort Study: Estimating the basic reproductive number from viral sequence data. Mol Biol Evol. 2012, 29: 347-357. 10.1093/molbev/msr217.

SECTION 3:
DERANGEMENTS AND THE EULER'S TOTIENT

8

A Note on Some Identities of Derangement Polynomials

Taekyun Kim[1,2] , Dae San Kim[3] , Gwan-Woo Jang[2] and Jongkyum Kwon[4]

[1]Department of Mathematics, College of Science, Tianjin Polytechnic University, Tianjin, China

[2]Department of Mathematics, Kwangwoon University, Seoul, Republic of Korea

[3]Department of Mathematics, Sogang University, Seoul, Republic of Korea

[4]Department of Mathematics Education and ERI, Gyeongsang National University, Jinju, Republic of Korea.

ABSTRACT

The problem of counting derangements was initiated by Pierre Rémond de Montmort in 1708 (see Carlitz in Fibonacci Q. 16(3):255–258, 1978, Clarke and Sved in Math. Mag. 66(5):299–303, 1993, Kim, Kim and Kwon in Adv. Stud. Contemp. Math. (Kyungshang) 28(1):1–11 2018. A derangement is a permutation that has no fixed points, and the derangement number d_n is the

Citation: Taekyun Kim, Dae San Kim, Gwan-Woo Jang and Jongkyum Kwon "A note on some identities of derangement polynomials" Journal of Inequalities and Applications 2018 2018:40. https://doi.org/10.1186/s13660-018-1636-8

number of fixed-point-free permutations on an n element set. In this paper, we study the derangement polynomials and investigate some interesting properties which are related to derangement numbers. Also, we study two generalizations of derangement polynomials, namely higher-order and r-derangement polynomials, and show some relations between them. In addition, we express several special polynomials in terms of the higher-order derangement polynomials by using umbral calculus.

Keywords: Derangement numbers, Derangement polynomials, r-derangement numbers, r-derangement polynomials, Umbral calculus

INTRODUCTION

Let $\mathbb{C}$ be the complex number field, and let $\mathcal{F}$ be the set of all formal power series in the variable t with coefficients in $\mathbb{C}$:

$$\mathcal{F} = \left\{ f(t) = \sum_{k=0}^{\infty} a_k \frac{t^k}{k!} \;\middle|\; a_k \in \mathbb{C} \right\}. \tag{1.1}$$

Let $\mathbb{P} = \mathbb{C}[x]$, and let $\mathbb{P}^*$ be the vector space of all linear functionals on $\mathbb{P}$. We denote the action of a linear functional L ∈ $\mathbb{P}^*$ on polynomials p(x) ∈ $\mathbb{P}$ by <L | p(x)>, and it is known that vector space operations on $\mathbb{P}^*$ are defined by

$$\langle L + M \mid p(x) \rangle = \langle L \mid p(x) \rangle + \langle M \mid p(x) \rangle, \qquad \langle cL \mid p(x) \rangle = c\langle L \mid p(x) \rangle, \tag{1.2}$$

where c is a complex constant (see [3–5]).

For $f(t) = \sum_{k=0}^{\infty} a_k \frac{t^k}{k!}$, we define a linear functional on $\mathbb{P}$ by setting

$$\langle f(t) \mid x^n \rangle = a_n \quad (n \geq 0) \text{ (see [6, 7])}. \tag{1.3}$$

From (1.3), we note that

$$\langle t^k \mid x^n \rangle = n!\delta_{n,k} \quad (n, k \geq 0) \text{ (see [8])}, \tag{1.4}$$

where $\delta_{n,k}$ is the Kronecker symbol.

The order o(f (t)) of a power series f (t)(≠ 0) ∈ F is the smallest integer k such that the coefficients of t^k do not vanish. For f (t), g(t) ∈ F, with o(f (t)) = 1 and o(g(t)) = 0, there exists a unique sequence $S_n(x)$ of polynomials such that $\langle g(t)f(t)^k | S_n(x) \rangle = n!\delta_{n,k}$ for n, k ≥ 0 (see [5, 8]). The sequence $S_n(x)$ is called the Sheffer sequence for (g(t), f (t)), which is denoted by $S_n(x) \sim (g(t), f(t))$. It is known that $S_n(x) \sim (g(t), f(t))$ if and only if

$$\frac{1}{g(\bar{f}(t))}e^{x\bar{f}(t)} = \sum_{n=0}^{\infty} S_n(x)\frac{t^n}{n!}, \tag{1.5}$$

where $\bar{f}(t)$ is the compositional inverse of $\bar{f}(t)$ with

$$f(\bar{f}(t)) = \bar{f}(f(t)) = t \quad \text{(see [8, 9])}. \tag{1.6}$$

For $f(t) \in F$ and $p(x) \in \mathbb{P}$, by (1.4), we get

$$f(t) = \sum_{k=0}^{\infty} \langle f(t)|x^k\rangle \frac{t^k}{k!}, \qquad p(x) = \sum_{k=0}^{\infty} \langle t^k|p(x)\rangle \frac{x^k}{k!} \quad \text{(see [9])}. \tag{1.7}$$

From (1.7), we note that

$$p^{(k)}(0) = \langle t^k|p(x)\rangle = \langle 1|p^{(k)}(x)\rangle \quad (k \geq 0), \tag{1.8}$$

where $p^{(k)}(x) = (\frac{d}{dx})^k p(x)$.

From (1.8), we easily get

$$t^k p(x) = p^{(k)}(x), \qquad e^{yt}p(x) = p(x+y), \qquad \langle e^{yt}|p(x)\rangle = p(y) \quad \text{(see [9])}. \tag{1.9}$$

Let $S_n(x) \sim (g(t), f(t))$ and $r_n(x) \sim (h(t), l(t))$ $(n \geq 0)$. Then we have

$$S_n(x) = \sum_{m=0}^{n} C_{n,m} r_m(x) \quad (n \geq 0) \text{ (see [8, 9])}, \tag{1.10}$$

where

$$C_{n,m} = \frac{1}{m!}\left\langle \frac{h(\bar{f}(t))}{g(\bar{f}(t))} l(\bar{f}(t))^m \middle| x^n \right\rangle \quad (n, m \geq 0). \tag{1.11}$$

For $u(\neq 1) \in \mathbb{C}$, the Frobenius-Euler numbers are defined by the generating function

$$\frac{1-u}{e^t - u} = \sum_{n=0}^{\infty} H_n(u)\frac{t^n}{n!} \quad \text{(see [10–12])}. \tag{1.12}$$

When $u = -1$, $H_n(-1) = E_n$ are the ordinary Euler numbers.

The Bernoulli polynomials are given by

$$\frac{t}{e^t - 1}e^{xt} = \sum_{n=0}^{\infty} B_n(x)\frac{t^n}{n!} \quad \text{(see [3, 12, 13])}. \tag{1.13}$$

When $x = 0$, $B_n = B_n(0)$ are the Bernoulli numbers.

We know that the Euler polynomials are defined by

$$\frac{2}{e^t + 1}e^{xt} = \sum_{n=0}^{\infty} E_n(x)\frac{t^n}{n!} \quad \text{(see [10, 11])}. \tag{1.14}$$

When $x = 0$, $E_n = E_n(0)$ are the Euler numbers.

The falling factorial sequence is defined as

$$(x)_0 = 1, (x)_n = x(x-1)\cdots(x-n+1) \quad (n \geq 1). \tag{1.15}$$

The Stirling numbers of the first kind are defined by

$$(x)_n = \sum_{l=0}^{n} S_1(n,l)x^l \quad (n \geq 0) \text{ (see [8])}, \tag{1.16}$$

and the Stirling numbers of the second kind are given by

$$x^n = \sum_{l=0}^{n} S_2(n,l)(x)_l \quad (n \geq 0) \text{ (see [8, 14, 15])}. \tag{1.17}$$

The Stirling numbers of the second kind are also given by the exponential generating function (see [8, p.59])

$$\frac{1}{k!}(e^t - 1)^k = \sum_{n=k}^{\infty} S_2(n,k)\frac{t^n}{n!}. \tag{1.18}$$

It is well known that the Bell polynomials are defined by the generating function

$$e^{x(e^t-1)} = \sum_{n=0}^{\infty} \mathrm{Bel}_n(x)\frac{t^n}{n!} \quad \text{(see [9])}. \tag{1.19}$$

When x = 1, $\mathrm{Bel}_n = \mathrm{Bel}_n(1)$ $(n \geq 0)$ are the Bell numbers.

From (1.19), we have

$$\mathrm{Bel}_n(x) = \sum_{k=0}^{n} S_2(n,k)x^k \quad (n \geq 0) \text{ (see [9])}. \tag{1.20}$$

A derangement is a permutation that has no fixed points. The derangement number d_n is the number of fixed-point-free permutations on an n element set (see [1–3]). The problem of counting derangements was initiated by Pierre Rémond de Montmort in 1708 (see [1–3]). The first few terms of the derangement number sequence $\{d_n\}_{n=0}^{\infty}$ are $d_0 = 1$, $d_1 = 0$, $d_2 = 1$, $d_3 = 2$, $d_4 = 9$, $d_5 = 44$, $d_6 = 265$, $d_7 = 1854$, ...

Indeed, d_n is given by the closed form formula:

$$d_n = n!\sum_{k=0}^{n} \frac{(-1)^k}{k!} \quad \text{(see [3])}. \tag{1.21}$$

From (1.21), we note that the generating function of derangement numbers is given by

$$\frac{1}{1-t}e^{-t} = \sum_{n=0}^{\infty} d_n \frac{t^n}{n!} \quad \text{(see [9]).} \tag{1.22}$$

By using (1.22), it is not difficult to show that

$$d_m = (m-1)(d_{m-1} + d_{m-2}) \quad (m \geq 2), \qquad d_0 = 1, \qquad d_1 = 0, \tag{1.23}$$

and

$$d_m = md_{m-1} + (-1)^m \quad (m \geq 1), \qquad d_0 = 1 \quad \text{(see [1–3]).} \tag{1.24}$$

Forr ∈ $\mathbb{N}$, the derangement numbers $d_n^{(r)}$ of orderr (n ≥ 0), are defined by the generating function

$$\left(\frac{1}{1-t}\right)^r e^{-t} = \sum_{n=0}^{\infty} d_n^{(r)} \frac{t^n}{n!} \quad \text{(see [3]).} \tag{1.25}$$

The umbral calculus comes under the heading of combinatorics, the calculus of finite differences, the theory of special functions, and formal solutions to differential equations. Also, formal power series play a predominant role in the umbral calculus. In this paper, we study the derangement polynomials and investigate some interesting properties which are related to derangement numbers. Further, we study two generalizations of derangement polynomials, namely higher-order and r-derangement polynomials, and show some relations between them. In addition, we express several special polynomials in terms of the higher-order derangement polynomials by using umbral calculus.

SOME IDENTITIES OF DERANGEMENT POLYNOMIALS ARISING FROM UMBRAL CALCULUS

Now, we define the derangement polynomials by

$$\sum_{n=0}^{\infty} d_n(x) \frac{t^n}{n!} = \frac{1}{1-t} e^{xt}. \tag{2.1}$$

We note here that, for x = –1, $d_n = d_n(-1)$ are the derangement numbers.

We observe that

$$\frac{1}{1-t}e^{xt} = e^{-\log(1-t)}e^{xt} = \left(\sum_{m=0}^{\infty}(-1)^m \frac{1}{m!}(\log(1-t))^m\right)\left(\sum_{l=0}^{\infty}x^l \frac{t^l}{l!}\right)$$

$$= \left(\sum_{k=0}^{\infty}\left(\sum_{m=0}^{k}(-1)^{k-m}S_1(k,m)\right)\frac{t^k}{k!}\right)\left(\sum_{l=0}^{\infty}x^l \frac{t^l}{l!}\right)$$

$$= \sum_{n=0}^{\infty}\left(\sum_{k=0}^{n}\sum_{m=0}^{k}\binom{n}{k}(-1)^{k-m}S_1(k,m)x^{n-k}\right)\frac{t^n}{n!}. \tag{2.2}$$

By (2.1) and (2.2), we get

$$d_n(x) = \sum_{k=0}^{n}\sum_{m=0}^{k}\binom{n}{k}(-1)^{k-m}S_1(k,m)x^{n-k} \quad (n \geq 0), \tag{2.3}$$

and

$$d_n = \sum_{k=0}^{n}\sum_{m=0}^{k}\binom{n}{k}(-1)^{n-m}S_1(k,m) \quad (n \geq 0). \tag{2.4}$$

Therefore we obtain the following lemma.

Lemma 2.1 For n ≥ 0, we have

$$d_n(x) = \sum_{k=0}^{n}\sum_{m=0}^{k}\binom{n}{k}(-1)^{k-m}S_1(k,m)x^{n-k}$$

and

$$d_n = \sum_{k=0}^{n}\sum_{m=0}^{k}\binom{n}{k}(-1)^{n-m}S_1(k,m).$$

From (2.1), we have

$$\sum_{n=0}^{\infty}d_n(x)\frac{t^n}{n!} = \left(\frac{1}{1-t}e^{-t}\right)e^{(x+1)t}$$

$$= \left(\sum_{m=0}^{\infty}d_m\frac{t^m}{m!}\right)\left(\sum_{l=0}^{\infty}(x+1)^l\frac{t^l}{l!}\right)$$

$$= \sum_{n=0}^{\infty}\left(\sum_{m=0}^{n}\binom{n}{m}d_m(x+1)^{n-m}\right)\frac{t^n}{n!}. \tag{2.5}$$

Therefore, we obtain the following proposition.

Proposition 2.2 For $n \geq 0$, we have

$$d_n(x) = \sum_{m=0}^{n} \binom{n}{m} d_m (x+1)^{n-m} = (d+x+1)^n,$$

with the usual convention about replacing d_n by d_n.

From Proposition 2.2, we have

$$\frac{d}{dx} d_n(x) = \frac{d}{dx}(d+x+1)^n = n(d+x+1)^{n-1} = nd_{n-1}(x) \quad (n \geq 1). \tag{2.6}$$

By (1.5) and (2.1), we get

$$d_n(x) \sim (1-t, t). \tag{2.7}$$

That is, $d_n(x)$ $(n \geq 0)$ is an Appell sequence.

Now, we note that

$$\begin{aligned} \sum_{n=0}^{\infty} d_n(x) \frac{t^n}{n!} &= \left(\frac{1}{1-t} e^t \right) e^{(x-1)t} \\ &= \sum_{n=0}^{\infty} \left(\sum_{l=0}^{n} \binom{n}{l} a_l (x-1)^{n-l} \right) \frac{t^n}{n!}, \end{aligned} \tag{2.8}$$

where a_n are the arrangement numbers defined by

$$\frac{1}{1-t} e^t = \sum_{n=0}^{\infty} a_n \frac{t^n}{n!}. \tag{2.9}$$

Replacing t by $e^t - 1$ in (2.1), we get

$$\begin{aligned} \sum_{k=0}^{\infty} d_k(x) \frac{1}{k!} (e^t - 1)^k &= \left(\frac{1}{2-e^t} \right) e^{x(e^t - 1)} \\ &= \left(\sum_{l=0}^{\infty} H_l(2) \frac{t^l}{l!} \right) \left(\sum_{m=0}^{\infty} \mathrm{Bel}_m(x) \frac{t^m}{m!} \right) \\ &= \sum_{n=0}^{\infty} \left(\sum_{l=0}^{n} \binom{n}{l} H_l(2) \mathrm{Bel}_{n-l}(x) \right) \frac{t^n}{n!}. \end{aligned} \tag{2.10}$$

On the other hand,

$$\begin{aligned} \sum_{k=0}^{\infty} d_k(x) \frac{1}{k!} (e^t - 1)^k &= \sum_{k=0}^{\infty} d_k(x) \sum_{n=k}^{\infty} S_2(n,k) \frac{t^n}{n!} \\ &= \sum_{n=0}^{\infty} \left(\sum_{k=0}^{n} S_2(n,k) d_k(x) \right) \frac{t^n}{n!}. \end{aligned} \tag{2.11}$$

Therefore, by (2.10) and (2.11), we obtain the following theorem.

Theorem 2.3 For n ≥ 0, we have

$$\sum_{k=0}^{n}\binom{n}{k}H_k(2)\mathrm{Bel}_{n-k}(x) = \sum_{k=0}^{n} S_2(n,k)d_k(x).$$

For $S_n(x) \sim (g(t),t)$, from (1.5) we have

$$\frac{1}{g(t)}e^{xt} = \sum_{n=0}^{\infty} S_n(x)\frac{t^n}{n!}. \tag{2.12}$$

Thus, by (2.12), we get

$$\frac{1}{g(t)}x^n = S_n(x) \quad (n \geq 0) \quad \Longleftrightarrow \quad S_n(x) \sim \big(g(t), t\big). \tag{2.13}$$

In (2.13), we take g(t)=1− t, then we have

$$\frac{1}{1-t}x^n = d_n(x) \quad (n \geq 0), \qquad t\, d_n(x) = n\, d_{n-1}(x) \quad (n \geq 1). \tag{2.14}$$

Now, we observe that

$$\begin{aligned} d_n(x+y) &= (d+x+y+1)^n = \sum_{l=0}^{n}\binom{n}{l}(d+x+1)^l y^{n-l} \\ &= \sum_{l=0}^{n}\binom{n}{l}d_l(x)y^{n-l} \quad (n \geq 0). \end{aligned} \tag{2.15}$$

From (2.15), we note that

$$\begin{aligned} \frac{1}{n+1}\big(d_{n+1}(x+y) - d_{n+1}(x)\big) &= \frac{1}{n+1}\sum_{k=1}^{n+1}\binom{n+1}{k}d_{n+1-k}(x)y^k \\ &= \sum_{k=1}^{n+1}\frac{n(n-1)\cdots(n-k+2)}{k!}d_{n+1-k}y^k \\ &= \sum_{k=1}^{n+1}\frac{y^k}{k!}t^{k-1}d_n(x). \end{aligned} \tag{2.16}$$

By (2.15) and (2.16), we get

$$\int_x^{x+y} d_n(u)\,du = \sum_{k=1}^{n+1} \frac{y^k}{k!} t^{k-1} d_n(x)$$
$$= \frac{1}{t}\sum_{k=1}^{n} \binom{n}{k} d_{n-k}(x) y^k = \frac{1}{t}\left(e^{yt} d_n(x) - d_n(x)\right)$$
$$= \frac{1}{t}\left(e^{yt} - 1\right) d_n(x) \quad (n \geq 0). \tag{2.17}$$

From (2.17), we can derive the following equation.

$$d_n(x) = \frac{t}{e^t - 1} \int_x^{x+1} d_n(u)\,du = \frac{1}{1-t} x^n \quad (n \geq 0). \tag{2.18}$$

Theorem 2.4 For n ≥ 0, we have

$$d_n(x) = \frac{t}{e^t - 1} \int_x^{x+1} d_n(u)\,du = \frac{1}{1-t} x^n.$$

From (1.10), we have

$$\left\langle \frac{e^{yt} - 1}{t} \,\middle|\, d_n(x) \right\rangle = \left\langle e^{yt} - 1 \,\middle|\, \frac{1}{n+1} d_{n+1}(x) \right\rangle$$
$$= \left\langle 1 \,\middle|\, \frac{1}{n+1}\left(d_{n+1}(x+y) - d_{n+1}(x)\right) \right\rangle$$
$$= \frac{1}{n+1}\left(d_{n+1}(y) - d_{n+1}(0)\right)$$
$$= \int_0^y d_n(u)\,du \quad (n \geq 0). \tag{2.19}$$

In particular

$$\sum_{n=0}^{\infty} d_n(0)\frac{t^n}{n!} = \left(\frac{1}{1-t}e^{-t}\right)e^{t}$$

$$= \left(\sum_{l=0}^{\infty} d_l \frac{t^l}{l!}\right)\left(\sum_{m=0}^{\infty} \frac{t^m}{m!}\right)$$

$$= \sum_{n=0}^{\infty}\left(\sum_{l=0}^{n}\binom{n}{l}d_l\right)\frac{t^n}{n!}. \quad (2.20)$$

Comparing the coefficients on both sides of (1.17), we have

$$d_n(0) = \sum_{l=0}^{n}\binom{n}{l}d_l \quad (n \geq 0). \quad (2.21)$$

Therefore, we obtain the following corollary.

Corollary 2.5 For $n \geq 0$, we have

$$d_n(0) = \sum_{l=0}^{n}\binom{n}{l}d_l$$

and

$$\left\langle \frac{e^{yt}-1}{t} \,\middle|\, d_n(x)\right\rangle = \int_0^y d_n(u)\,du.$$

For $r \in \mathbb{N}$, we define the derangement polynomials of order r by

$$\left(\frac{1}{1-t}\right)^r e^{xt} = \sum_{n=0}^{\infty} d_n^{(r)}(x)\frac{t^n}{n!}. \quad (2.22)$$

When $x = -1$, $d_n^{(r)}(-1) = d_n^{(r)}$ are the derangement numbers of order r.

For $0 \leq r \leq n$, the r-derangement numbers, denoted by $D_n^{(r)}$, are the number of derangements on n+r elements under the restriction that the first r-elements are in disjoint cycles. It is known that the generating function of the r-derangement numbers is given by

$$\sum_{n=0}^{\infty} D_n^{(r)}\frac{t^n}{n!} = \frac{t^r}{(1-t)^{r+1}}e^{-t}. \quad (2.23)$$

We consider the r-derangement polynomials given by

$$\frac{t^r}{(1-t)^{r+1}}e^{xt} = \sum_{n=0}^{\infty} D_n^{(r)}(x)\frac{t^n}{n!} \quad (0 \leq r \leq n). \tag{2.24}$$

From (2.24), we note that $D_n^{(r)}(-1) = D_n^{(r)}$ are the r-derangement numbers. By (2.13) and (2.22), we easily get

$$d_n^{(r)}(x) \sim \left((1-t)^r, t\right) \quad (n \geq 0) \tag{2.25}$$

and

$$t^r d_n^{(r)}(x) = (n)_r d_{n-r}^{(r)}(x) = r!\binom{n}{r} d_{n-r}^{(r)}(x). \tag{2.26}$$

From (2.22) and (2.24), we have

$$\sum_{n=0}^{\infty} D_n^{(r)}(x)\frac{t^n}{n!} = t^r \sum_{n=0}^{\infty} d_n^{(r+1)}(x)\frac{t^n}{n!}$$

$$= \sum_{n=r}^{\infty} \binom{n}{r} r! d_{n-r}^{(r+1)}(x) \tag{2.27}$$

Comparing the coefficients on both sides of (2.27), we get

$$D_n^{(r)}(x) = \binom{n}{r} r! d_{n-r}^{(r+1)}(x) \quad (n \geq r). \tag{2.28}$$

From (2.22), we have

$$\sum_{n=0}^{\infty} d_n^{(r)}(0)\frac{t^n}{n!} = \left(\frac{1}{1-t}\right)^r = \sum_{n=0}^{\infty}\binom{n+r-1}{n} t^n = \sum_{n=0}^{\infty} (n+r-1)_n \frac{t^n}{n!}. \tag{2.29}$$

Thus, by (2.29), we get

$$d_n^{(r)}(0) = (n+r-1)_n \quad (n \geq 0). \tag{2.30}$$

From (2.22) and (2.24), we have

$$\begin{aligned} t^r \sum_{n=0}^{\infty} d_n^{(r+1)}(x)\frac{t^n}{n!} &= \frac{t^r}{(1-t)^{r+1}} e^{-t} e^{(x+1)t} \\ &= \left(\sum_{l=0}^{\infty} D_l^{(r)}\frac{t^l}{l!}\right)\left(\sum_{m=0}^{\infty} (x+1)^m \frac{t^m}{m!}\right) \\ &= \sum_{n=0}^{\infty}\left(\sum_{l=0}^{n}\binom{n}{l} D_l^{(r)}(x+1)^{n-l}\right)\frac{t^n}{n!}. \end{aligned} \tag{2.31}$$

Therefore, by (2.27) and (2.31), we obtain the following theorem.

Theorem 2.6 For $n \geq r$, we have

$$\sum_{l=0}^{n}\binom{n}{l}D_l^{(r)}(x+1)^{n-l}=\binom{n}{r}r!d_{n-r}^{(r+1)}(x).$$

Now, we observe that

$$\begin{aligned}\frac{t^{r-1}}{(1-t)^r}e^{xt}+\frac{t^r}{(1-t)^{r+1}}e^{xt}&=\frac{t^{r-1}}{(1-t)^{r+1}}e^{xt}\\&=\left(\frac{t^{r-1}}{(1-t)^r}e^{-t}\right)\left(\frac{1}{1-t}e^{-t}\right)e^{(x+2)t}\\&=\sum_{n=0}^{\infty}\left(\sum_{k=0}^{n}\sum_{l=0}^{k}\binom{n}{k}\binom{k}{l}D_l^{(r-1)}d_{k-l}(x+2)^{n-k}\right)\frac{t^n}{n!}.\end{aligned} \tag{2.32}$$

On the other hand, by (2.24), we get

$$\frac{t^{r-1}}{(1-t)^r}e^{xt}+\frac{t^r}{(1-t)^{r+1}}e^{xt}=\sum_{n=0}^{\infty}\left(D_n^{(r-1)}(x)+D_n^{(r)}(x)\right)\frac{t^n}{n!}. \tag{2.33}$$

From (2.32) and (2.33), we have

$$D_n^{(r-1)}(x)+D_n^{(r)}(x)=\sum_{k=0}^{n}\sum_{l=0}^{k}\binom{n}{k}\binom{k}{l}D_l^{(r-1)}d_{k-l}(x+2)^{n-k}. \tag{2.34}$$

In particular, for $x = -1$, we get

$$D_n^{(r-1)}+D_n^{(r)}=\sum_{k=0}^{n}\sum_{l=0}^{k}\binom{n}{k}\binom{k}{l}D_l^{(r-1)}d_{k-l}. \tag{2.35}$$

Therefore, by (2.34) and (2.35), we obtain the following theorem.

Theorem 2.7 For $n \geq 0$, we have

$$D_n^{(r-1)}(x)+D_n^{(r)}(x)=\sum_{k=0}^{n}\sum_{l=0}^{k}\binom{n}{k}\binom{k}{l}D_l^{(r-1)}d_{k-l}(x+2)^{n-k}.$$

Moreover,

$$D_n^{(r-1)}+D_n^{(r)}=\sum_{k=0}^{n}\sum_{l=0}^{k}\binom{n}{k}\binom{k}{l}D_l^{(r-1)}d_{k-l}.$$

By (2.22), we easily get

$$\sum_{n=0}^{\infty} d_n^{(r)}(x)\frac{t^n}{n!} = \left(\frac{1}{1-t}\right)^r e^{-t}e^{(x+1)t} = \left(\sum_{l=0}^{\infty} d_l^{(r)}\frac{t^l}{l!}\right)\left(\sum_{m=0}^{\infty}(x+1)^m\frac{t^m}{m!}\right)$$
$$= \sum_{n=0}^{\infty}\left(\sum_{l=0}^{n}\binom{n}{l}d_l^{(r)}(x+1)^{n-l}\right)\frac{t^n}{n!}. \quad (2.36)$$

Comparing the coefficients on both sides of (2.36), we have

$$d_n^{(r)}(x) = \sum_{l=0}^{n}\binom{n}{l}d_l^{(r)}(x+1)^{n-l}, \quad (2.37)$$

with the usual convention about replacing $(d^{(r)})^l$ by $d_l^{(r)}$. Thus, by (2.37), we get

$$d_n^{(r)}(x+y) = \left(d^{(r)}+x+y+1\right)^n = \left(d^{(r)}+x+1+y\right)^n$$
$$= \sum_{l=0}^{n}\binom{n}{l}\left(d^{(r)}+x+1\right)^l y^{n-l} = \sum_{l=0}^{n}\binom{n}{l}d_l^{(r)}(x)y^{n-l} \quad (n \geq 0). \quad (2.38)$$

From (2.22), we can derive the following equation:

$$\sum_{n=0}^{\infty} d_n^{(r)}(x)\frac{t^n}{n!} = \left(\frac{1}{1-t}\right)^r e^{xt} = \left(\frac{1}{1-t}\right)^{r-1} e^{-t}\left(\frac{1}{1-t}\right)e^{(x+1)t}$$
$$= \left(\sum_{l=0}^{\infty} d_l^{(r-1)}\frac{t^l}{l!}\right)\left(\sum_{m=0}^{\infty} d_m(x+1)\frac{t^m}{m!}\right)$$
$$= \sum_{n=0}^{\infty}\left(\sum_{l=0}^{n}\binom{n}{l}d_l^{(r-1)}d_{n-l}(x+1)\right)\frac{t^n}{n!}. \quad (2.39)$$

Thus, by (2.39), we get

$$d_n^{(r)}(x) = \sum_{l=0}^{n}\binom{n}{l}d_l^{(r-1)}d_{n-l}(x+1) \quad (n \geq 0). \quad (2.40)$$

For x = –2, from (2.37) and (2.40) we have

$$d_n^{(r)}(-2) = \sum_{l=0}^{n}\binom{n}{l}d_l^{(r-1)}d_{n-l}$$
$$= \sum_{l=0}^{n}\binom{n}{l}d_l^{(r)}(-1)^{n-l} \quad (n \geq 0). \quad (2.41)$$

From (2.17), we have

$$\frac{e^t-1}{t}d_n^{(r)}(x)=\int_x^{x+1}d_n^{(r)}(u)\,du=\frac{1}{n+1}\{d_{n+1}^{(r)}(x+1)-d_{n+1}^{(r)}(x)\}$$
$$=\frac{1}{n+1}\left\{\sum_{l=0}^{n+1}\binom{n+1}{l}d_l^{(r)}(x)-d_{n+1}^{(r)}(x)\right\}$$
$$=\frac{1}{n+1}\sum_{l=0}^{n}\binom{n+1}{l}d_l^{(r)}(x)$$
$$=\frac{1}{n+1}\sum_{l=1}^{n+1}\binom{n+1}{l}d_{n+1-l}^{(r)}(x)\quad(n\geq 0).\tag{2.42}$$

By (2.37) and (2.42), we get

$$d_n^{(r)}(x)=\frac{1}{n+1}\sum_{l=1}^{n+1}\sum_{m=0}^{n+1-l}\binom{n+1}{l}\binom{n+1-l}{m}d_m^{(r)}\frac{t}{e^t-1}(x+1)^{n+1-l-m}$$
$$=\frac{1}{n+1}\sum_{l=1}^{n+1}\sum_{m=0}^{n+1-l}\binom{n+1}{l}\binom{n+1-l}{m}d_m^{(r)}B_{m+1-l-m}(x+1).\tag{2.43}$$

Therefore, by (2.43), we obtain the following theorem.

Theorem 2.8 For n ≥ 0, we have

$$d_n^{(r)}(x)=\frac{1}{n+1}\sum_{l=1}^{n+1}\sum_{m=0}^{n+1-l}\binom{n+1}{l}\binom{n+1-l}{m}d_m^{(r)}B_{m+1-l-m}(x+1).$$

For n ≥ 0, let

$$\mathbb{P}_n=\{p(x)\in\mathbb{C}[x]\,|\,\deg p(x)\leq n\}.$$

Then $\mathbb{P}_n$ is an (n + 1)-dimensional vector space over $\mathbb{C}$.
For p(x) ∈ $\mathbb{P}_n$, we let

$$p(x)=\sum_{k=0}^{n}C_k d_k(x).\tag{2.44}$$

From (1.4), we have

$$\langle(1-t)t^k|p(x)\rangle=\sum_{l=0}^{n}C_l\langle(1-t)t^k|d_l(x)\rangle$$
$$=\sum_{l=0}^{n}C_l l!\delta_{k,l}=C_k k!\quad(k\geq 0).\tag{2.45}$$

Thus, we have

$$C_k = \frac{1}{k!}\left\langle (1-t)t^k \middle| p(x)\right\rangle = \frac{1}{k!}\left\langle (1-t) \middle| p^{(k)}(x)\right\rangle. \tag{2.46}$$

Therefore, by (2.44) and (2.46), we obtain the following theorem.

Theorem 2.9 For p(x) ∈ $\mathbb{P}_n$, we have

$$p(x) = \sum_{k=0}^{n} C_k d_k(x),$$

Where $C_k = \frac{1}{k!}\langle (1-t)t^k | p(x)\rangle = \frac{1}{k!}\langle (1-t) | p^{(k)}(x)\rangle$.

Let us take $p(x) = d_n^{(r)}(x) \in \mathbb{P}_n$. Then we have

$$p(x) = \sum_{l=0}^{n} C_l d_l(x), \tag{2.47}$$

where

$$\begin{aligned} C_l &= \frac{1}{l!}\left\langle 1-t \middle| p^{(l)}(x)\right\rangle = \frac{1}{l!}\left\langle 1-t \middle| (n)_l d_{n-l}^{(r)}(x)\right\rangle \\ &= \binom{n}{l}\left\langle 1-t \middle| d_{n-l}^{(r)}(x)\right\rangle = \binom{n}{l} d_{n-l}^{(r)}(0) - \binom{n}{l+1}(l+1)d_{n-l-1}^{(r)}(0). \end{aligned} \tag{2.48}$$

Hence, by (2.47) and (2.48), we get

$$d_n^{(r)}(x) = \sum_{l=0}^{n}\left\{\binom{n}{l} d_{n-l}^{(r)}(0) - \binom{n}{l+1}(l+1)d_{n-l-1}^{(r)}(0)\right\} d_l(x).$$

Assume that $p(x) = \sum_{k=0}^{n} C_k^{(r)} d_k^{(r)}(x) \in \mathbb{P}_n$. Then, by (2.25), we get

$$\begin{aligned} \left\langle (1-t)^r t^k \middle| p(x)\right\rangle &= \sum_{l=0}^{n} C_l^{(r)} \left\langle (!-t)^r t^k \middle| d_l^{(r)}(x)\right\rangle \\ &= \sum_{l=0}^{n} C_l^{(r)} l! \delta_{l,k} = C_k^{(r)} k! \quad (k \geq 0). \end{aligned} \tag{2.49}$$

Thus, from (2.49), we note that

$$C_k^{(r)} = \frac{1}{k!}\left\langle (1-t)^r t^k \middle| p(x)\right\rangle = \frac{1}{k!}\left\langle (1-t)^r \middle| p^{(k)}(x)\right\rangle. \tag{2.50}$$

Therefore, we obtain the following theorem.

Theorem 2.10 For n ≥ 0, we have

$$p(x) = \sum_{k=0}^{n} C_k^{(r)} d_k^{(r)}(x) \in \mathbb{P}_n,$$

where

$$C_k = \frac{1}{k!}\langle (1-t)^r t^k | p(x) \rangle = \frac{1}{k!}\langle (1-t)^r | p^{(k)}(x) \rangle.$$

Example 1 For p(x) = $d_n(x) \in \mathbb{P}_n$, we have

$$p(x) = \sum_{k=0}^{n} C_k^{(r)} d_k^{(r)}(x),$$

where

$$C_k^{(r)} = \frac{1}{k!}\langle (1-t)^r | p^{(k)}(x) \rangle = \binom{n}{k}\langle (1-t)^r | d_{n-k}(x) \rangle$$

$$= \binom{n}{k}\sum_{j=0}^{r}\binom{r}{j}(-1)^j \langle t^j | d_{n-k}(x) \rangle = \binom{n}{k}\sum_{j=0}^{r}\binom{r}{j}(-1)^j (n-k)_j d_{n-k-j}(0)$$

$$= \binom{n}{k}\sum_{j=0}^{r}\binom{r}{j}(-1)^j (n-k)_j (n-k-j)! = \binom{n}{k}\sum_{j=0}^{r}\binom{r}{j}(-1)^j (n-k)!.$$

Thus, we note that

$$d_n(x) = \sum_{k=0}^{n}\left(\sum_{j=0}^{r}\binom{n}{k}(n-k)!\binom{r}{j}(-1)^j\right) d_k^{(r)}(x).$$

Example 2 For p(x) = $B_n(x)$ ($n \geq 0$), we have

$$B_n(x) = \sum_{k=0}^{n} C_k^{(r)} d_k^{(r)}(x),$$

where

$$C_k^{(r)} = \frac{1}{k!}\langle (1-t)^r t^k | B_n(x) \rangle = \binom{n}{k}\langle (1-t)^r | B_{n-k}(x) \rangle$$

$$= \binom{n}{k}\sum_{j=0}^{r}\binom{r}{j}(-1)^j \langle t^j | B_{n-k}(x) \rangle$$

$$= \binom{n}{k}\sum_{j=0}^{r}\binom{r}{j}(-1)^j (n-k)_j \langle 1 | B_{n-k-j}(x) \rangle$$

$$= \binom{n}{k}\sum_{j=0}^{r}\binom{r}{j}(-1)^j \binom{n-k}{j} j! B_{n-k-j}.$$

Hence

$$B_n(x) = \sum_{k=0}^{n}\left(\binom{n}{k}\sum_{j=0}^{r}\binom{r}{j}(-1)^j\binom{n-k}{j}j!B_{n-k-j}\right)d_k^{(r)}(x).$$

Example 3 For p(x) = $E_n(x) \in \mathbb{P}_n$ $(n \geq 0)$, we have

$$E_n(x) = \sum_{k=0}^{n} C_k^{(r)} d_k^{(r)}(x),$$

where

$$C_k^{(r)} = \frac{1}{k!}\langle (1-t)t^k | E_n(x)\rangle = \binom{n}{k}\langle (1-t)|E_{n-k}(x)\rangle$$

$$= \binom{n}{k}E_{n-k} - \binom{n}{k}(n-k)E_{n-k-1}.$$

Thus, we get

$$E_n(x) = \sum_{k=0}^{n}\left\{\binom{n}{k}E_{n-k} - \binom{n}{k}(n-k)E_{n-k-1}\right\}d_k^{(r)}(x).$$

Example 4 For p(x) = $\mathrm{Bel}_n(x) \in \mathbb{P}_n$, we have

$$\mathrm{Bel}_n(x) = \sum_{k=0}^{n} C_k^{(r)} d_k^{(r)}(x),$$

where

$$C_k^{(r)} = \frac{1}{k!}\langle (1-t)t^k | \mathrm{Bel}_n(x)\rangle = \frac{1}{k!}\left\langle (1-t)t^k \middle| \sum_{m=0}^{n} S_2(n,m)x^m \right\rangle$$

$$= \sum_{m=k}^{n} S_2(n,m)\binom{m}{k}\langle (1-t)|x^{m-k}\rangle$$

$$= S_2(n,k) - \sum_{m=k}^{n} S_2(n,m)\binom{m}{k}(m-k)0^{m-k-1}$$

$$= S_2(n,k) - S_2(n,k+1)(k+1) = 2S_2(n,k) - S_2(n+1,k+1).$$

Hence

$$\mathrm{Bel}_n(x) = \sum_{k=0}^{n}\left(2S_2(n,k) - S_2(n+1,k+1)\right)d_k^{(r)}(x).$$

The ordered Bell polynomials are defined by the generating function

$$\frac{1}{2-e^t}e^{xt} = \sum_{n=0}^{\infty} b_n(x)\frac{t^n}{n!}. \tag{2.51}$$

When $x = 0$, $b_n = b_n(0)$ $(n \geq 0)$ are the ordered Bell numbers. From (2.12) and (2.51), we note that $b_n(x) \sim (2 - e^t, t)$ $(n \geq 0)$. For $b_n(x) \sim (2 - e^t, t)$, $d_n(x) \sim (1 - t, t)$, by (2.7) and (2.13), we get

$$b_n(x) = \sum_{m=0}^{n} C_{n,m} d_m(x) \quad (n \geq 0), \tag{2.52}$$

where

$$\begin{aligned} C_{n,m} &= \frac{1}{m!}\left\langle \frac{1-t}{2-e^t} t^m \middle| x^n \right\rangle = \binom{n}{m}\left\langle \frac{1-t}{2-e^t} \middle| x^{n-m} \right\rangle \\ &= \binom{n}{m}\left\langle 1-t \middle| \frac{1}{2-e^t} x^{n-m} \right\rangle = \binom{n}{m}\langle 1-t | b_{n-m}(x)\rangle \\ &= \binom{n}{m}\{b_{n-m} - (n-m)b_{n-m-1}\}. \end{aligned} \tag{2.53}$$

Therefore, we obtain the following theorem.

Theorem 2.11 For $n \geq 0$, we have

$$b_n(x) = \sum_{m=0}^{n} \binom{n}{m}(b_{n-m} - (n-m)b_{n-m-1})d_m(x).$$

For $d_n(x) \sim (1-t, t)$, $(x)_n \sim (1, e^t - 1)$, we have

$$d_n(x) = \sum_{m=0}^{n} C_{n,m}(x)_m \quad (n \geq 0), \tag{2.54}$$

where

$$\begin{aligned} C_{n,m} &= \frac{1}{m!}\left\langle \frac{1}{1-t}(e^t-1)^m \middle| x^n \right\rangle = \sum_{l=m}^{n} S_2(l,m)\frac{1}{l!}\left\langle \frac{t^l}{1-t} \middle| x^n \right\rangle \\ &= \sum_{l=m}^{n} S_2(l,m)\binom{n}{l}\left\langle \frac{1}{1-t} \middle| x^{n-l} \right\rangle = \sum_{l=m}^{n} S_2(l,m)\binom{n}{l}\langle 1 | d_{n-l}(x)\rangle \\ &= \sum_{l=m}^{n} S_2(l,m)\binom{n}{l} d_{n-l}(0). \end{aligned} \tag{2.55}$$

Therefore, by (2.54) and (2.55), we obtain the following theorem

Theorem 2.12 For $n \geq 0$, we have

$$d_n(x) = \sum_{m=0}^{n}\left(\sum_{l=m}^{n} S_2(l,m)\binom{n}{l} d_{n-l}(0)\right)(x)_m.$$

RESULTS AND DISCUSSION

In this paper, as a natural companion to derangement numbers, we have investigated derangement polynomials and derived several interesting properties on them which are related to derangement numbers. Also, we have considered two generalizations of derangement polynomials, namely the higher-order and r-derangement polynomials, and showed some relations between them and also with some other special polynomials. In addition, by using umbral calculus, we derived a formula expressing any polynomials as linear combinations of higher-order derangement polynomials and illustrated this with several special polynomials.

CONCLUSION

The introduction of derangement numbers goes back to as early as 1708 when Pierre Rémond de Montmort considered some counting problem on derangements. However, it seems that the umbral calculus approach to the derangement polynomials and their generalizations has not yet been done. In this paper, we have used umbral calculus in order to study some interesting properties on them, certain relations between them, and some connections with several other special polynomials.

ACKNOWLEDGEMENTS

This work was supported by the National Research Foundation of Korea (NRF) grant funded by the Korea government (MEST) (No. 2017R1E1A1A03070882).

AUTHORS' CONTRIBUTIONS

All authors contributed equally to the manuscript, read, and approved the final manuscript.

REFERENCES

1. Carlitz, L.: The number of derangements of a sequence with given specification. Fibonacci Q. 16(3), 255–258 (1978)
2. Clarke, R.J., Sved, M.: Derangements and Bell numbers. Math. Mag. 66(5), 299–303 (1993)
3. Kim, D.S., Kim, T., Kwon, H.-I.: Fourier series for r-derangement and higher-order derangement functions. Adv. Stud. Contemp. Math. (Kyungshang) 28(1), 1–11 (2018)
4. Dere, R., Simsek, Y.: Applications of umbral algebra to some special polynomials. Adv. Stud. Contemp. Math. (Kyungshang) 22(3), 433–438 (2012)
5. Kim, D.S., Kim, T., Ryoo, C.S.: Sheffer sequences for the powers of Sheffer pairs under umbral composition. Adv. Stud. Contemp. Math. (Kyungshang) 23(2), 275–285 (2013)
6. Kim, T., Kim, D.S.: On -Bell polynomials associated with umbral calculus. Russ. J. Math. Phys. 24(1), 69–78 (2017)
7. Kim, T., Kim, D.S., Jang, G.-W., Jang, L.-C.: Degenerate ordered Bell numbers and polynomials associated with umbral calculus. J. Nonlinear Sci. Appl. 10(10), 5142–5155 (2017)
8. Roman, S.: The Umbral Calculus. Pure and Applied Mathematics, vol. 111, Academic Press, New York (1984) x+193 pp.
9. Kim, T., Kim, D.S., Jang, G.-W.: Some formulas of ordered Bell numbers and polynomials arising from umbral calculus. Proc. Jangjeon Math. Soc. 20(4), 659–670 (2017)
10. Bayad, A., Kim, T.: Identities for the Bernoulli, the Euler and the Genocchi numbers and polynomials. Adv. Stud. Contemp. Math. (Kyungshang) 20(2), 247–253 (2010)
11. Ding, D., Yang, J.: Some identities related to the Apostol–Euler and Apostol–Bernoulli polynomials. Adv. Stud. Contemp. Math. (Kyungshang) 20(1), 7–21 (2010)
12. He, Y., Kim, T.: General convolution identities for Apostol–Bernoulli, Euler and Genocchi polynomials. J. Nonlinear Sci. Appl. 9(6), 4780–4797 (2016)
13. Simsek, Y.: Generating functions of the twisted Bernoulli numbers and polynomials associated with their interpolation functions. Adv. Stud. Contemp. Math. (Kyungshang) 16(2), 251–278 (2008)

14. Kwasniewski, A.K.: On Ψ-umbral extensions of Stirling numbers and Dobinski-like formulas. Adv. Stud. Contemp. Math. (Kyungshang) 12(1), 73–100 (2006)
15. Kim, T.: Identities involving Laguerre polynomials derived from umbral calculus. Russ. J. Math. Phys. 21(3), 584–589 (2014)

SECTION 4:
PARTITIONS AND GENERATING FUNCTIONS

9

A Rademacher Type Formula for Partitions and Overpartitions

Andrew V. Sills

Department of Mathematical Sciences, Georgia Southern University, Statesboro, GA, 30460-8093, USA

ABSTRACT

A Rademacher-type convergent series formula which generalizes the Hardy-Ramanujan Rademacher formula for the number of partitions of n and the Zuckerman formula for the Fourier coefficients of $\vartheta_4(0 \mid \tau)^{-1}$ is presented.

BACKGROUND

Partitions

A partition of an integer n is a representation of n as a sum of positive integers, where the order of the summands (called parts) is considered irrelevant. It is customary to write the parts in nonincreasing order. For example, there are

Citation: Andrew V. Sills, "A Rademacher Type Formula for Partitions and Overpartitions," International Journal of Mathematics and Mathematical Sciences, vol. 2010, Article ID 630458, 21 pages, 2010. https://doi.org/10.1155/2010/630458

three partitions of the integer 3, namely, 3, $2+1$, and $1+1+1$. Let $p(n)$. Let $p(n)$ denote the number of partitions of , with the convention that $p(0)=1$, and let $f(x)$ denote the generating function of $p(n)$, that is, let

$$f(x) := \sum_{n=0}^{\infty} p(n)x^n. \tag{1.1}$$

Euler [1] was the first to systematically study partitions. He showed that

$$f(x) = \prod_{m=1}^{\infty} \frac{1}{1-x^m}. \tag{1.2}$$

Euler also showed that

$$\frac{1}{f(x)} = \sum_{n=-\infty}^{\infty} (-1)^n x^{n(3n-1)/2}, \tag{1.3}$$

and since the exponents appearing on the right side of (1.3) are the pentagonal numbers, (1.3) is often called "Euler's pentagonal number theorem."

Although Euler's results can all be treated from the point of view of formal power series, the series and infinite products (above and indeed all the series and infinite products mentioned in this paper) converge absolutely when |x| < 1, which is important for analytic study of these series and products.

Hardy and Ramanujan were the first to study p(n) analytically and produced an incredibly accurate asymptotic formula [2, page 85, equation (1.74)], namely,

$$p(n) = \frac{1}{2\pi\sqrt{2}} \sum_{k=1}^{[\alpha\sqrt{n}]} \sqrt{k} \sum_{\substack{0\le h<k \\ (h,k)=1}} \omega(h,k) e^{-2\pi i n h/k} \frac{d}{dn}\left(\frac{\exp\left((\pi/k)\sqrt{(2/3)(n-1/24)}\right)}{\sqrt{n-1/24}} \right) + O\left(n^{-1/4}\right), \tag{1.4}$$

wher

$$\omega(h,k) = \exp\left(\pi i \sum_{r=1}^{k-1} \frac{r}{k} \left(\frac{hr}{k} - \left\lfloor \frac{hr}{k} \right\rfloor - \frac{1}{2} \right) \right). \tag{1.5}$$

α is an arbitrary constant, and here and throughout (h, k) is an abbreviation for gcd (h, k).

Later Rademacher [3] improved upon (1.4) by finding the following convergent series representation for p(n):

$$p(n) = \frac{1}{\pi\sqrt{2}} \sum_{k=1}^{\infty} \sqrt{k} \sum_{\substack{0 \le h < k \\ (h,k)=1}} \omega(h,k) e^{-2\pi i n h/k} \frac{d}{dn}\left(\frac{\sinh\left((\pi/k)\sqrt{(2/3)(n-1/24)} \right)}{\sqrt{n-1/24}} \right). \tag{1.6}$$

Rademacher's method was used extensively by many practitioners, including Grosswald [4, 5], Haberzetle [6], Hagis [7–15], Hua [16], Iseki [17–19], Lehner [20], Livingood [21], Niven [22], and Subramanyasastri [23] to study various restricted partitions functions.

Recently, Bringmann and Ono [24] have given exact formulas for the coefficients of all harmonic Maass forms of weight 1/2. The generating functions considered herein are weakly holomorphic modular forms of weight −1/2, and thus they are harmonic Maass forms of weight 1/2. Accordingly, the results of this present paper could be derived from the general theorem in [24]. However, here we opt to derive the results via classical method of Rademacher.

Overpartitions

Overpartitions were introduced by Corteel and Lovejoy in [25] and have been studied extensively by them and others including Bringmann, Chen, Fu, Goh, Hirschhorn, Hitczenko, Lascoux, Mahlburg, Robbins, Rødseth, Sellers, Yee, and Zhao [25–43].

An overpartition of n is a representation of n as a sum of positive integers with summands in nonincreasing order, where the last occurrence of a given summand may or may not be overlined. Thus the eight overpartitions of 3 are $3, \overline{3}, 2+1, \overline{2}+1, 2+\overline{1}, \overline{2}+\overline{1}, 1+1+1, 1+1+\overline{1}$.

Let $\overline{p}(n)$ denote the number of over partitions of n and let $\overline{f}(x)$ denote the generating function $\sum_{n=0}^{\infty} \overline{p}(n)x^n$ of $\overline{p}(n)$. Elementary techniques are sufficient to show that

$$\overline{f}(x) = \prod_{m=1}^{\infty} \frac{1+x^m}{1-x^m} = \frac{f(x)^2}{f(x^2)}. \tag{1.7}$$

Note that

$$\frac{1}{\overline{f}(x)} = \sum_{n=-\infty}^{\infty} (-1)^n x^{n^2}. \tag{1.8}$$

via an identity of Gauss (see equation (2.2.12), page 23 in [44, 45]), so that the reciprocal of the generating function for overpartitions is a series wherein a coefficient is nonzero if and only if the exponent of x is a perfect square, just as the reciprocal of the generating function for partitions is a series wherein a coefficient is nonzero if and only if the exponent of x is a pentagonal number.

Hardy and Ramanujan, writing more than 80 years before the coining of the term "overpartition," stated [2, page 109-110] that the function which we are calling $\overline{p}(n)$ "has no very simple arithmetical interpretation; but the series is none the less, as the direct reciprocal of a simple ϑ-function, of particular interest." They went on to state that

$$\overline{p}(n) = \frac{1}{4\pi}\frac{d}{dn}\left(\frac{e^{\pi\sqrt{n}}}{\sqrt{n}}\right) + \frac{\sqrt{3}}{2\pi}\cos\left(\frac{2}{3}n\pi - \frac{1}{6}\pi\right)\frac{d}{dn}\left(e^{\pi\sqrt{n}/3}\right) + \cdots + O\left(n^{-1/4}\right). \tag{1.9}$$

In fact, (1.9) was improved to the following Rademacher-type convergent series by Zuckerman [46, page 321, equation (8.53)]:

$$\overline{p}(n) = \frac{1}{2\pi}\sum_{\substack{k\geq 1\\ 2\nmid k}}\sqrt{k}\sum_{\substack{0\leq h<k\\ (h,k)=1}}\frac{\omega(h,k)^2}{\omega(2h,k)}e^{-2\pi inh/k}\frac{d}{dn}\left(\frac{\sinh(\pi\sqrt{n}/k)}{\sqrt{n}}\right). \tag{1.10}$$

A simplified proof of (1.10) was given by Goldberg in his Ph.D. thesis [47].

Partitions Where No Odd Part Is Repeated

Let pod(n) denote the number of partitions of n where no odd part appears more than once. Let g(x) denote the generating function of pod(n), so we have

$$g(x) = \sum_{n=0}^{\infty} pod(n)x^n = \prod_{m=1}^{\infty}\frac{1+x^{2m-1}}{1-x^{2m}} = \frac{f(x)f(x^4)}{f(x^2)}. \tag{1.11}$$

Via another identity of Gauss (see equation (2.2.13), page 23 in [44, 45]), it turns out that

$$\frac{1}{g(x)} = \sum_{n=0}^{\infty}(-x)^{n(n+1)/2} = \sum_{n=-\infty}^{\infty}(-1)^n x^{2n^2-n}; \tag{1.12}$$

so in this case the reciprocal of the generating function under consideration has nonzero coefficients at the exponents which are triangular (or equivalently, hexagonal) numbers.

The analogous Rademacher-type formula for pod(n) is as follows:

$$pod(n) = \frac{2}{\pi}\sum_{k\geq 1}\sqrt{k\left(1-(-1)^k+\left\lfloor\frac{(k,4)}{4}\right\rfloor\right)}$$

$$\times \sum_{\substack{0\leq h<k \\ (h,k)=1}} \frac{\omega(h,k)\ \omega(4h/(k,4),k/(k,4))}{\omega(2h/(k,2),k/(k,2))} e^{-2\pi i n h/k}$$

$$\times \frac{d}{dn}\left(\frac{\sinh\left(\pi\sqrt{(k,4)(8n-1)}/4k\right)}{\sqrt{8n-1}}\right). \tag{1.13}$$

Equation *1.13) is the case r = 2 of Theorem 2.1 to be presented in the next section

A COMMON GENERALIZATION

Let us define

$$f_r(x) := \frac{f(x)f(x^{2^r})}{f(x^2)}, \tag{2.1}$$

where r is a nonnegative integer. Thus,

$$f_0(x) = \overline{f}(x) = \sum_{n=0}^{\infty}\overline{p}(n)x^n,$$

$$f_1(x) = f(x) = \sum_{n=0}^{\infty}p(n)x^n,$$

$$f_2(x) = g(x) = \sum_{n=0}^{\infty}pod(n)x^n. \tag{2.2}$$

Let p_r (n) denote the coefficient of x^n in the expansion of f_r (n), that is,

$$f_r(x) = \sum_{n=0}^{\infty}p_r(n)x^n. \tag{2.3}$$

Notice that f_r (x) can be represented by several forms of equivalent infinite products, each of which has a natural combinatorial interpretation:

$$f_r(x) = \prod_{m=1}^{\infty}\frac{1+x^m}{1-x^{2^r m}} \tag{2.4}$$

$$= \prod_{m=1}^{\infty} \frac{1}{(1 - x^{2m-1})(1 - x^{2^r m})} \tag{2.5}$$

$$= \prod_{m=1}^{\infty} \frac{1}{1 - x^{2^{r-1} m}} \prod_{\lambda=1}^{2^{r-1}-1} \left(1 + x^{2^{r-1} m + \lambda}\right). \tag{2.6}$$

Thus, pr (n) equals each of the following:

(i)the number of overpartitions of n where nonoverlined parts are multiples of 2^n(by (2.4));

(ii) the number of partitions of n where all parts are either odd or multiples of 2^r (by (2.5)), provided $r \geqq 1$;

(iii) the number of partitions of n where nonmultiples of 2^{r-1} are distinct (by (2.6)), provided $r \geqq 1$.

Theorem 2.1. For r = 0, 1, 2, 3, 4,

$$p_r(n) = \frac{2^{(r+1)/2}\sqrt{3}}{\pi} \sum_{\substack{k \geq 1 \\ (k, 2^{\max(r,1)})=1}} \sqrt{k}$$

$$\times \sum_{\substack{0 \leq h < k \\ (h,k)=1}} e^{-2\pi i n h/k} \frac{\omega(h,k)\omega(2^r h, k)}{\omega(2h,k)}$$

$$\times \frac{d}{dn} \left\{ \frac{\sinh\left(\pi\sqrt{(24n - 2^r + 1)(1 + 2^{r-1})} / (2^{r/2} \cdot 6k)\right)}{\sqrt{24n - 2^r + 1}} \right\}$$

$$+ \frac{\sqrt{3}}{\pi} \sum_{j=1+\lfloor r/2 \rfloor}^{r} 2^{(2-j+r)/2} \sum_{\substack{k \geq 1 \\ (k,2^r)=2^j}} \sqrt{k}$$

$$\times \sum_{\substack{0 \leq h < k \\ (h,k)=1}} e^{-2\pi i n h/k} \frac{\omega(h,k)\omega(2^{r-j}h, 2^{-j}k)}{\omega(h,k/2)}$$

$$\times \frac{d}{dn} \left\{ \frac{\sinh\left(\pi\sqrt{(24n - 2^r + 1)(-1 + 2^{2j-r})} / 6k\right)}{\sqrt{24n - 2^r + 1}} \right\}. \tag{2.7}$$

A PROOF OF THEOREM 2.1

The method of proof is based on Rademacher's proof of (1.6) in [48] with the necessary modifications. Additional details of Rademacher's proof of (1.6) are provided in [49], [50, Chapter 14], and [51, Chapter 5]. Of fundamental importance is the path of integration to be used. In [48], Rademacher improved upon his original proof of (1.6) given in [3], by altering his path of integration from a carefully chosen circle to a more complicated path based on Ford circles, which in turn led to considerable simplifications later in the proof.

Farey Fractions

The sequence $\mathcal{F}_N$ of proper Farey fractions of order N is the set of all h/k with (h, k)= 1 and $0 \leq h/k < 1$, arranged in increasing order. Thus, for example, $\mathcal{F}_4$ ={0/1, 1/4, 1/3, 1/2, 2/3, 3/4}.

For a given N, let h_p, h_s, k_p, and k_s be such that h_p/k_p is the immediate predecessor of h/k and h_s/k_s is the immediate successor of h/k in $\mathcal{F}_N$. It will be convenient to view each $\mathcal{F}_N$ cyclically, that is, to view 0/1 as the immediate successor of (N – 1)/N.

Ford Circles and the Rademacher Path

Let h and k be integers with $(h,k) = 1$ and $0 \leq h < k$. The Ford circle [52] $C(h,k)$ is the circle in $\mathbb{C}$ of radius $1/2k^2$ centered at the point

$$\frac{h}{k} + \frac{1}{2k^2}i. \tag{3.1}$$

The upper arc γ(h, k) of the Ford circle C(h, k) is those points of C(h, k) from the initial point

$$\alpha_I(h,k) := \frac{h}{k} - \frac{k_p}{k(k^2+k_p^2)} + \frac{1}{k^2+k_p^2}i \tag{3.2}$$

to the terminal point

$$\alpha_T(h,k) := \frac{h}{k} + \frac{k_s}{k(k^2+k_s^2)} + \frac{1}{k^2+k_s^2}i, \tag{3.3}$$

traversed in the clockwise direction.

Note that we have

$$\alpha_I(0,1) = \alpha_T(N-1,N). \tag{3.4}$$

Every Ford circle is in the upper half plane. For h_1/k_1, $h_2/k_2 \in$ FN, $C(h_1, k_1)$ and $C(h_2, k_2)$ are either tangent or do not intersect.

The Rademacher path P(N) of order N is the path in the upper half of the τ-plane from i to i + 1 consisting of

$$\bigcup_{h/k \in \mathfrak{F}_N} \gamma(h,k) \tag{3.5}$$

traversed left to right and clockwise. In particular, we consider the left half of the Ford circle C(0, 1) and the corresponding upper arc γ(0, 1) to be translated to the right by 1 unit. This is legal given then periodicity of the function which is to be integrated over P(N).

Set Up the Integral

Let n and r be fixed, with $n > (2^{r-1})/24$.

Since

$$f_r(x) = \sum_{n=0}^{\infty} p_r(n) x^n, \tag{2.6}$$

Cauchy's residue theorem implies that

$$p_r(n) = \frac{1}{2\pi i} \int_C \frac{f_r(x)}{x^{n+1}} dx, \tag{3.7}$$

where C is any simply closed contour enclosing the origin and inside the unit circle. We introduce the change of variable

$$x = e^{2\pi i \tau} \tag{3.8}$$

so that the unit disk |x| 1 in the x-plane maps to the infinitely tall, unit-wide strip in the τ-plane where $0 \le \Re\tau \le 1$ and $\Im\tau \ge 0$.

The contour C is then taken to be the preimage of P(N) under the map $x \mapsto e^{2\pi i \tau}$.

$$\begin{aligned} p_r(n) &= \int_{P(N)} f_r\left(e^{2\pi i \tau}\right) e^{-2\pi i n \tau} d\tau \\ &= \sum_{h/k \in \mathfrak{F}_N} \int_{\gamma(h,k)} f_r\left(e^{2\pi i \tau}\right) e^{-2\pi i n \tau} d\tau \\ &= \sum_{k=1}^{N} \sum_{\substack{0 \le h < k \\ (h,k)=1}} \int_{\gamma(h,k)} f_r\left(e^{2\pi i \tau}\right) e^{-2\pi i n \tau} d\tau. \end{aligned} \tag{3.9}$$

Another Change of Variable

Next, we change variables again, taking

$$\tau = \frac{iz + h}{k}, \tag{3.10}$$

so that

$$z = -ik\left(\tau - \frac{h}{k}\right). \tag{3.11}$$

Thus C(h, k) (in the τ-plane) maps to the clockwise-oriented circle $K_k^{(-)}$ (in the z-plane) centered at 1/2k with radius 1/2k.

$$p_r(n) = i\sum_{k=1}^{N} k^{-1} \sum_{\substack{0 \le h < k \\ (h,k)=1}} e^{-2\pi i n h/k} \int_{\substack{z_I(h,k) \\ \text{arc of } K_k^{(-)}}}^{z_T(h,k)} e^{2n\pi z/k} f_r\left(e^{2\pi i h/k - 2\pi z/k}\right) dz, \tag{3.12}$$

where z_I (h, k(resp., z_T (h, k)) is the image of α_I (h, k) see (3.2)) (resp. αT (h, k)[see 3.2]) under the transformation (3.11).
So the transformation (3.11) maps the upper arc γ(h, k) of C(h, k) in the τ-plane to the arc on $K_k^{(-)}$ which initiates at

$$z_I(h,k) = \frac{k}{k^2 + k_p^2} + \frac{k_p}{k^2 + k_p^2} i \tag{3.13}$$

and terminates at

$$z_T(h,k) = \frac{k}{k^2 + k_s^2} - \frac{k_s}{k^2 + k_s^2} i. \tag{3.14}$$

Exploiting a Modular Transformation

From the theory of modular forms, we have the transformation formula [2, page 93, Lemma 4.31]:

$$f\left(\exp\left(\frac{2\pi i h}{k} - \frac{2\pi z}{k}\right)\right) = \omega(h,k) \exp\left(\frac{\pi(z^{-1} - z)}{12k}\right) \sqrt{z} f\left(\exp\left(2\pi i \frac{iz^{-1} + H}{k}\right)\right). \tag{3.15}$$

where $\sqrt{z}$ is the principal branch, (h, k) = 1, and H is a solution to the congruence

$$hH \equiv -1 \pmod{k}. \tag{3.16}$$

From (3.15), we deduce the analogous transformation for f_r (x).
The transformation formula is a piecewise defined function with r + 1 cases corresponding to j = 0, 1, 2,...,r, where (k, 2^r)= 2^j :

$$f_r\left(\exp\left(\frac{2\pi ih}{k}-\frac{2\pi z}{k}\right)\right)$$

$$=\frac{\omega(h,k)\omega(2^{r-j}h,2^{-j}k)}{\omega(2h/(2-\delta_{j0}),k/(2-\delta_{j0}))}\exp\left(\frac{\pi\left(2+2^{2j-r+1}-(2-\delta_{j0})^2\right)}{24kz}+\frac{\pi(1-2^r)z}{12k}\right)$$

$$\times\sqrt{z\ 2^{r-j-1}(2-\delta_{j0})}$$

$$\times\frac{f(\exp(-2\pi/kz+2H_j\pi i/k))f(\exp(-2^{2j-r+1}\pi/kz+2^{2j-r+1}H_j\pi i/k))}{f\left(\exp\left(-\pi(2-\delta_{j0})^2/kz+H_j\pi(2-\delta_{j0})^2 i/k\right)\right)},\qquad(3.17)$$

where H_j is divisible by 2^{r-j} and is a solution to the congruence $hH_j \equiv -1(\mathrm{mod}\,k)$, and

$$\delta_{j0}=\begin{cases}1 & \text{if } j=0,\\ 0 & \text{if } j\neq 0\end{cases}\qquad(3.18)$$

is the Kronecker δ-function.

Notice that in particular, for $\lfloor r/2\rfloor \leqq j \leqq r$, (3.17) simplifies to

$$f_r\left(\exp\left(\frac{2\pi ih}{k}-\frac{2\pi z}{k}\right)\right)$$

$$=\frac{\omega(h,k)\omega(2^{r-j}h,2^{-j}k)}{\omega(h,k/2)}\exp\left(\frac{\pi}{12k}\left[\left(2^{2j-r}-1\right)z^{-1}+(1-2^r)z\right]\right)$$

$$\times\sqrt{z2^{r-j}}\,f_{2j-r}\left(\exp\left(\frac{-2\pi}{kz}+\frac{2H_j\pi i}{k}\right)\right).\qquad(3.19)$$

Since the r = 0 case was established by Zuckerman, and the r = 1 case by Rademacher, we will proceed with the assumption that r > 1.

Apply (3.17) to (3.12) to obtain

$$p_r(n)=i\sum_{j=0}^{r}\sum_{\substack{k=1\\(k,2^r)=j}}^{N}k^{-1}\sum_{\substack{0\leq h<k\\(h,k)=1}}e^{-2\pi inh/k}$$

$$\times\int_{\substack{z_I(h,k)\\ \text{arc of } K_k^{(-1)}}}^{z_T(h,k)}\frac{\omega(h,k)\omega(2^{r-j}h,2^{-j}k)}{\omega(2h/(2-\delta_{j0}),k/(2-\delta_{j0}))}$$

$$\times\exp\left(\frac{\pi\left(2+2^{2j-r+1}-(2-\delta_{j0})^2\right)}{24kz}+\frac{\pi(24n+1-2^r)z}{12k}\right)$$

$$\times\sqrt{z2^{r-j-1}(2-\delta_{j0})}$$

$$\times\frac{f(\exp(-2\pi/kz+2H_j\pi i/k))f(\exp(-2^{2j-r+1}\pi/kz+2^{2j-r+1}H_j\pi i/k))}{f\left(\exp\left(-\pi(2-\delta_{j0})^2/kz+H_j\pi(2-\delta_{j0})^2 i/k\right)\right)}dz.\qquad(3.20)$$

Normalization

Next, introduce a normalization $\zeta = zk$ (this is not strictly necessary, but it will allow us in the sequel to quote various useful results directly from the literature):

$$p_r(n) = i\sum_{j=0}^{r} \sum_{\substack{k=1 \\ (k,2^r)=j}}^{N} k^{-5/2} \sum_{\substack{0 \leq h < k \\ (h,k)=1}} e^{-2\pi inh/k}$$
$$\times \int_{\substack{\zeta_I(h,k) \\ \text{arc of } K^{(-)}}}^{\zeta_T(h,k)} \frac{\omega(h,k)\omega(2^{r-j}h, 2^{-j}k)}{\omega(2h/(2-\delta_{j0}), k/(2-\delta_{j0}))}$$
$$\times \exp\left(\frac{\pi\left(2 + 2^{2j-r+1} - (2-\delta_{j0})^2\right)}{24\zeta} + \frac{\pi(24n + 1 - 2^r)\zeta}{12k^2}\right)$$
$$\times \sqrt{\zeta 2^{r-j-1}(2-\delta_{j0})}$$
$$\times \frac{f(\exp(-2\pi/\zeta + 2H_j\pi i/k)) f(\exp(-2^{2j-r+1}\pi/\zeta + 2^{2j-r+1}H_j\pi i/k))}{f\left(\exp\left(-\pi(2-\delta_{j0})^2/\zeta + H_j\pi(2-\delta_{j0})^2 i/k\right)\right)} d\zeta, \tag{3.21}$$

where

$$\zeta_I(h,k) = \frac{k^2}{k^2 + k_p^2} + \frac{kk_p}{k^2 + k_p^2} i,$$
$$\zeta_T(h,k) = \frac{k^2}{k^2 + k_s^2} - \frac{kk_s}{k^2 + k_s^2} i. \tag{3.22}$$

Let us now rewrite (3.21) as

$$p_r(n) = i\sum_{j=0}^{r} \sum_{\substack{k=1 \\ (k,2^{\max(r,1)})=2^j}}^{N} k^{-5/2} \sum_{\substack{0 \leq h < k \\ (h,k)=1}} e^{-2\pi inh/k} \frac{\omega(h,k)\omega(2^{r-j}h, 2^{-j}k)}{\omega(2h/(2-\delta_{j0}), k/(2-\delta_{j0}))}$$
$$\times (\mathcal{I}_{j,1} + \mathcal{I}_{j,2}), \tag{3.23}$$

where

$$\mathcal{I}_{j,1} := \int_{\substack{\zeta_I(h,k) \\ \text{arc}}}^{\zeta_T(h,k)} \exp\left(\frac{\pi\left(2 + 2^{2j-r+1} - (2-\delta_{j0})^2\right)}{24\zeta} + \frac{\pi(24n + 1 - 2^r)\zeta}{12k^2}\right)$$
$$\times \sqrt{\zeta 2^{r-j-1}(2-\delta_{j0})}$$
$$\times \left\{-1 + \frac{f(\exp(-2\pi/\zeta + 2H_j\pi i/k)) f(\exp(-2^{2j-r+1}\pi/\zeta + 2^{2j-r+1}H_j\pi i/k))}{f\left(\exp\left(-\pi(2-\delta_{j0})^2/\zeta + H_j\pi(2-\delta_{j0})^2 i/k\right)\right)}\right\} d\zeta,$$

$$\mathcal{I}_{j,2} := \int_{\substack{\zeta_I(h,k)\\ \text{arc}}}^{\zeta_T(h,k)} \exp\left(\frac{\pi\left(2 + 2^{2j-r+1} - (2-\delta_{j0})^2\right)}{24\zeta} + \frac{\pi(24n+1-2^r)\zeta}{12k^2} \right) \times \sqrt{\zeta 2^{r-j-1}(2-\delta_{j0})}\,d\zeta. \tag{3.24}$$

Estimation

It will turn out that as N → ∞, only $\mathcal{I}_{j,2}$ for j = 0 and $\lfloor r/2 \rfloor < j \leqq r$ ultimately make a contribution. Note that all the integrations in the ζ-plane occur on arcs and chords of the circle K of radius 1/2 centered at the point 1/2. So, inside and on K, $0 < \Re\zeta \leqq 1$ and $\Re(1/\zeta) \geqq 1$.

Estimation of $\mathcal{I}_{j,2}$ *for* $1 \leqq j \leqq \lfloor r/2 \rfloor$

The regularity of the integrand allows us to alter the path of integration from the arc connecting ζI (h, k) and ζT (h, k) to the directed segment. By [51, page 104, Theorem 5.9], the length of the path of integration does not exceed $2\sqrt{2}k/N$, and on the segment connecting $\zeta_I(h,k)$ to $\zeta_T(h,k)$, $|\zeta| < \sqrt{2}k/N$. Thus, the absolute value of the integrand is

$$\begin{aligned} &\left| \exp\left(\frac{\pi(2^{2j-r}-1)}{12\zeta} + \frac{\pi(24n+1-2^r)\zeta}{12k^2} \right) \sqrt{\zeta 2^{r-j}} \right| \\ &= |\zeta|^{1/2} 2^{(r-j)/2} \exp\left(\frac{(24n+1-2^r)\pi\Re\zeta}{12k^2} \right) \exp\left(\frac{\pi(2^{2j-r}-1)}{12} \Re\frac{1}{\zeta} \right) \\ &\leqq |\zeta|^{1/2} 2^{r/2} \exp(2\pi n). \end{aligned} \tag{3.25}$$

Thus, for $1 \leqq j \leqq \lfloor r/2 \rfloor$

$$|\mathcal{I}_{j,2}| \leqq \frac{2\sqrt{2}k}{N} \left(\frac{\sqrt{2}k}{N} \right)^{1/2} 2^{r/2} e^{2\pi n} \leqq C_j' k^{3/2} N^{-3/2} \tag{3.26}$$

for a constant C_j' (recalling that n and r are fixed).

Estimation of $\mathcal{I}_{j,1}$ *for* $1 \leqq j \leqq \lfloor r/2 \rfloor$

We have the absolute value of the integrand:

$$\left| \sqrt{\zeta 2^{r-j}} \exp\left(\frac{\pi(2^{2j-r}-1)}{12\zeta} + \frac{\pi(24n+1-2^r)\zeta}{12k^2} \right) \right| \times \left| -1 + \frac{f(\exp(-2\pi/\zeta + 2H_j\pi i/k)) f(\exp(-2^{2j-r+1}\pi/\zeta + 2^{2j-r+1}H_j\pi i/k))}{f\left(\exp\left(\frac{-4\pi}{\zeta} + \frac{4H_j\pi i}{k}\right)\right)} \right|$$

$$= \left| \sqrt{\zeta 2^{r-j}} \exp\left(\frac{\pi(2^{2j-r}-1)}{12\zeta} + \frac{\pi(24n+1-2^r)\zeta}{12k^2} \right) \right|$$

$$\times \left| -1 + f_{2j-r}\left(\exp\left(\frac{2\pi}{\zeta} + \frac{2H_j\pi i}{k} \right) \right) \right|$$

$$= |\zeta|^{1/2} 2^{(r-j)/2} \exp\left(\frac{(24n+1-2^r)\pi \Re\zeta}{12k^2} \right) \exp\left(\frac{(2^{2j-r}-1)\pi}{12} \Re\frac{1}{\zeta} \right)$$

$$\times \left| \sum_{m=1}^{\infty} p_{2j-r}(m) \exp\left(\frac{-2\pi m}{\zeta} + \frac{2H_j\pi i m}{k} \right) \right|$$

$$\leq |\zeta|^{1/2} 2^{(r-j)/2} \exp\left(\frac{(24n+1-2^r)\pi}{12} \right) \sum_{m=1}^{\infty} p_{2j-r}(m) \exp\left(-\frac{\pi}{12}\left(24m - 2^{2j-r}+1\right) \right)$$

$$\leq |\zeta|^{1/2} 2^{r/2} e^{2\pi n} \sum_{m=1}^{\infty} p_0(m) e^{-2\pi m}$$

$$= c_j |\zeta|^{1/2} \tag{3.27}$$

for a constant c_j. So, for $1 \leq j \leq \lfloor r/2 \rfloor$

$$|I_{j,1}| \leq \frac{2\sqrt{2}k}{N} \left(\frac{\sqrt{2}k}{N} \right)^{1/2} c_j < C_j k^{3/2} N^{-3/2} \tag{3.28}$$

for a constant C_j.

Estimation of $I_{0,1}$

Let $p_r^*(x)$ be defined by

$$\sum_{n=0}^{\infty} p_r^*(n) x^n = \frac{f(x^{2^r}) f(x)}{f(x^{2^{r-1}})}. \tag{3.29}$$

Again, the regularity of the integrand allows us to alter the path of integration from the arc connecting ζ_I (h, k) and ζ_T (h, k) to the directed segment.

With this in mind, we estimate the absolute value of the integrand:

$$\left| \exp\left(\frac{\pi(1+2^{1-r})}{24\zeta} + \frac{\pi(24n+1-2^r)\zeta}{12k^2} \right) \sqrt{\zeta 2^{r-1}} \right|$$

$$\times \left| -1 + \frac{f(\exp(-2\pi/\zeta + 2H_0\pi i/k)) f(\exp(-2^{1-r}\pi/\zeta + 2^{1-r}H_0\pi i/k))}{f(\exp(-\pi/\zeta + H_0\pi i/k))} \right|$$

$$= \left| \exp\left(\frac{\pi(1+2^{1-r})}{24\zeta} + \frac{\pi(24n+1-2^r)\zeta}{12k^2} \right) \sqrt{\zeta 2^{r-1}} \right|$$

$$\times \left| \sum_{m=1}^{\infty} p^*(m) \exp\left(\frac{-2^{1-r}\pi m}{\zeta} + \frac{2^{1-r}H_0\pi i m}{k} \right) \right|$$

$$= \exp\left(\frac{\pi(1+2^{1-r})}{24} \Re\frac{1}{\zeta} \right) \exp\left(\frac{\pi(24n+1-2^r)\Re\zeta}{12k^2} \right) |\zeta|^{1/2} 2^{(r-1)/2}$$

$$\times \left| \sum_{m=1}^{\infty} p^*(m) \exp\left(\frac{-2^{1-r}\pi m}{\zeta} \right) \exp\left(\frac{2^{1-r}H_0\pi i m}{k} \right) \right|$$

$$\leq e^{2\pi n} |\zeta|^{1/2} 2^{(r-1)/2}$$

$$\times \sum_{m=1}^{\infty} |p^*(m)| \exp\left(\frac{\pi(1+2^{1-r})}{24}\Re\frac{1}{\zeta} - 2^{1-r}\pi m \Re\frac{1}{\zeta}\right)$$

$$= e^{2\pi n}|\zeta|^{1/2}2^{(r-1)/2}\sum_{m=1}^{\infty}|p^*(m)|\exp\left(-\frac{\pi}{24\cdot 2^{r-1}}\Re\frac{1}{\zeta}\left(24m-1-2^{r-1}\right)\right)$$

$$\leq e^{2\pi n}|\zeta|^{1/2}2^{(r-1)/2}\sum_{m=1}^{\infty}|p^*(m)|\exp\left(-\frac{\pi}{24}\left(24m-1-2^{r-1}\right)\right)$$

$$< e^{2\pi n}|\zeta|^{1/2}2^{(r-1)/2}\sum_{m=1}^{\infty}\left|p^*\left(24m-1-2^{r-1}\right)\right|y^{24m-1-2^{r-1}} \quad \left(\text{where } y = e^{-\pi/24}\right)$$

$$= c_0|\zeta|^{1/2} \tag{3.30}$$

for a constant c_0. So,

$$|\mathfrak{I}_{0,1}| \leq \frac{2\sqrt{2}k}{N}\left(\frac{\sqrt{2}k}{N}\right)^{1/2} c_0 < C_0 k^{3/2} N^{-3/2} \tag{3.31}$$

for a constant C_0.

Estimation of $\mathfrak{I}_{j,1}$ *for* $1 + \lfloor r/2 \rfloor \leq j \leq r \leq 4$

Again, the regularity of the integrand allows us to alter the path of integration from the arc connecting ζ_I (h, k) and ζ_T (h, k) to the directed segment. With this in mind,

$$\left|\exp\left(\frac{\pi(2^{2j-r}-1)}{12\zeta} + \frac{\pi(24n+1-2^r)\zeta}{12k^2}\right)\sqrt{\zeta 2^{r-j}}\right|$$

$$\times \left|-1 + f_{2j-r}\left(\exp\left(\frac{-2\pi}{\zeta} + \frac{2H_j\pi i}{k}\right)\right)\right|$$

$$= \left|\exp\left(\frac{\pi(2^{2j-r}-1)}{12\zeta}\right)\exp\left(\frac{\pi(24n+1-2^r)\zeta}{12k^2}\right)\sqrt{\zeta 2^{r-j}}\right|$$

$$\times \left|\sum_{m=1}^{\infty} p_{2j-r}(m)\exp\left(\frac{-2\pi m}{\zeta}\right)\exp\left(\frac{2H_j\pi i m}{k}\right)\right|$$

$$= \exp\left(\frac{\pi(2^{2j-r}-1)}{12}\Re\frac{1}{\zeta}\right)\exp\left(\frac{\pi(24n+1-2^r)\Re\zeta}{12k^2}\right)|\zeta|^{1/2}2^{(r-j)/2}$$

$$\times \left|\sum_{m=1}^{\infty} p_{2j-r}(m)\exp\left(-2\pi m\Re\frac{1}{\zeta}\right)\exp\left(\frac{2H_j\pi i m}{k}\right)\right|$$

$$\leq e^{2\pi n}|\zeta|^{1/2}2^{r/2}\sum_{m=1}^{\infty}p_{2j-r}(m)\exp\left(-\frac{\pi}{12}\Re\frac{1}{\zeta}\left(24m-2^{2j-r}+1\right)\right)$$

$$\leq e^{2\pi n}|\zeta|^{1/2}2^{r/2}\sum_{m=1}^{\infty}p_0(m)\exp\left(-\frac{\pi}{12}\Re\frac{1}{\zeta}(24m-2^r+1)\right)$$

$$= c_j|\zeta|^{1/2} \tag{3.32}$$

for a constant c_j. So,

$$|\mathfrak{I}_{j,1}| \leq \frac{2\sqrt{2}k}{N}\left(\frac{\sqrt{2}k}{N}\right)^{1/2} c_j < C_j k^{3/2} N^{-3/2} \tag{3.33}$$

for a constant C_j, when $1 + \lfloor r/2 \rfloor \leq j \leq r$.

Combining the Estimates

One has

$$\left| i \sum_{\substack{k=1 \\ (k,2^r)=1}}^{N} k^{-5/2} \sum_{\substack{0 \leq h < k \\ (h,k)=1}} e^{-2\pi inh/k} \frac{\omega(h,k)\omega(2^r h,k)}{\omega(2h,k)} \mathcal{I}_{0,1} \right.$$
$$+ i \sum_{j=1}^{\lfloor r/2 \rfloor} \sum_{\substack{k=1 \\ (k,2^r)=2^j}}^{N} k^{-5/2} \sum_{\substack{0 \leq h < k \\ (h,k)=1}} e^{-2\pi inh/k} \frac{\omega(h,k)\omega(2^{r-j}h, 2^{-j}k)}{\omega(h,k/2)} (\mathcal{I}_{j,1} + \mathcal{I}_{j,2})$$
$$\left. + i \sum_{j=1+\lfloor r/2 \rfloor}^{r} \sum_{\substack{k=1 \\ (k,2^r)=2^j}}^{N} k^{-5/2} \sum_{\substack{0 \leq h < k \\ (h,k)=1}} e^{-2\pi inh/k} \frac{\omega(h,k)\omega(2^{r-j}h, 2^{-j}k)}{\omega(h,k/2)} \mathcal{I}_{j,1} \right|$$
$$< \sum_{j=0}^{r} \sum_{k=1}^{N} \sum_{h=0}^{k-1} C_j k^{-1} N^{-3/2} + \sum_{j=1}^{\lfloor r/2 \rfloor} \sum_{k=1}^{N} \sum_{h=0}^{k-1} C'_j k^{-1} N^{-3/2}$$
$$\leq C'' N^{-\frac{3}{2}} \sum_{k=1}^{N} 1, \quad \left(\text{where } C'' = \sum_{j=0}^{r} C_j + \sum_{j=1}^{\lfloor r/2 \rfloor} C'_j \right)$$
$$= O\left(N^{-1/2}\right). \tag{3.34}$$

Thus, we may revise (3.21) to

$$p_r(n) = i \sum_{k=1}^{N} k^{-5/2} \sum_{\substack{0 \leq h < k \\ (h,k)=1}} e^{-2\pi inh/k} \frac{\omega(h,k)\omega(2^r h,k)}{\omega(2h,k)} \mathcal{I}_{0,2}$$
$$+ i \sum_{j=1+\lfloor r/2 \rfloor}^{r} \sum_{k=1}^{N} k^{-5/2} \sum_{\substack{0 \leq h < k \\ (h,k)=1}} e^{-2\pi inh/k} \frac{\omega(h,k)\omega(2^{r-j}h, 2^{-j}k)}{\omega(h,k/2)} \mathcal{I}_{j,2} + O\left(N^{-1/2}\right). \tag{3.35}$$

Evaluation of $\mathcal{I}_{j,2}$ *for* $j = 0$ *and* $1 + \lfloor r/2 \rfloor \leq j \leq r$

Write $\mathcal{I}_{j,2}$ as

$$\mathcal{I}_{j,2} = \int_{K^{(-)}} \exp\left(\frac{\pi(2^{2j-r} - 1)}{12\zeta} + \frac{\pi(24n + 1 - 2^r)\zeta}{12k^2} \right) \sqrt{\zeta 2^{r-j}}\, d\zeta - \mathcal{I}_{j,3} - \mathcal{I}_{j,4}, \tag{3.36}$$

where

$$\mathcal{I}_{j,3} := \int_0^{\zeta_l(h,k)}, \quad \mathcal{I}_{j,4} := \int_{\zeta_r(h,k)}^0,$$

(3.37)

and $\mathcal{I}_{j,3}$ and $\mathcal{I}_{j,4}$ have the same integrand as (3.36) (and analogously for $I_{0,2}$).

Estimation of $\mathcal{I}_{j,3}$ *and* $\mathcal{I}_{j,4}$

We note that the length of the arc of integration in $\mathcal{I}_{j,3}$ is less than $\pi k/\sqrt{2}N$, and on this arc $|\zeta| < \sqrt{2}k/N$. [50, page 272]. Also, $\Re 1/\zeta = 1$ on K [50, page 271, equation (120.2)]. Further, $0 < \Re\zeta < 2k^2/N^2$ [50, page 271, equation (119.6)]. The absolute value of the integrand is thus

$$\left|2^{r-j}\zeta\right|^{1/2} \exp\left(\frac{(24n+1-2^r)\pi\Re\zeta}{12k^2} + \frac{(2^{2j-r}-1)\pi}{12}\Re\frac{1}{\zeta}\right)$$
$$< 2^{(r-j)/2}2^{1/4}k^{1/2}N^{-1/2}\exp\left(\frac{(24n+1-2^r)\pi}{6N^2} + \frac{(2^{2j-r}-1)\pi}{12}\right)$$

(3.38)

so that

$$|\mathcal{I}_{j,2}| < \pi k 2^{-1/2}N^{-1}2^{(r-j)/2}2^{1/4}k^{1/2}N^{-1/2}\exp\left(\frac{(24n+1-2^r)\pi}{6N^2} + \frac{(2^{2j-r}-1)\pi}{12}\right)$$
$$= \pi k^{3/2}N^{-3/2}2^{(2r-2j-1)/4}\exp\left(\frac{(24n+1-2^r)\pi}{6N^2} + \frac{(2^{2j-r}-1)\pi}{12}\right)$$
$$= O\left(k^{3/2}N^{-3/2}\exp\left(\frac{(24n+1-2^r)\pi}{6N^2}\right)\right).$$

(3.39)

By the same reasoning, $|\mathcal{I}_{j,3}| = O(k^{3/2}N^{-3/2}\exp((24n+1-2^r)\pi/6N^2))$
We may therefore revise (3.35) to

$$p_r(n) = i\sum_{k=1}^{N} k^{-5/2} \sum_{\substack{0\leq h<k \\ (h,k)=1}} e^{-2\pi inh/k}\frac{\omega(h,k)\omega(2^r h,k)}{\omega(2h,k)}$$
$$\times \int_{K^{(-1)}} \sqrt{\zeta 2^{r-1}}\exp\left\{\frac{\pi(2^{1-r}+1)}{24\zeta} + \frac{\pi(24n+1-2^r)\zeta}{12k^2}\right\}d\zeta$$
$$+ i\sum_{j=1+\lfloor r/2\rfloor}^{r}\sum_{k=1}^{N} k^{-5/2} \sum_{\substack{0\leq h<k \\ (h,k)=1}} e^{-2\pi inh/k}\frac{\omega(h,k)\omega(2^{r-j}h,2^{-j}k)}{\omega(h,k/2)}$$
$$\times \int_{K^{(-j)}} \sqrt{\zeta 2^{r-j}}\exp\left\{\frac{\pi(2^{2j-r}-1)}{12\zeta} + \frac{\pi(24n+1-2^r)\zeta}{12k^2}\right\}d\zeta + O\left(N^{-1/2}\right)$$

(3.40)

and upon letting N tend to infinity, obtain

$$p_r(n) = i\sum_{k=1}^{\infty} k^{-5/2} \sum_{\substack{0 \le h < k \\ (h,k)=1}} e^{-2\pi inh/k} \frac{\omega(h,k)\omega(2^r h,k)}{\omega(2h,k)}$$

$$\times \int_{K^{(-)}} \sqrt{\zeta 2^{r-1}} \exp\left\{ \frac{\pi(2^{1-r}+1)}{24\zeta} + \frac{\pi(24n+1-2^r)\zeta}{12k^2} \right\} d\zeta$$

$$+ i \sum_{j=1+\lfloor r/2 \rfloor}^{r} \sum_{k=1}^{\infty} k^{-5/2} \sum_{\substack{0 \le h < k \\ (h,k)=1}} e^{-2\pi inh/k} \frac{\omega(h,k)\omega(2^{r-j}h, 2^{-j}k)}{\omega(h,k/2)}$$

$$\times \int_{K^{(-)}} \sqrt{\zeta 2^{r-j}} \exp\left\{ \frac{\pi(2^{2j-r}-1)}{12\zeta} + \frac{\pi(24n+1-2^r)\zeta}{12k^2} \right\} d\zeta. \tag{3.41}$$

The Final Form

We may now introduce the change of variable

$$\zeta = \frac{\pi\left(2 + 2^{2j-r+1} - (2-\delta_{j0})^2\right)}{24t} \tag{3.42}$$

(where the first summation in (3.41) is the j=0 term separated out for clarity), which allows the integral to be evaluated in terms of $I_{3/2}$, the Bessel function of the first kind of order 3/2 with purely imaginary argument [53, page 372, §17.7] when we bear in mind that a "bent" path of integration is allowable according to the remark preceding Equation (8) on page 177 of [54]. See also [51, page 109]. The final form of the formula is then obtained by using the fact that Bessel functions of half-odd integer order can be expressed in terms of elementary functions.

We therefore have, for $c = \pi(2 + 2^{2j-r+1} - (\delta_{j0})^2)/24$,

$$p_r(n) = i\sum_{k=1}^{\infty} k^{-5/2} \sum_{\substack{0 \le h < k \\ (h,k)=1}} e^{-2\pi inh/k} \frac{\omega(h,k)\omega(2^r h,k)}{\omega(2h,k)} 2^{(r-1)/2} \left(\pi \frac{2^{1-r}+1}{24}\right)^{3/2}$$

$$\times \int_{c-i\infty}^{c+i\infty} t^{-5/2} \exp\left\{ t + \frac{(1+2^{1-r})(24n+1-2^r)\pi^2}{288k^2 t} \right\} dt$$

$$+ i \sum_{j=1+\lfloor r/2 \rfloor}^{r} \sum_{k=1}^{\infty} k^{-5/2}$$

$$\times \sum_{\substack{0 \le h < k \\ (h,k)=1}} e^{-2\pi inh/k} \frac{\omega(h,k)\omega(2^{r-j}h, 2^{-j}k)}{\omega(h,k/2)} \times 2^{(r-j)/2} \left(\pi \frac{2^{2j-r}-1}{12}\right)^{3/2}$$

$$\times \int_{c-i\infty}^{c+i\infty} t^{-5/2} \exp\left\{ t + \frac{(2^{2j-r}-1)(24n-2^r+1)\pi^2}{144k^2 t} \right\} dt$$

$$
\begin{aligned}
&= \frac{\pi}{(24n - 2^r + 1)^{3/4}} \\
&\times \Bigg\{ \frac{(1+2^{r-1})^{3/4}}{2^{(r-2)/4}} \sum_{\substack{k \geq 1 \\ (k,\, 2^{\max(r,1)})=1}} k^{-1} \\
&\quad \times \sum_{\substack{0 \leq h < k \\ (h,k)=1}} e^{-2\pi i n h/k} \frac{\omega(h,k)\omega(2^r h,k)}{\omega(2h,k)} I_{3/2}\left(\frac{\pi\sqrt{(24n - 2^r + 1)(1+2^{r-1})}}{2^{r/2} \cdot 6k} \right) \\
&\quad + \sum_{j=1+\lfloor r/2 \rfloor}^{r} \left(2^{2j-r} - 1\right)^{3/4} 2^{(2-j+r)/2} \sum_{\substack{k \geq 1 \\ (k,2^r)=2^j}} k^{-1} \\
&\quad \times \sum_{\substack{0 \leq h < k \\ (h,k)=1}} e^{-2\pi i n h/k} \frac{\omega(h,k)\omega(2^{r-j}h, 2^{-j}k)}{\omega(h,k/2)} I_{3/2}\left(\frac{\pi\sqrt{(24n - 2^r + 1)(1+2^{2j-r})}}{6k} \right) \Bigg\},
\end{aligned}
\tag{3.43}
$$

which, after application of the formula [51, page 110]

$$
I_{3/2}(z) = \sqrt{\frac{2z}{\pi}} \frac{d}{dz} \left(\frac{\sinh z}{z} \right),
\tag{3.44}
$$

is equivalent to Theorem 2.1.

ACKNOWLEDGMENTS

The author thanks the anonymous referee for bringing the work of Zuckerman [46] and Goldberg [47] to his attention. This, in turn, led the author to seek the more general result presented here in this final version of the paper.

REFERENCES

1. L. Euler, Introductio in Analysin Infinatorum, Marcum-Michaelem Bousquet, Lausanne, Switzerland, 1748.
2. G. H. Hardy and S. Ramanujan, "Asymptotic formulae in combinatory analysis," Proceedings of the London Mathematical Society, vol. 17, pp. 75–115, 1918.
3. H. Rademacher, "On the partition function p(n)," Proceedings of the London Mathematical Society, vol. 43, pp. 241–254, 1937.
4. E. Grosswald, "Some theorems concerning partitions," Transactions of the American Mathematical Society, vol. 89, pp. 113–128, 1958.
5. E. Grosswald, "Partitions into prime powers," The Michigan Mathematical Journal, vol. 7, pp. 97–122, 1960.
6. M. Haberzetle, "On some partition functions," American Journal of Mathematics, vol. 63, pp. 589–599, 1941.
7. P. Hagis Jr., "A problem on partitions with a prime modulus p≥3," Transactions of the American Mathematical Society, vol. 102, pp. 30–62, 1962.
8. P. Hagis Jr., "Partitions into odd summands," American Journal of Mathematics, vol. 85, pp. 213–222, 1963.
9. P. Hagis Jr., "On a class of partitions with distinct summands," Transactions of the American Mathematical Society, vol. 112, pp. 401–415, 1964.
10. P. Hagis Jr., "Partitions into odd and unequal parts," American Journal of Mathematics, vol. 86, pp. 317–324, 1964.
11. P. Hagis Jr., "Partitions with odd summands—some comments and corrections," American Journal of Mathematics, vol. 87, pp. 218–220, 1965.
12. P. Hagis Jr., "A correction of some theorems on partitions," Transactions of the American Mathematical Society, vol. 118, p. 550, 1965.
13. P. Hagis Jr., "On the partitions of an integer into distinct odd summands," American Journal of Mathematics, vol. 87, pp. 867–873, 1965.
14. P. Hagis Jr., "Some theorems concerning partitions into odd summands," American Journal of Mathematics, vol. 88, pp. 664–681, 1966.
15. P. Hagis Jr., "Partitions with a restriction on the multiplicity of the summands," Transactions of the American Mathematical Society, vol.

155, pp. 375–384, 1971.

16. L.-K. Hua, "On the number of partitions of a number into unequal parts," Transactions of the American Mathematical Society, vol. 51, pp. 194–201, 1942.
17. S. Iseki, "A partition function with some congruence condition," American Journal of Mathematics, vol. 81, pp. 939–961, 1959.
18. S. Iseki, "On some partition functions," Journal of the Mathematical Society of Japan, vol. 12, pp. 81–88, 1960.
19. S. Iseki, "Partitions in certain arithmetic progressions," American Journal of Mathematics, vol. 83, pp. 243–264, 1961.
20. J. Lehner, "A partition function connected with the modulus five," Duke Mathematical Journal, vol. 8, pp. 631–655, 1941.
21. J. Livingood, "A partition function with the prime modulus p>3," American Journal of Mathematics, vol. 67, pp. 194–208, 1945.
22. I. Niven, "On a certain partition function," American Journal of Mathematics, vol. 62, pp. 353–364, 1940.
23. V. V. Subramanyasastri, "Partitions with congruence conditions," The Journal of the Indian Mathematical Society, vol. 11, pp. 55–80, 1972.
24. K. Bringmann and K. Ono, "Coefficients of harmonic Maass forms," in Proceedings of the University of Florida Conference on Partitions, q-Series, and Modular Forms, 2008.
25. S. Corteel and J. Lovejoy, "Overpartitions," Transactions of the American Mathematical Society, vol. 356, no. 4, pp. 1623–1635, 2004.
26. K. Bringmann and J. Lovejoy, "Dyson's rank, overpartitions, and weak Maass forms," International Mathematics Research Notices, vol. 2007, no. 19, Article ID rnm063, 34 pages, 2007.
27. W. Y. C. Chen and J. J. Y. Zhao, "The Gaussian coefficients and overpartitions," Discrete Mathematics, vol. 305, no. 1–3, pp. 350–353, 2005.
28. S. Corteel, W. M. Y. Goh, and P. Hitczenko, "A local limit theorem in the theory of overpartitions," Algorithmica, vol. 46, no. 3-4, pp. 329–343, 2006.
29. S. Corteel and P. Hitczenko, "Multiplicity and number of parts in overpartitions," Annals of Combinatorics, vol. 8, no. 3, pp. 287–301,

2004.

30. S. Corteel, J. Lovejoy, and A. J. Yee, "Overpartitions and generating functions for generalized Frobenius partitions," in Mathematics and Computer Science. III, Trends in Mathematics, pp. 15–24, Birkhäuser, Basel, Switzerland, 2004.
31. S. Corteel and O. Mallet, "Overpartitions, lattice paths, and Rogers-Ramanujan identities," Journal of Combinatorial Theory. Series A, vol. 114, no. 8, pp. 1407–1437, 2007.
32. A. M. Fu and A. Lascoux, "q-identities related to overpartitions and divisor functions," Electronic Journal of Combinatorics, vol. 12, pp. 1–7, 2005, research paper 38.
33. M. D. Hirschhorn and J. A. Sellers, "Arithmetic relations for overpartitions," Journal of Combinatorial Mathematics and Combinatorial Computing, vol. 53, pp. 65–73, 2005.
34. M. D. Hirschhorn and J. A. Sellers, "Arithmetic properties of overpartitions into odd parts," Annals of Combinatorics, vol. 10, no. 3, pp. 353–367, 2006.
35. J. Lovejoy, "Gordon's theorem for overpartitions," Journal of Combinatorial Theory. Series A, vol. 103, no. 2, pp. 393–401, 2003.
36. J. Lovejoy, "Overpartitions and real quadratic fields," Journal of Number Theory, vol. 106, no. 1, pp. 178–186, 2004.
37. J. Lovejoy, "Overpartition theorems of the Rogers-Ramanujan type," Journal of the London Mathematical Society, vol. 69, no. 3, pp. 562–574, 2004.
38. J. Lovejoy, "A theorem on seven-colored overpartitions and its applications," International Journal of Number Theory, vol. 1, no. 2, pp. 215–224, 2005.
39. J. Lovejoy, "Rank and conjugation for the Frobenius representation of an overpartition," Annals of Combinatorics, vol. 9, no. 3, pp. 321–334, 2005.
40. J. Lovejoy, "Partitions and overpartitions with attached parts," Archiv der Mathematik, vol. 88, no. 4, pp. 316–322, 2007.
41. K. Mahlburg, "The overpartition function modulo small powers of 2," Discrete Mathematics, vol. 286, no. 3, pp. 263–267, 2004.
42. N. Robbins, "Some properties of overpartitions," JP Journal of Algebra, Number Theory and Applications, vol. 3, no. 3, pp. 395–404, 2003.

43. Ø. J. Rødseth and J. A. Sellers, "On m-ary overpartitions," Annals of Combinatorics, vol. 9, no. 3, pp. 345–353, 2005.
44. G. E. Andrews, The Theory of Partitions, vol. 2 of Encyclopedia of Mathematics and Its Applications, Addison-Wesley, Reading, Mass, USA, 1976.
45. G. E. Andrews, The Theory of Partitions, Cambridge Mathematical Library, Cambridge University Press, Cambridge, UK, 1998.
46. H. S. Zuckerman, "On the coefficients of certain modular forms belonging to subgroups of the modular group," Transactions of the American Mathematical Society, vol. 45, no. 2, pp. 298–321, 1939.
47. L. Goldberg, Transformation of theta-functions and analogues of Dedekind sums, Ph.D. thesis, University of Illinois, Urbana, Ill, USA, 1981.
48. H. Rademacher, "On the expansion of the partition function in a series," Annals of Mathematics, vol. 44, pp. 416–422, 1943.
49. H. Rademacher, Lectures on Analytic Number Theory, Tata Institute, Bombay, India, 1954-1955.
50. H. Rademacher, Topics in Analytic Number Theory, vol. 169 of Die Grundlehren der mathematischen Wissenschaften, Springer, New York, NY, USA, 1973.
51. T. M. Apostol, Modular Functions and Dirichlet Series in Number Theory, vol. 41 of Graduate Texts in Mathematics, Springer, New York, NY, USA, 2nd edition, 1990.
52. L. R. Ford, "Fractions," The American Mathematical Monthly, vol. 45, no. 9, pp. 586–601, 1938.
53. E. T. Whittaker and G. N. Watson, A Course of Modern Analysis, Cambridge University Press, Cambridge, UK, 4th edition, 1927.
54. G. N. Watson, A Treatise on the Theory of Bessel Functions, Cambridge University Press, Cambridge, UK, 2nd edition, 1944.

10

On The Exponential Generating Function For Non-Backtracking Walks

FrancescaArrigo[a], PeterGrindrod[b], Desmond J.Higham[a], VanniNoferini[c]

[a] Department of Mathematics and Statistics, University of Strathclyde, Glasgow, G1 1XH, UK

[b] Mathematical Institute, University of Oxford, Andrew Wiles Building, Radcliffe Observatory Quarter, Woodstock Road, Oxford, OX2 6GG, UK

[c] Department of Mathematical Sciences, University of Essex, Wivenhoe Park, Colchester, CO4 3SQ, UK

ABSTRACT

We derive an explicit formula for the exponential generating function associated with non-backtracking walks around a graph. We study both undirected and directed graphs. Our results allow us to derive computable expressions for non-backtracking versions of network centrality measures based on the matrix exponential. We find that eliminating backtracking walks in this context does not significantly increase the computational expense. We show how the new measures may be interpreted in terms of standard exponential centrality computation on a certain multilayer network. Insights

Citation: Arrigo, Francesca and Grindrod, Peter and Higham, Desmond J. and Noferini, Vanni (2018) "On the exponential generating function for non-backtracking walks". Linear Algebra and its Applications, 556. pp. 381-399. https://doi.org/10.1016/j.laa.2018.07.010

from this block matrix interpretation also allow us to characterize centrality measures arising from general matrix functions. Rigorous analysis on the star graph illustrates the effect of non-backtracking and shows that unwanted localization effects can be eliminated when we restrict to non-backtracking walks. We also investigate the localization issue on synthetic networks.

Keywords. Generating function, inverse participation ratio, Katz, localization, loops, multilayer, graph spectrum, backtracking walks.

INTRODUCTION

Motivation

The concept of a walk on a graph is very natural—on arriving at a node, the walker may continue by traversing any edge pointing out of that node. If, at any stage, the edge along which the walker continues is the reverse of the edge on which they arrived, then the walk is said to be backtracking. Non-backtracking walks have been analyzed in a number of fields. They play a key role in the study of zeta functions on graphs [24], with applications in spectral graph theory [4, 23], number theory [44], discrete mathematics [12, 41], quantum chaos [39], random matrix theory [40], stochastic analysis [3], applied linear algebra [42] and computer science [38, 45]. Within network science, constraining walks to be non-backtracking has been shown to offer benefits in community detection [26, 28], centrality measurement [6, 20, 29, 36], network comparison [30] and in the study of related issues concerning optimal percolation [31, 32].

A natural task across all these fields is to count the number of distinct nonbacktracking walks of a given length between pairs of nodes, and to form a compact expression for the associated generating function. For network centrality, in the case of standard walk counts it has been argued that an exponential-style power series gives a useful alternative to the standard resolvent-style version [15]. For example, based on an analogy with a physical oscillator, it may be argued that moving from the resolvent to the exponential takes us from classical to quantum physics [16]. Further, there are effective and reliable tools for computing the action of the matrix exponential [2, 21, 22]. This provides the initial motivation for our article, where we study the exponential generating function associated with non-backtracking walks. We also note that more general matrix functions have been proposed in this walk-counting context [17]. Hence, we then extend the analysis to cover generating functions based on arbitrary power series.

Background

We let $A = (a_{ij}) \in \mathbb{R}^{n\times n}$ denote the adjacency matrix for an unweighted graph with n nodes; that is, $a_{ij} = 1$ if there is an edge from node i to node j, and $a_{ij} = 0$ otherwise. We will assume the graph to be loopless, so that $a_{ii} = 0$ for all i, and to have no multiple edges. If we further assume the graph represented by A to be undirected, i.e., if $A = A^T$, then the graph will be said to be simple. We use D to denote the diagonal matrix whose diagonal entries are $D_{ii} = (A^2)_{ii}$. So D_{ii} counts the number of reciprocal neighbours of node i, that is, the number of nodes j such that $a_{ij} = a_{ji} = 1$. If $A = A^T$, so that all edges are reciprocated, then D_{ii} reduces to the vector of degrees and any node i for which $D_{ii} = 1$ will be called a leaf.

If A $6\neq$ A^T the network is said to be directed and we will denote by $S = (s_{ij}) \in \mathbb{R}^{n\times n}$ the matrix associated with the graph obtained by removing all edges that are not reciprocated, so that $s_{ij} = a_{ij}a_{ji}$ for all i, j = 1, 2, . . . , n.

A walk of length r from node i to node j is characterized by a sequence of edges $i = i_1 \mapsto i_2$, i_2 $7\mapsto$ i_3, . . . , $i_r \mapsto i_{r+1} = j$. Note that edges and nodes may be revisited in the course of a walk. It follows from the definition of matrix multiplication that A^r has (i, j)th element that counts the number of distinct walks of length r from i to j; see, for example, [13, Theorem 2.2.1]. Given a parameter t, the ordinary generating function [46] associated with this sequence of walk counts may be represented by a resolvent function:

$$\sum_{r=0}^{\infty} t^r A^r = (I - tA)^{-1},$$

where I is the identity matrix. If t is interpreted as a formal algebraic variable, this expression holds within the algebra of formal power series. Analytically, if t is interpreted as a real (or complex) valued parameter, it is valid for $|t| < 1/\rho(A)$, where $\rho(\cdot)$ denotes the spectral radius. Similarly, the associated exponential generating function [46] takes the form of a matrix exponential:

$$\sum_{r=0}^{\infty} \frac{t^r A^r}{r!} = \exp(tA),$$

analytically valid for all $t \in \mathbb{R}$ (or in $\mathbb{C}$).

Power series expansions that combine walk counts of different lengths have become widely used in the field of network science [17]. Here, the idea is to assign a weight to each node that summarizes its "importance" in the network. If we interpret importance as the ability of a node to initiate messages along the edges in the graph, then we see that $(I - tA)^{-1}1$, for $0 < t < 1/\rho(A)$, and $\exp(tA)1$, for $t > 0$, are candidates for measuring and comparing nodes, where $1 \in \mathbb{R}^n$ is the vector of ones. Here, the ith component arises by summing the number of distinct walks from node i to all nodes in the graph, with walks of length r discounted by t^r in the resolvent case and by $t^r/r!$ in the exponential case. The resolvent version dates back to the work of Katz [25] and the alternative based on the exponential function, called the total (node) communicability, was introduced and studied in [9]. We also note that both versions are related to eigenvector centrality [10, 11], which uses the Perron-Frobenius eigenvector of A when the underlying graph is strongly connected [8].

In recent work [29], Martin et al. argued that traditional centrality measures may suffer from unwanted localization effects when the network under study has a scale-free degree distribution (see, e.g., [7, 14]). In this type of network, there are a few nodes with a high number of neighbours, that are thus assigned most of the centrality, and many nodes with a very small number of connections, that hence share the remainder of the mass of the centrality vector. This unbalanced partitioning makes it difficult to distinguish between the importance of the nodes with few connections. The authors therefore proposed a variation of eigenvector centrality motivated by the concept of non-backtracking walks. This idea was pursued in [6, 20] in the context of resolvent-based centrality. Our aim here is to develop and study the analogous exponential version of non-backtracking walk centrality, and also to consider general matrix functions.

Formally, we say that a walk is backtracking if it contains at least one pair of successive edges of the form $i \mapsto j, j \mapsto i$; that is, having left some node i it immediately returns to that node. We say the walk is non-backtracking otherwise. We use the abbreviation NBTW to mean non-backtracking walk. We will denote by $p_r(A) \in \mathbb{R}^{n\times n}$ the matrix whose (i, j)th entry counts the number of NBTWs of length r from node i to node j in the graph represented by the matrix A. Our first task is therefore to obtain useful expressions for the corresponding exponential generating function

$$F(t) := \sum_{r=0}^{\infty} \frac{t^r p_r(A)}{r!} . \qquad (1.1)$$

and centrality measure

$$\mathbf{b} := F(t)\mathbf{1}. \tag{1.2}$$

We do this in Sections 2 and 3 for the case of undirected and directed graphs, respectively. Although, of course, an undirected graph is a special case of a directed graph, we treat the two cases separately for two reasons. First, undirected graphs are common in network science, and hence it is useful to have results tailored to this case. Second, in our context it seems more straightforward to study undirected graphs directly than to simplify the more general expression for directed graphs. In Section 4, we briefly describe how to compute the proposed centrality measures and discuss computational costs. In Section 5, we interpret the new expressions from the viewpoint of traversals around multilayer networks with possibly negative interlayer weights. This block matrix framework also allows us to derive expressions for the case where the exponential is replaced by a general matrix function. In Section 6 we give a concrete example of the benefit of non-backtracking, and further illustrations on realistic model networks appear in Section 7. We end with a short summary in Section 8.

The key analytical steps used here are to derive and solve an ODE—this is a widely used approach in the study of exponential generating functions [46]. To this end, we note from (1.1) that

$$F'(t) = \sum_{r=0}^{\infty} \frac{t^r p_{r+1}(A)}{r!} \quad \text{and} \quad F''(t) = \sum_{r=0}^{\infty} \frac{t^r p_{r+2}(A)}{r!}. \tag{1.3}$$

We also require a lemma on matrix-valued ODEs. The following result arises from column wise application of a standard result on vector-valued linear systems; see, for example, [21, Section 2.1]. The statement of the result makes use of the shifted exponential function $\psi_1(z) = (e^z - 1)/z$, which may be defined formally via the power series expansion

$$\psi_1(z) = \sum_{r=0}^{\infty} \frac{z^r}{(r+1)!}. \tag{1.4}$$

Lemma 1.1. Let $C \in \mathbb{R}^{m \times m}$ and $P, X_0 \in \mathbb{R}^{m \times \ell}$. Then

$$X(t) = t\psi_1(tC)(P + CX_0) + X_0 \tag{1.5}$$

is the unique solution to the linear, constant coefficient, inhomogeneous, autonomous, initial value ODE system

$$X'(t) = CX(t) + P, \qquad X(0) = X_0.$$

Moreover, in the homogeneous case P = 0, the expression (1.5) simplifies to

$$X(t) = \exp(tC)X_0. \quad (1.6)$$

Remark 1.2. We note that Lemma 1.1 does not require the Jacobian matrix C to be nonsingular.

EXPONENTIAL GENERATING FUNCTION FOR UNDIRECTED GRAPHS.

We assume throughout this section that the graph is undirected. We also assume for convenience that the graph is connected—in the disconnected case, each component may be studied separately. We note that each node is therefore assumed to have at least one neighbour.

As we mentioned in Section 1.2, the matrix Ar counts walks of length r. Hence the trivial relationship $A^{r+1} = AA^r$ gives a recurrence between walk counts for length r and r + 1. In the case of NBTWs on undirected graphs, the following two term recurrence relates walk counts of successive lengths. We have $p_1(A) = A$, $p_2(A) = A^2 - D$, and

$$p_{r+2}(A) = Ap_{r+1}(A) - (D - I)p_r(A), \quad \text{for } r \geq 1. \quad (2.1)$$

This recursion was proved in [12] for general directed graphs. For the undirected case considered here, the recurrence was re-derived independently in [41].

We now give the main statement of this section: an explicit formula for the generating function (1.1) and the corresponding centrality measure (1.2).

Theorem 2.1. For a simple graph with associated adjacency matrix A, the exponential generating function F(t) in (1.1) and total communicability vector b in (1.2) satisfy

$$F(t) = \begin{bmatrix} I & 0 \end{bmatrix} t\psi_1(tY) \begin{bmatrix} A \\ A^2 - D \end{bmatrix} + I,$$

$$\mathbf{b} = \begin{bmatrix} I & 0 \end{bmatrix} t\psi_1(tY) \begin{bmatrix} A\mathbf{1} \\ (A^2 - D)\mathbf{1} \end{bmatrix} + \mathbf{1},$$

where

$$Y := \begin{bmatrix} 0 & I \\ I - D & A \end{bmatrix} \in \mathbb{R}^{2n \times 2n}. \quad (2.2)$$

4

In the case where the graph has no leaves, these expressions may be written as

$$F(t) = \begin{bmatrix} I & 0 \end{bmatrix} \exp(tY) \begin{bmatrix} I + (D-I)^{-1} \\ A \end{bmatrix} - (D-I)^{-1},$$

$$\mathbf{b} = \begin{bmatrix} I & 0 \end{bmatrix} \exp(tY) \begin{bmatrix} \mathbf{1} + (D-I)^{-1}\mathbf{1} \\ A\mathbf{1} \end{bmatrix} - (D-I)^{-1}\mathbf{1}.$$

Proof. To obtain the exponential generating function F(t) in (1.1), we multiply by $t^r/r!$ in (2.1) and then sum from r = 1 to ∞, to obtain

$$\sum_{r=1}^{\infty} \frac{t^r p_{r+2}(A)}{r!} = A \sum_{r=1}^{\infty} \frac{t^r p_{r+1}(A)}{r!} - (D-I) \sum_{r=1}^{\infty} \frac{t^r p_r(A)}{r!}.$$

From (1.3), this shows

$$F''(t) - p_2(A) = A(F'(t) - p_1(A)) - (D-I)(F(t) - p_0(A)),$$

which simplifies to

$$F''(t) - AF'(t) + (D-I)F(t) = -I. \tag{2.3}$$

To convert to a first order system, we introduce

$$G(t) = \begin{bmatrix} G_1(t) \\ G_2(t) \end{bmatrix} := \begin{bmatrix} F(t) \\ F'(t) \end{bmatrix} \in \mathbb{R}^{2n \times n},$$

so that (2.3) becomes

$$G'(t) = YG(t) + \begin{bmatrix} 0 \\ -I \end{bmatrix}, \qquad \text{where } G(0) = \begin{bmatrix} I \\ A \end{bmatrix},$$

where Y is defined in (2.2). It then follows from Lemma 1.1 that

$$G(t) = t\psi_1(tY) \left(\begin{bmatrix} 0 \\ -I \end{bmatrix} + Y \begin{bmatrix} I \\ A \end{bmatrix} \right) + \begin{bmatrix} I \\ A \end{bmatrix}. \tag{2.4}$$

Finally, we note that the Jacobian matrix Y in (2.2) is invertible if and only if $(D - I)^{-1}$ exists, that is, if the graph does not contain leaves, i.e., nodes of degree one. In this case

$$Y^{-1} = \begin{bmatrix} (D-I)^{-1}A & -(D-I)^{-1} \\ I & 0 \end{bmatrix}$$

and (2.4) simplifies to

$$G(t) = \exp(tY) \begin{bmatrix} I + (D-I)^{-1} \\ A \end{bmatrix} - \begin{bmatrix} (D-I)^{-1} \\ 0 \end{bmatrix}. \tag{2.5}$$

Noting that F(t) = I 0 G(t), we see that (2.4) and (2.5) yield the statement. We discuss these results further in Sections 4 and 5.

EXPONENTIAL GENERATING FUNCTION FOR DIRECTED GRAPHS.

For directed graphs the matrices $p_r(A)$ that count NBTWs satisfy a three term recurrence. This was shown in [12], with a more linear algebra oriented derivation given in [42]. With $p_0(A) = I$, $p_1(A) = A$, $p_2(A) = A^2 - D$, the recurrence takes the form

$$p_{r+3}(A) = Ap_{r+2}(A) - (D-I)p_{r+1}(A) - (A-S)p_r(A), \quad \text{for } r \geq 0. \tag{3.1}$$

Recall that $S \in \mathbb{R}^{n \times n}$ is the adjacency matrix of the graph obtained by only considering reciprocated links in the original network; so $s_{ij} = a_{ij}a_{ji}$. We emphasize that unlike the undirected case (2.1), this recurrence is valid from r = 0. Here is the analogue for directed graphs of Theorem 2.1.

Theorem 3.1. Let A be the adjacency matrix of an unweighted directed graph with no self-loops nor multiple edges. The exponential generating function F(t) in (1.1) and total communicability vector b in (1.2) satisfy

$$F(t) = \begin{bmatrix} I & 0 & 0 \end{bmatrix} \exp(tZ) \begin{bmatrix} I \\ A \\ A^2 - D \end{bmatrix},$$

$$\mathbf{b} = \begin{bmatrix} I & 0 & 0 \end{bmatrix} \exp(tZ) \begin{bmatrix} \mathbf{1} \\ A\mathbf{1} \\ (A^2 - D)\mathbf{1} \end{bmatrix},$$

Where

$$Z := \begin{bmatrix} 0 & I & 0 \\ 0 & 0 & I \\ -(A-S) & -(D-I) & A \end{bmatrix} \in \mathbb{R}^{3n \times 3n}. \tag{3.2}$$

Proof. Multiplying (3.1) by $t^r/r!$ and summing from r = 0 to ∞ we arrive at

$$F'''(t) = AF''(t) - (D-I)F'(t) - (A-S)F(t). \tag{3.3}$$

We note that, in contrast to (2.3), this ODE does not have an inhomogeneous term. To convert to a first order system, introduce

$$H(t) = \begin{bmatrix} H_1(t) \\ H_2(t) \\ H_3(t) \end{bmatrix} = \begin{bmatrix} F(t) \\ F'(t) \\ F''(t) \end{bmatrix},$$

so that

$$H'(t) = ZH(t), \qquad \text{where } H(0) = \begin{bmatrix} I \\ A \\ A^2 - D \end{bmatrix};$$

where Z is defined in (3.2).

Using (1.6) in Lemma 1.1 and noting that F(t) = I 0 0 H(t), we arrive at the statement.

It is of interest to note that the removal of backtracking walks may be interpreted as the introduction of damping terms −(D − I)F ′ (t) and −(A − S)F(t) in (3.3). Further, Theorem 3.1 shows that, in general, as t increases the total communicability vector b aligns with the first n components of the dominant eigenvector of Z.

COMPUTING THE CENTRALITY VECTORS.

In this section we consider computational issues. First, we note that the standard expression for the total (node) communicability, $e^{tA}1$, requires the action of the exponential of a sparse n by n matrix on a vector in $\mathbb{R}^n$. It is reasonable to assume that a suitable approximation may be obtained iteratively, using a small number of sparse matrix-vector multiplications [2]. Assuming that A has O(n) nonzeros (edges), then each sparse matrix-vector multiplication will cost O(n). Hence, using k to denote the number of iterations, the overall cost will be O(nk). In practice, it is reasonable to regard k as finite and independent from n, so that this expression becomes O(n).

In Theorem 2.1, we see that for an undirected graph with no leaves, the cost of computing the non-backtracking total communicability vector b is dominated by the action of e^{tY} on a vector in $\mathbb{R}^{2n}$, where the 2n by 2n sparse matrix Y is defined in (2.2). We note that Y has only 2n more nonzeros than the underlying adjacency matrix A, and hence, under the assumptions above, if the same number of iterations are required, then the same O(n) cost applies.

In the directed case, the expression for b in Theorem 3.1 requires the action of e^{tZ} on a vector in $\mathbb{R}^{3n}$, where the 3n by 3n sparse matrix Z is defined in (3.2). Since Z also has O(n) nonzeros, we again recover the O(n) cost under our assumptions.

It remains to consider the case of an undirected graph with leaves. In Theorem 2.1, the expression for b involves the shifted exponential, ψ_1, from (1.4). We may proceed by reducing the key computation to the approximation of the action of a matrix exponential in a slightly higher dimension. Let

$$\widehat{Y} := \begin{bmatrix} tY & \mathrm{v} \\ \mathbf{0}^T & 0 \end{bmatrix} \in \mathbb{R}^{(2n+1)\times(2n+1)},$$

where $\mathrm{v} \in \mathbb{R}^{2n}$ and $0 \in \mathbb{R}^{2n}$ is a vector of all zeros. It was shown in [37,

Section 2.3] that

$$\exp(\widehat{Y}) = \begin{bmatrix} e^{tY} & \psi_1(tY)\mathbf{v} \\ \mathbf{0}^T & 1 \end{bmatrix}.$$

So, if we let

$$\mathbf{v} = \begin{bmatrix} A\mathbf{1} \\ (A^2 - D)\mathbf{1} \end{bmatrix}$$

then b in Theorem 2.1 may be computed by first approximating the vector $(\widehat{Y})\mathbf{e}_{2n+1}$, where e_{2n+1} is the (2n + 1)th vector of the standard basis of $\mathbb{R}^{2n+1}$. We then extract the top n entries of the resulting vector, scale by t and shift by 1. This again will have an O(n) cost under our assumptions.

Remark 4.1. It is worth emphasizing that the final steps of scaling and shifting can be avoided if we are only interested, as is customary in network science, in the rankings of the nodes rather than their actual centrality scores.

BLOCK MATRIX INTERPRETATIONS

Theorems 2.1 and 3.1 show that, loosely, for exponential-based centrality the operation of eliminating backtracking walks is equivalent to replacing the original adjacency matrix A with a suitable block matrix. In the undirected case, the 2 by 2 version Y in (2.2) is used, whereas the directed case has the 3 by 3 version Z in (3.2). These block matrices were constructed indirectly, as a consequence of the generating function approach, [46]. In this section, we add insight by showing how these block matrices may be interpreted directly. We also show that they remain relevant when the exponential is replaced by a general matrix function.

We begin with an algebraic interpretation. The matrices Z and Y are the first companion matrix linearization [18] of the matrix polynomials $(A - S) + (D - I)\lambda - A\lambda^2 + I\lambda^3$ and, respectively, $(D - I) - A\lambda + I\lambda^2$. These are in turn the reversals of, respectively, the directed and undirected version of the deformed graph Laplacian, a matrix polynomial that plays a crucial role in the theory of NBTW combinatorics, see [6, 20]. Algebraically, the companion matrix is the multiplication-by-λI operator in the module obtained by forming quotients in the ring of matrix polynomials using the left ideal generated by the linearized matrix polynomial [34]. Linear-algebraic functions of the companion matrix also have an analogous interpretation; for example, exp(tZ) represents multiplication by $\exp(t\lambda)I$ modulo $(A-S)+$

$(D-I)\lambda - A\lambda^2 + I\lambda^3$. For more details on this algebraic view of companion matrices see, e.g., [33, Sec. 2], [34, Sec. 5], [35, Sec. 9], and the references therein.

There is also a graph-theoretical interpretation that we now describe in detail. We focus first on the directed case, involving Z in (3.2). A useful connection is that Z represents the adjacency matrix for a three layer weighted network. This network is node-aligned [27], so that the general ith nodes from each layer may be regarded as copies of the same node. Looking at the diagonal blocks of Z, we see that the within-layer connections correspond to the empty graph for layers one and two, and the original graph for layer three. Next, we consider the off-diagonal blocks. The identity matrix in the (1, 2) block and zero matrix in the (1, 3) block show that the only edge out of a node in layer one points to the corresponding node in layer two. Similarly, we may only leave layer two by moving to the corresponding node in layer three. Within layer three, in addition to moving within the layer using edges from A, traversals are possible to layers two and one. The (3, 2) block shows that for $D_{ii} > 1$ a traversal may move from node i in layer three to node i in layer two, with the negative weight $1 - D_{ii}$ assigned to that edge. The (3, 1) block shows that a move from node i in layer three to node j in layer one is allowed if and only if the original graph has a nonreciprocal edge from node i to node j, in which case the inter-layer edge carries weight −1.

The expression for b in Theorem 3.1 shows that exponential NBTW centrality arises from a post-processed version of the "standard" exponential centrality on this multilayer network, where negative inter-layer weights have the effect of negating the contributions of backtracking walks in the underlying graph.

The two lemmas and the theorem below show that Z is also relevant for more general matrix functions.

Lemma 5.1. For all $r \geq 0$ we have

$$Z^r \widetilde{Z} \begin{bmatrix} 0 \\ 0 \\ I \end{bmatrix} = \begin{bmatrix} p_r(A) \\ p_{r+1}(A) \\ p_{r+2}(A) \end{bmatrix}$$

Proof. For $r < 2$, the identity may be verified directly. For $r \geq 2$ the result follows by straightforward induction using the three-term recurrence (3.1).

Theorem 5.2 (Directed graph). *Let* $\{\alpha_r\}_{r=0}^{\infty}$ be a sequence such that $f0(Z) = \sum_{r=0}^{\infty} \alpha_r Z^r$ converges. Then $\sum_{r=0}^{\infty} \alpha_r p_r(A)$ converges, and

$$\sum_{r=0}^{\infty} \alpha_r p_r(A) = [I\ 0\ 0] f_0(Z) \begin{bmatrix} I \\ A \\ A^2 - D \end{bmatrix}$$

Moreover, $\sum_{r=0}^{\infty} \alpha_r p_r(A)$ is equal to the (3, 3) block in $f_0(Z) - f_2(Z)$, where

$$f_k(Z) = \sum_{r=0}^{\infty} \alpha_{r+k} Z^r.$$

Proof. By Lemma 5.1 it follows that

$$\sum_{r=0}^{\infty} \alpha_r p_r(A) = [I\ 0\ 0] \sum_{r=0}^{\infty} \alpha_r Z^r \widetilde{Z} \begin{bmatrix} 0 \\ 0 \\ I \end{bmatrix}$$

and, since

$$\widetilde{Z} \begin{bmatrix} 0 \\ 0 \\ I \end{bmatrix} = \begin{bmatrix} I \\ A \\ A^2 - D \end{bmatrix}$$

the first part of the statement follows.

For the second part, note first that $\sum_{r=0}^{\infty} \alpha_{r+2} Z^r$ also converges under the assumptions in the statement. Then

$$\begin{aligned} \sum_{r=0}^{\infty} \alpha_r p_r(A) &= \alpha_0 I + \alpha_1 A + \sum_{r=2}^{\infty} \alpha_r p_r(A) \\ &= \alpha_0 I + \alpha_1 A + [0\ 0\ I] \sum_{r=2}^{\infty} \alpha_r Z^{r-2} \widetilde{Z} \begin{bmatrix} 0 \\ 0 \\ I \end{bmatrix} \\ &= \alpha_0 I + \alpha_1 A + [0\ 0\ I] \sum_{r=2}^{\infty} \alpha_r Z^r \begin{bmatrix} 0 \\ 0 \\ I \end{bmatrix} - [0\ 0\ I] \sum_{r=2}^{\infty} \alpha_r Z^{r-2} \begin{bmatrix} 0 \\ 0 \\ I \end{bmatrix} \\ &= [0\ 0\ I] \sum_{r=0}^{\infty} (\alpha_r - \alpha_{r+2}) Z^r \begin{bmatrix} 0 \\ 0 \\ I \end{bmatrix}, \end{aligned}$$

as required.

Theorem 5.2 shows that we can accumulate combinations of NBTW counts in the original network by looking at the (3, 3) block of an appropriate function of the matrix Z. From a multilayer perspective, the (3, 3) block of Z is associated with the layer containing the adjacency matrix A of the original network. Theorem 5.2 therefore concerns walks of any length that start and end at that layer. This includes not only walks that take place within layer three, but also those that make one or more intermediate visits to the other layers.

The contribution coming from $f_0(Z)$ counts walks in the multilayer network using the original weights. The correction term involving $f_2(Z)$ removes walks from the count by weighting walks of length r − 2 as if they were of length r.

In more detail, the term $f_2(Z)$ is used to correct the appearance of an identity matrix in Z_2 that is then propagated at higher powers of Z. This identity matrix at level r = 2 originates from the one appearing in the (3, 2) block. This latter is used in the recursion (3.1) defining pr(A) when r > 2 to correct the penalization of walks of the type

$$i \to \cdots \to k \to j \to k \to j \qquad \text{of length } r$$

which are backtracking at length $(r-1)^*$ and thus need not be removed by the factor $p_{r-2}(A)D$. Walks of this type take place entirely in the third layer and are balanced by the existence of walks of the form

$$\begin{array}{lccccccccccc} \text{node:} & i & \to & \cdots & \to & k & \to & j & \to & j & \to & j \\ \text{layer:} & [3] & \to & \cdots & \to & [3] & \to & [3] & \to & [2] & \to & [3] \end{array}$$

provided that r > 2. When r = 2, walks of the form

$$\begin{array}{lccccc} \text{node:} & j & \xrightarrow{1-d_j} & j & \xrightarrow{1} & j \\ \text{layer:} & [3] & \to & [2] & \to & [3] \end{array}$$

are improperly weighted, as the first hop contains a +1 that is used to balance the removal of walks that have never even existed. This introduces a "delay" that is then propagated when longer walks are counted. To counteract it, the matrix function $f_2(Z)$ removes walks of length r−2 that are however weighted as if they were of length r; this type of weighting precisely targets those walks obtained from the propagation of the identity matrix introduced at r = 2 in $f_0(Z)$.

Theorem 5.2 can be used to recover the known characterisation of the ordinary generating function. To see this, we let $\alpha r = t^r$ for all r, with $|t| < \rho(Z)^{-1}$. Then we have

$$\sum_{r=0}^{\infty} t^r p_r(A) = [0\ 0\ I]\left[(I-tZ)^{-1} - t^2(I-tZ)^{-1}\right]\begin{bmatrix}0\\0\\I\end{bmatrix}$$

$$= (1-t^2)[0\ 0\ I](I-tZ)^{-1}\begin{bmatrix}0\\0\\I\end{bmatrix}.$$

This is equivalent to the result in [6, Section 3].

Similarly, setting $\alpha r = t^r/r!$ for some t > 0, Theorem 5.2 gives

$$\sum_{r=0}^{\infty} \frac{t^r p_r(A)}{r!} = [0\ 0\ I]\left(e^{tZ} - t^2\psi_2(tZ)\right) \begin{bmatrix} 0 \\ 0 \\ I \end{bmatrix}, \tag{5.1}$$

where $\psi_2(x) = \sum_{r=0}^{\infty} x^r/(r+2)!$. This description of F(t) from (1.1) is equivalent to that given in Theorem 3.1. However, (5.1) casts the result entirely in terms of the (3, 3) block, which, as we have argued above, has a natural multilayer interpretation

We finish this section by briefly discussing the undirected case. Here, the block matrix Y in (2.2) may be associated with a two-layer network. In layer one, the only possible transition is to the equivalent node in layer two. We may move within layer two using any edge that exists in the original graph, or, for $D_{ii} > 1$, we may move from node i in layer two to node i in layer one along an edge with negative weight $1 - D_{ii}$. The analogues of Lemma 5.1 and Theorem 5.2 are given below, with proofs following in a similar manner

Lemma 5.3. For all $r \geq 0$ we have

$$Y^r \widetilde{Y} \begin{bmatrix} 0 \\ I \end{bmatrix} = \begin{bmatrix} p_{r+1}(A) \\ p_{r+2}(A) \end{bmatrix}.$$

Theorem 5.4 (Undirected graph). Let $\{\alpha_r\}_{r=0}^{\infty}$ be a sequence of nonnegative weights such that $f_k(Y) := \sum_{r=0}^{\infty} \alpha_{r+k} Y^r$ converges for k = 0. Then

$$\sum_{r=0}^{\infty} \alpha_r p_r(A) = [0\ I]\left(f_0(Y) - f_2(Y)\right) \begin{bmatrix} 0 \\ I \end{bmatrix}.$$

Theorem 5.4 quantifies in the undirected case how NBTWs on the original network may also be counted by considering walks on the associated multilayer network that start and end in the layer containing A.

STAR GRAPH ANALYSIS

In this section, we give further insights into the effect of restricting to NBTWs in a total communicability centrality measure. For eigenvector centrality, the authors in [29] argued that non-backtracking is a means to avoid localization; that is, the concentration of weight on a small number of nodes. Following [19], they characterized localization through an asymptotic limit. Given $v = (v_i) \in \mathbb{R}^n$ the inverse participation ratio is defined as

$$S(\mathbf{v}) = \frac{1}{\|\mathbf{v}\|_2^4} \sum_{i=1}^{n} v_i^4 = \frac{\|\mathbf{v}\|_4^4}{\|\mathbf{v}\|_2^4}. \tag{6.1}$$

This quantity is of order one in the case where a small number of elements in v are responsible for most of the weight. More formally, we say that the measure is localized if S(v) = O(1) and nonlocalized if S(v) = o(1), in the asymptotic limit $n \to \infty$.

We will analyze this property on the undirected star graph with n nodes. Here the adjacency matrix takes the form

$$A = \left[\begin{array}{c|c} & \mathbf{1}_{n-1}^T \\ \hline \mathbf{1}_{n-1} & \end{array}\right], \tag{6.1}$$

where $1_s \in \mathbb{R}^s$ is the vector of all ones. Thus node 1 is a hub with n − 1 connections to nodes 2, 3, . . . , n, which are only connected to node 1. Intuitively, the effect of backtracking should be significant in this example. The hub node has n−1 neighbours, and hence n − 1 walks of length 1, but cannot initiate walks of length two or more without backtracking. Each peripheral node, in contrast, has a single neighbour, but n − 2 NBTWs of length two. Hence, in the non-backtracking regime, if t is taken sufficiently large that degree does not dominate, then having one high quality link (to a hub) may be comparable in importance to having many low quality links (to leaves).

It is straightforward to show for (6.2) that (see, for example, [5, 21])

$$e^{tA} = \left[\begin{array}{c|c} \cosh(t\sqrt{n-1}) & \sinh(t\sqrt{n-1})(\sqrt{n-1})^{-1}\mathbf{1}_{n-1}^T \\ \hline \sinh(t\sqrt{n-1})(\sqrt{n-1})^{-1}\mathbf{1}_{n-1} & \cosh(t\sqrt{\mathbf{1}_{n-1}\mathbf{1}_{n-1}^T}) \end{array}\right].$$

Hence, the standard, backtracking walk version of total communicability, x = e ^{tA1}, satisfies

$$\mathbf{x} = \left[\begin{array}{c} \cosh(t\sqrt{n-1}) + \sqrt{n-1}\sinh(t\sqrt{n-1}) \\ \cosh(t\sqrt{n-1})\mathbf{1}_{n-1} + (\sqrt{n-1})^{-1}\sinh(t\sqrt{n-1})\mathbf{1}_{n-1} \end{array}\right].$$

So the values x_1 for the hub node and x_2 for a typical leaf node satisfy

$$x_1 = \cosh(t\sqrt{n-1}) + \sqrt{n-1}\sinh(t\sqrt{n-1}),$$
$$x_2 = \cosh(t\sqrt{n-1}) + (\sqrt{n-1})^{-1}\sinh(t\sqrt{n-1}).$$

It follows immediately that for any $t > 0$ we have $x_1 > x_2$ when the graph has at least three nodes. Hence, the standard total communicability measure always ranks the hub node above the leaves. We also find that, for fixed $t > 0$,

$$(x_1)^4 = \frac{n^2}{16}e^{4t\sqrt{n-1}} + o\left(n^2e^{4t\sqrt{n-1}}\right),$$

$$(x_i)^4 = \frac{n^2}{16(n-1)^2}e^{4t\sqrt{n-1}} + o\left(e^{4t\sqrt{n-1}}\right), \quad \text{for all } i = 2, \ldots, n,$$

$$\|\mathbf{x}\|_2^4 = \frac{n^2}{4}e^{4t\sqrt{n-1}} + o\left(n^2e^{4t\sqrt{n-1}}\right).$$

It follows that the inverse participation ratio (6.1) satisfies

$$\lim_{n\to\infty} S(\mathbf{x}) = \frac{1}{4}$$

and hence the measure is localized.

We may evaluate the non-backtracking version, $b = F(t)1$, from first principles. The hub node has $n - 1$ NBTWs of length one and no others, and each leaf node has one NBTW of length one, $n - 2$ NBTWs of length two and no others. So

$$b_1 = 1 + t(n-1) \quad \text{and} \quad b_2 = 1 + t + \frac{t^2}{2}(n-2).$$

Straightforward computation shows that $b_1 > b_2$ for $0 < t < 2$ and $b_1 < b_2$ for $t > 2$. Therefore, we observe a transition that is not present with standard exponential centrality, albeit in a parameter regime where t is greater than unity.

More tellingly, for fixed $t > 0$

$$(b_1)^4 = t^4n^4 + o(n^4),$$

$$(b_i)^4 = \frac{t^8}{16}n^4 + o(n^4), \quad \text{for all } i = 2, \ldots, n,$$

$$\|\mathbf{b}\|_2^4 = \frac{t^8}{16}n^6 + o(n^6).$$

Hence,

$$\lim_{n\to\infty} S(\mathbf{b}) = 0;$$

so the measure is nonlocalized.

NUMERICAL TESTS

In this section we study the localization effect in a realistic setting. We used a random graph model rather than a fixed data set so that we can control the dimension, n, and study the asymptotics. We used an implementation of the Barab´asi-Albert preferential attachment model for undirected networks in the MATLAB toolbox CONTEST [43]. In this model, nodes are added sequentially until the desired size is reached. Each node is given, upon arrival, d links to current nodes.

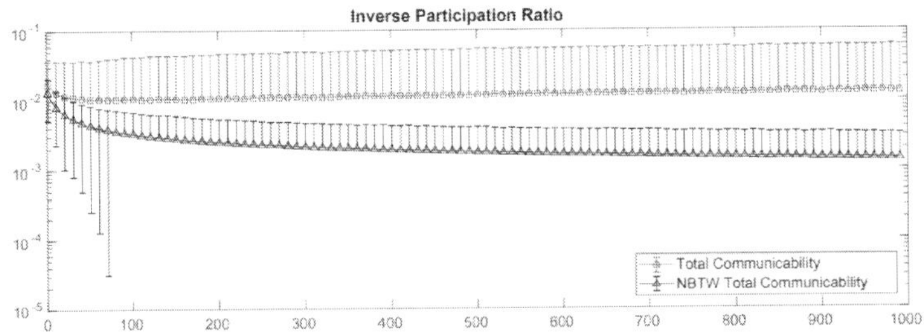

Figure 1. Inverse participation ratio of vectors F(t)1 and $e^{tA}1$ as the number of nodes, n, varies.

These target nodes are chosen independently with a probability proportional to their current degree. The resulting network will have a scale-free degree distribution, where a few nodes (hubs) have a high degree and many nodes have a very small degree. In this case, the vector of degrees is localized by construction. We have $\rho(A) \leq \|A\|_\infty$, and when this inequality is sharp the presence of a high degree node forces Katz to use a small parameter α, producing results close to degree centrality [8].

We used the default setting of the toolbox, giving each node d = 2 links upon arrival to the network, and we built networks of increasing size from n = 10 to n = 1000. For each network, we computed the inverse participation ratio (6.1) for the normalized total communicability vector $e^{tA}1$ and its nonbactracking analogue b = F(t)1 for t = 1 fixed. In order to compute both centrality vectors, we used the MATLAB toolbox funm kryl [1], which computes the action of a matrix function over a vector using a restarted Arnoldi algorithm. We repeated our tests 1000 times.

The results are displayed in Figure 1, where the sample average of the inverse participation ratio is plotted against n in a semilogarithmic scale. The error bar is used to display the standard error around the average values. The results are in line with the analytical results presented in Section 6—the inverse participation ratio decays towards zero for the non-backtracking version, but appears to be bounded away from zero when backtracking walks are included.

SUMMARY

The main contributions of this work were

1. to derive analytical expressions for the non-backtracking walk exponential generating function for both undirected and directed graphs,
2. to show that it is feasible to compute the resulting network centrality measures: the key matrix involved in the computations has twice (three times) the dimension of the analogous backtracking version for undirected (directed) graphs, and has a comparable level of sparsity,
3. to interpret the new measures in the context of traditional centrality on a multilayer network,
4. to use insights from the multilayer/block matrix setting in order to derive expressions for general matrix function-based centrality measures,
5. to study how non-backtracking within exponential centrality reduces localization effects on a star graph,
6. to give further illustrative results on a controllable test network.

Our overall message is that it is both analytically and computationally attractive to study non-backtracking walk counts in a general matrix function setting that includes the matrix exponential. We therefore hope to have paved the way for future work that, for example, considers

- physical interpretations of the new measures,
- applications to specific fields of network science,
- further links with areas of discrete mathematics, stochastic processes and quantum physics,
- the development and analysis of customized numerical methods for approximating the action of a matrix function on the type of large, sparse unstructured matrix arising in network science.

ACKNOWLEDGEMENTS

The work of FA and DJH was supported by grant EP/M00158X/1 from the EPSRC/RCUK Digital Economy Programme.

REFERENCES

1. Martin Afanasjew, Michael Eiermann, Oliver G Ernst, and Stefan G¨uttel. Implementation of a restarted Krylov subspace method for the evaluation of matrix functions. Linear Algebra and its applications, 429(10):2293–2314, 2008.
2. A. H. Al-Mohy and N. J. Higham. Computing the action of the matrix exponential, with an application to exponential integrators. SIAM J. Sci. Comp., 33:488–511, 2011.
3. Noga Alon, Itai Benjamini, Eyal Lubetzky, and Sasha Sodin. Non-backtracking random walks mix faster. Communications in Contemporary Mathematics, 09:585–603, 2007.
4. O. Angel, J. Friedman, and S Hoory. The non-backtracking spectrum of the universal cover of a graph. Transactions of the American Mathematical Society, 326:4287–4318, 2015.
5. Francesca Arrigo, Michele Benzi, and Caterina Fenu. Computation of generalized matrix functions. SIAM Journal on Matrix Analysis and Applications, 37(3):836–860, 2016.
6. Francesca Arrigo, Peter Grindrod, Desmond J. Higham, and Vanni Noferini. Non-backtracking walk centrality for directed networks. Journal of Complex Networks, (published online ahead of print), 2017.
7. A.-L. Barab´asi and R. Albert. Emergence of scaling in random networks. Science, 286(5439):509–12, 1999.
8. M. Benzi and C. Klymko. On the limiting behavior of parameter-dependent network centrality measures. SIAM J. Matrix Anal. Appl., 36:686–706, 2015.
9. Michele Benzi and Christine Klymko. Total communicability as a centrality measure. Journal of Complex Networks, 1(2):124–149, 2013.
10. P. Bonacich. Factoring and weighting approaches to status scores and clique identification. Journal of Mathematical Sociology, 2:113–120, 1972.
11. P. Bonacich. Power and centrality: a family of measures. American Journal of Sociology, 92:1170–1182, 1987.
12. R. Bowen and O. E. Lanford. Zeta functions of restrictions of the shift transformation. In ShiingShen Chern and Stephen Smale, editors, Global Analysis: Proceedings of the Symposium in Pure Mathematics of the Americal Mathematical Society, University of California,

Berkely, 1968, pages 43–49. American Mathematical Society, 1970.

13. D. Cvetkov´c, P. Rowlinson, and S. Simi´c. Eigenspaces of Graphs. Cambridge University Press, Cambridge, 1997.
14. E. Estrada. The Structure of Complex Networks. Oxford University Press, Oxford, 2011.
15. Ernesto Estrada and Naomichi Hatano. Communicability in complex networks. Phys. Rev. E, 77(3):036111, 2008.
16. Ernesto Estrada, Naomichi Hatano, and Michele Benzi. The physics of communicability in complex networks. Physics Reports, 514:89–119, 2011.
17. Ernesto Estrada and Desmond J. Higham. Network properties revealed through matrix functions. SIAM Review, 52:696–671, 2010.
18. I. Gohberg, P. Lancaster, and L. Rodman. Matrix Polynomials. SIAM, Philadelphia, PA, 2009.
19. A. V. Goltsev, S. N. Dorogovtsev, J. G. Oliveira, and J. F. F. Mendes. Localization and spreading of diseases in complex networks. Phys. Rev. Lett., 109:128702, 2012.
20. Peter Grindrod, Desmond J. Higham, and Vanni Noferini. The deformed graph Laplacian and its applications to network centrality analysis. University of Essex Research Repository 18919, University of Essex, UK, 2017.
21. Nicholas J. Higham. Functions of Matrices: Theory and Computation. Society for Industrial and Applied Mathematics, Philadelphia, PA, USA, 2008.
22. Nicholas J. Higham and Edvin Deadman. A catalogue of software for matrix functions. Version 2.0. MIMS EPrint 2016.3, Manchester Institute for Mathematical Sciences, The University of Manchester, UK, January 2016. Updated March 2016.
23. Matthew D. Horton. Ihara zeta functions on digraphs. Linear Algebra and its Applications, 425:130–142, 2007.
24. Matthew D. Horton, H. M. Stark, and Audrey A. Terras. What are zeta functions of graphs and what are they good for? In G. Bertolaiko, R. Carlson, S. A. Fulling, and P. Kuchment, editors, Quantum graphs and their applications, volume 415 of Contemp. Math., pages 173–190. 2006.
25. L. Katz. A new index derived from sociometric data analysis. Psychometrika, 18:39–43, 1953.

26. Tatsuro Kawamoto. Localized eigenvectors of the non-backtracking matrix. Journal of Statistical Mechanics: Theory and Experiment, 2016:023404, 2016.
27. Mikko Kivel¨a, Alex Arenas, Marc Barthelemy, James P. Gleeson, Yamir Moreno, and Mason A. Porter. Multilayer networks. Journal of Complex Networks, 2:203–271, 2014.
28. Florent Krzakala, Cristopher Moore, Elchanan Mossel, Joe Neeman, Allan Sly, Lenka Zdeborov´a, and Pan Zhang. Spectral redemption: clustering sparse networks. Proceedings of the National Academy of Sciences, 110:20935–20940, 2013.
29. Travis Martin, Xiao Zhang, and M. E. J. Newman. Localization and centrality in networks. Phys. Rev. E, 90:052808, 2014.
30. Pierre-Andr´e G. Maugis, Sofia C. Olhede, and Patrick J. Wolfe. Topology reveals universal features for network comparison. arXiv:1705.05677
31. stat.ME., 2017.
32. Flaviano Morone and Hern´an A. Makse. Influence maximization in complex networks through optimal percolation. Nature, 524:65–68, 2015.
33. Flaviano Morone, Byungjoon Min, Lin Bo, Romain Mari, and Hern´an A. Makse. Collective influence algorithm to find influencers via optimal percolation in massively large social media. Scientific Reports, 6:30062, 2016.
34. Y. Nakatsukasa and V. Noferini. On the stability of computing polynomial roots via confederate linearizations. Mathematics of Computation, 85:2391–2425, 2016.
35. Y. Nakatsukasa, V. Noferini, and A. Townsend. Vector spaces of linearizations of matrix polynomials: a bivariate polynomial approach. SIAM J. Matrix Anal. Appl., 38(1):1–29, 2016.
36. V. Noferini and F. Poloni. Duality of matrix pencils, Wong chains and linearizations. Linear Algebra Appl., 471:730–767, 2015.
37. Romualdo Pastor-Satorras and Claudio Castellano. Distinct types of eigenvector localization in networks. Scientific Reports, 6:18847, 2016.
38. Yousef Saad. Analysis of some Krylov subspace approximations to the matrix exponential operator. SIAM Journal on Numerical Analysis, 29(1):209–228, 1992.

39. Alaa Saade, Florent Krzakala, and Lenka Zdeborov´a. Spectral clustering of graphs with the Bethe Hessian. In Z. Ghahramani, M. Welling, C. Cortes, N. D. Lawrence, and K. Q. Weinberger, editors, Advances in Neural Information Processing Systems 27, pages 406–414. 2014.
40. Uzy Smilansky. Quantum chaos on discrete graphs. Journal of Physics A: Mathematical and Theoretical, 40:F621, 2007.
41. Sasha Sodin. Random matrices, non-backtracking walks, and the orthogonal polynomials. J. Math. Phys, 48:123503, 2007.
42. H.M. Stark and A.A. Terras. Zeta functions of finite graphs and coverings. Advances in Mathematics, 121(1):124–165, 1996.
43. Andrei Tarfulea and Robert Perlis. An Ihara formula for partially directed graphs. Linear Algebra and its Applications, 431:73–85, 2009.
44. Alan Taylor and Desmond J. Higham. CONTEST: A controllable test matrix toolbox for MATLAB. ACM Trans. Math. Softw., 35(4):26:1–26:17, 2009.
45. Audrey Terras. Harmonic Analysis on Symmetric Spaces — Euclidean Space, the Sphere, and the Poincar´e Upper Half-Plane. Springer, New York, 2nd edition, 2013.
46. Yusuke Watanabe and Kenji Fukumizu. Graph zeta function in the Bethe free energy and loopy belief propagation. In Y. Bengio, D. Schuurmans, J. Lafferty, C. Williams, and A. Culotta, editors, Advances in Neural Information Processing Systems 22, pages 2017–2025. 2009.
47. Herbert S. Wilf. Generatingfunctionology. A. K. Peters, Ltd., Natick, MA, USA, 2006.

SECTION 5:
LINEAR RECURRENCES AND THE FIBONACCI NUMBERS

11

On Sequences of Numbers and Polynomials Defined by Linear Recurrence Relations of Order 2

Tian-Xiao He[1] and Peter J.-S. Shiue[2]

[1]Department of Mathematics and Computer Science, Illinois Wesleyan University, Bloomington, IL 61702, USA

[2]Department of Mathematical Sciences, University of Nevada Las Vegas, Las Vegas, NV 89154, USA

ABSTRACT

Here we present a new method to construct the explicit formula of a sequence of numbers and polynomials generated by a linear recurrence relation of order 2. The applications of the method to the Fibonacci and Lucas numbers, Chebyshev polynomials, the generalized Gegenbauer-Humbert polynomials are also discussed. The derived idea provides a general method to construct identities of number or polynomial sequences defined by linear recurrence relations. The applications using the method to solve some algebraic and ordinary differential equations are presented.

Citation: Tian-Xiao He and Peter J.-S. Shiue, "On Sequences of Numbers and Polynomials Defined by Linear Recurrence Relations of Order 2," International Journal of Mathematics and Mathematical Sciences, vol. 2009, Article ID 709386, 21 pages, 2009. https://doi.org/10.1155/2009/709386

INTRODUCTION

Many number and polynomial sequences can be defined, characterized, evaluated, and classified by linear recurrence relations with certain orders. A number sequence $\{a_n\}$ is called sequence of order 2 if it satisfies the linear recurrence relation of order 2:

$$a_n = pa_{n-1} + qa_{n-2}, \quad n \geq 2, \tag{1.1}$$

for some nonzero constants p and q and initial conditions a_0 and a_1. In Mansour [1], the sequence $\{an\}_{n\geq 0}$ defined by [1.1] is called Horadam's sequence, which was introduced in 1965 by Horadam [2]. The work in [1] also obtained the generating functions for powers of Horadam's sequence. To construct an explicit formula of its general term, one may use a generating function, characteristic equation, or a matrix method (see Comtet [3], Hsu [4], Strang [5], Wilf [6], etc). In [7], Benjamin and Quinn presented many elegant combinatorial meanings of the sequence defined by recurrence relation [1.1]. For instance, a_n counts the number of ways to tile an n-board (i.e., board of length n) with squares (representing 1ss) and dominoes (representing 2s) where each tile, except the initial one has a color. In addition, there are p colors for squares and q colors for dominoes. In this paper, we will present a new method to construct an explicit formula of $\{an\}$ generated by (1.1). The key idea of our method is to reduce the relation (1.1) of order 2 to a linear recurrence relation of order 1:

$$a_n = ca_{n-1} + d, \quad n \geq 1, \tag{1.2}$$

for some constants $c \neq 0$ and d and initial condition a_0 via geometric sequence. Then, the expression of the general term of the sequence of order 2 can be obtained from the formula of the general term of the sequence of order 1:

$$a_n = \begin{cases} a_0 c^n + d\dfrac{c^n - 1}{c - 1}, & \text{if } c \neq 1; \\ a_0 + nd, & \text{if } c = 1. \end{cases} \tag{1.3}$$

The method and some related results on the generalized Gegenbauer-Humbert polynomial sequence of order 2 as well as a few examples will be given in Section 2. Section 3 will discuss the application of the method to the construction of the identities of sequences of order 2. There is an extension of the above results to higher order cases. In Section 4, we will discuss the applications of the method to the solution of algebraic equations and initial value problems of second-order ordinary differential equations.

MAIN RESULTS AND EXAMPLES

Let α and β be two roots of quadratic cquation $x^2 - px - q$ 0. We may write (1.1) as

$$a_n = (\alpha + \beta)a_{n-1} - \alpha\beta a_{n-2}, \tag{2.1}$$

where α and β satisfy α + β= p and αβ =−q. Therefore, from (2.1), we have

$$a_n - \alpha a_{n-1} = \beta(a_{n-1} - \alpha a_{n-2}), \tag{2.2}$$

which implies that $\{an - \alpha a_{n-1}\}_{n\geq 1}$ is a geometric sequence with common ratio β. Hence,

$$a_n - \alpha a_{n-1} = (a_1 - \alpha a_0)\beta^{n-1},$$

$$a_n = \alpha a_{n-1} + (a_1 - \alpha a_0)\beta^{n-1}. \tag{2.3}$$

Consequently,

$$\frac{a_n}{\beta^n} = \frac{\alpha}{\beta}\left(\frac{a_{n-1}}{\beta^{n-1}}\right) + \frac{a_1 - \alpha a_0}{\beta}. \tag{2.4}$$

Let b_n: a_n/β^n. We may write (2.4) as

$$b_n = \frac{\alpha}{\beta}b_{n-1} + \frac{a_1 - \alpha a_0}{\beta}. \tag{2.5}$$

If α ≠ β, by using (1.3), we immediately obtain

$$\frac{a_n}{\beta^n} = a_0\left(\frac{\alpha}{\beta}\right)^n + \frac{a_1 - \alpha a_0}{\beta}\frac{(\alpha/\beta)^n - 1}{(\alpha/\beta) - 1}$$
$$= \frac{1}{\beta^n}\left[\alpha^n a_0 + \frac{a_1 - \alpha a_0}{\alpha - \beta}(\alpha^n - \beta^n)\right], \tag{2.6}$$

which yields

$$a_n = \left(\frac{a_1 - \beta a_0}{\alpha - \beta}\right)\alpha^n - \left(\frac{a_1 - \alpha a_0}{\alpha - \beta}\right)\beta^n. \tag{2.7}$$

Similarly, if α β, then (1.3) implies

$$a_n = a_0\alpha^n + n\alpha^{n-1}(a_1 - \alpha a_0) = na_1\alpha^{n-1} - (n-1)a_0\alpha^n. \tag{2.8}$$

We may summarize the above result as follows

Proposition 2.1. Let {an} be a sequence of order 2 satisfying linear recurrence relation (2.1). Then

$$a_n = \begin{cases} \left(\frac{a_1 - \beta a_0}{\alpha - \beta}\right)\alpha^n - \left(\frac{a_1 - \alpha a_0}{\alpha - \beta}\right)\beta^n, & if\ \alpha \neq \beta; \\ n a_1 \alpha^{n-1} - (n-1) a_0 \alpha^n, & if\ \alpha = \beta. \end{cases} \tag{2.9}$$

In particular, if $\{a_n\}$ satisfies the linear recurrence relation (1.1) with q 1, namely,

$$a_n = pa_{n-1} + a_{n-2}, \tag{2.10}$$

then the equation $x^2 - px - 1 = 0$ has two solutions:

$$\alpha = \frac{p + \sqrt{p^2+4}}{2}, \qquad \beta = -\frac{1}{\alpha} = \frac{p - \sqrt{p^2+4}}{2}. \tag{2.11}$$

From Proposition 2.1, we have the following corollary

Corollary 2.2. Let {an} be a sequence of order 2 satisfying the linear recurrence relation a_n pa_{n-1}+ an−2. Then

$$a_n = \frac{2a_1 - \left(p - \sqrt{p^2+4}\right)a_0}{2\sqrt{p^2+4}}\alpha^n - \frac{2a_1 - \left(p + \sqrt{p^2+4}\right)a_0}{2\sqrt{p^2+4}}\left(-\frac{1}{\alpha}\right)^n, \tag{2.12}$$

where α is defined by (2.11).

Similarly, let {an} be a sequence of order 2 satisfying the linear recurrence relation a_n a_{n-1} + qa_{n-2}. Then

$$a_n = \begin{cases} \frac{2a_1 - (1 - \sqrt{4q+1})a_0}{2\sqrt{4q+1}}\alpha_1^n - \frac{2a_1 - (1 + \sqrt{4q+1})a_0}{2\sqrt{4q+1}}\alpha_2^n, & if\ q \neq -\frac{1}{4}; \\ \frac{1}{2^n}(2na_1 - (n-1)a_0), & if\ q = -\frac{1}{4}, \end{cases} \tag{2.13}$$

where $\alpha_1 = (1/2)(1+\sqrt{4q+1})\ and\ \alpha_2 = (1/2)(1-\sqrt{4q+1})$ are solutions of the equation $x^2{-}x{-}q$ 0.

The first special case (2.12) was studied by Falbo in [8]. If p 1, the sequence is clearly the Fibonacci sequence0/mm/. If p 2 (q 1), the corresponding sequence is the sequence of numerators (when two initial conditions are 1 and 3) or denominators (when two initial conditions are 1 and 2) of the convergent of a continued fraction to √2: {1/1, 3/2, 7/5, 17/12, 41/29,...}, called the closest rational approximation sequence to √2. The second special case is also a corollary of Proposition 2.1. If q 2 (p 1), {an} is the Jacobsthal sequence/mm/ (see Bergum et al. [9]).

Remark 2.3. Proposition 2.1 can be extended to the linear recurrence relations of order 2 with more general form: $a_n = pa_{n-1} + qa_{n-2} + \ell$ for $p+q \neq 1$. It can be seen that the above recurrence relation is equivalent to the form (1.1)

$b_n = pb_{n-1}+qb_{n-2}$, where $b_n = a_n - k$ and $k = \ell/(1-p-q)$.

Remark 2.4. Denote

$$u_n = \begin{bmatrix} a_{n+1} \\ a_n \end{bmatrix}, \quad A = \begin{bmatrix} p & q \\ 1 & 0 \end{bmatrix}. \tag{2.14}$$

We may write relation $a_n = pa_{n-1} + qa_{n-2}$ and $a_{n-1} = a_{n-1}$ into a matrix form un−1 Aun−2 with respect to 2 × 2 matrix A defined above. Thus u_{n-1} $A^{n-1}u_0$. To find explicit expression of u_{n-1}, the real problem is to calculate A^{n-1}. The key lies in the eigenvalues and eigenvectors. The eigenvalues of A are precisely α and β, which are two roots of the characteristic equation x^2-px-q 0 for the matrix A. However, an obvious identity can be obtained from (u_n, u_{n-1}) A^{n-1}(u_1, u_0) by taking determinants on the both sides: $a_{n+1}a_{n-1} - a_n^2 = (-q)^{n-1}(a_2a_0 - a_1^2)$ (see, e.g., [5] for more details).

Example 2.5. Let $\{Fn\}_{n\geq 0}$ be the Fibonacci sequences with the linear recurrence relation $F_n = F_{n-1} + F_{n-2}$, where F_0 and F_1 are assumed to be 0 and 1, respectively. Thus, the recurrence relation is a special case of (1.1) with p= q =1 and the special case of the sequence in Corollary 2.2, which can be written as (2.1) with

$$\alpha = \frac{1+\sqrt{5}}{2}, \qquad \beta = \frac{1-\sqrt{5}}{2}. \tag{2.15}$$

Since α − β $\sqrt{5}$, from (2.12) we have the expression of F_n as follows:

$$F_n = \frac{1}{\sqrt{5}}\left\{\left(\frac{1+\sqrt{5}}{2}\right)^n - \left(\frac{1-\sqrt{5}}{2}\right)^n\right\}. \tag{2.16}$$

Example 2.6. We have mentioned above that the denominators of the closest rational approximation to $\sqrt{2}$ form a sequence satisfying the recurrence relation an $a_n = 2a_{n-1} + a_{n-2}$. With an additional initial condition 0, the sequence becomes the Pell number sequence: {p_n 0, 1, 2, 5, 12, 29,...}, which also satisfies the recurrence relation $p_n = 2p_{n-1} + p_{n-2}$. Using formula (2.23) in Corollary 2.2, we obtain the general term of the Pell number sequence:

$$p_n = \frac{1}{2\sqrt{2}}\left\{\left(1+\sqrt{2}\right)^n - \left(1-\sqrt{2}\right)^n\right\}. \tag{2.17}$$

The numerators of the closest rational approximation to $\sqrt{2}$ are half the companion Pell numbers or Pell-Lucas numbers. By adding in initial condition 2, we obtain the Pell-Lucas number sequence {c_n 2, 2, 6, 14, 34, 82,...}, which satisfies c_n $2c_{n-1} + c_{n-2}$. Similarly, Corollary 2.2 gives

$$c_n = \left(1+\sqrt{2}\right)^n + \left(1-\sqrt{2}\right)^n. \tag{2.18}$$

We now consider the sequence of the sums of Pell number: $\{\sigma_n$ 0, 1, 3, 8, 20,...}, which satisfies the recurrence relation

$$\sigma_n = 2\sigma_{n-1} + \sigma_{n-2} + 1. \tag{2.19}$$

From Remark 2.3, the above expression can be transfered to an equivalent form

$$b_n = 2b_{n-1} + b_{n-2}, \tag{2.20}$$

where b_n σ_n + 1/2. Using Corollary 2.2, one easily obtain

$$b_n = \frac{1}{4}\left\{\left(1+\sqrt{2}\right)^{n+1} + \left(1-\sqrt{2}\right)^{n+1}\right\}. \tag{2.21}$$

Thus,

$$\sigma_n = \frac{1}{4}\left\{\left(1+\sqrt{2}\right)^{n+1} + \left(1-\sqrt{2}\right)^{n+1}\right\} - \frac{1}{2}. \tag{2.22}$$

If the coefficients of the linear recurrence relation of a function sequence {an(x)} of order 2 are real or complex-value functions of variable x, that is,

$$a_n(x) = p(x)a_{n-1}(x) + q(x)a_{n-2}(x), \tag{2.23}$$

we obtain a function sequence of order 2 with initial conditions $a_0(x)$ and $a_1(x)$. In particular, if all of p(x), q(x), $a_0(x)$, and $a_1(x)$ are polynomials, then the corresponding sequence $\{a_n(x)\}$ is a polynomial sequence of order 2. Denote the solutions of

$$t^2 - p(x)t - q(x) = 0 \tag{2.24}$$

by α(x) and β(x). Then

$$\alpha(x) = \frac{1}{2}\left(p(x) + \sqrt{p^2(x) + 4q(x)}\right), \quad \beta(x) = \frac{1}{2}\left(p(x) - \sqrt{p^2(x) + 4q(x)}\right). \tag{2.25}$$

Similar to Proposition 2.1, we have

Proposition 2.7. Let $\{a_n\}$ be a sequence of order 2 satisfying the linear recurrence relation (2.23). Then

$$a_n(x) = \begin{cases} \left(\frac{a_1(x) - \beta(x)a_0(x)}{\alpha(x) - \beta(x)}\right)\alpha^n(x) - \left(\frac{a_1(x) - \alpha(x)a_0(x)}{\alpha(x) - \beta(x)}\right)\beta^n(x), & \text{if } \alpha(x) \neq \beta(x); \\ na_1(x)\alpha^{n-1}(x) - (n-1)a_0(x)\alpha^n(x), & \text{if } \alpha(x) = \beta(x), \end{cases} \tag{2.26}$$

where α(x) and β(x) are shown in (2.25).

Example 2.8. Consider the Chebyshev polynomials of the first kind, $T_n(x)$, defined by

$$T_n(x) = \cos(n \text{ arc} \cos x), \tag{2.27}$$

which satisfies the recurrence relation

$$T_n(x) = 2xT_{n-1}(x) - T_{n-2}(x) \tag{2.28}$$

with $T_0(x) = 1$ and $T_1(x) = x$. Thus the corresponding p, q, α, and β are, respectively, 2x, −1, x+$\sqrt{x^2-1}$, and x− $\sqrt{x^2-1}$, which yields $T_1(x)-\beta T_0(x) = \sqrt{x^2-1}, T_1(x)-\alpha T_0(x) = -\sqrt{x^2-1}$ and $\alpha-\beta = 2\sqrt{x^2-1}$. Substituting the quantities into (2.7) yields

$$T_n(x) = \frac{1}{2}\left\{\left(x+\sqrt{x^2-1}\right)^n + \left(x-\sqrt{x^2-1}\right)^n\right\}. \tag{2.29}$$

All the Chebyshev polynomials of the second kind, third kind, and fourth kind satisfy the same recurrence relationship as the Chebyshev polynomials of the first kind with the same constant initial term 1. However, they possess different linear initial terms, which are 2x, 2x−1, and 2x+1, respectively (see, e.g., Mason and Handscomb [10] and Rivlin [11]). We will give the expression of the Chebyshev polynomials of the second kind later by sorting them into the class of the generalized Gegenbauer-Humbert polynomials. As for the Chebyshev polynomials of the third kind, $T_n^{(3)}(x)$, and the Chebyshev polynomials of the fourth kind, $T_n^{(4)}(x)$when $x^2 \neq 1$, we clearly have the following expressions using a similar argument presented for the Chebyshev polynomials of the first kind:

$$T_n^{(3)}(x) = \frac{\sqrt{x^2-1}+x-1}{2\sqrt{x^2-1}}\left(x+\sqrt{x^2-1}\right)^n + \frac{\sqrt{x^2-1}-x+1}{2\sqrt{x^2-1}}\left(x-\sqrt{x^2-1}\right)^n,$$

$$T_n^{(4)}(x) = \frac{\sqrt{x^2-1}+x+1}{2\sqrt{x^2-1}}\left(x+\sqrt{x^2-1}\right)^n + \frac{\sqrt{x^2-1}-x-1}{2\sqrt{x^2-1}}\left(x-\sqrt{x^2-1}\right)^n. \tag{2.30}$$

Example 2.9. In [12], André-Jeannin studied the generalized Fibonacci and Lucas polynomials defined, respectively, by

$$U_n(x;a,b) = U_n = (x+a)U_{n-1} - bU_{n-2}, \qquad U_0(x) = 0, \qquad U_1(x) = 1,$$

$$V_n(x;a,b) = V_n = (x+a)V_{n-1} - bV_{n-2}, \qquad V_0(x) = 2, \qquad V_1(x) = x+a, \tag{2.31}$$

where a and b are real parameters. Clearly, $V_n(2x;0,1) = 2T_n(x)$. Using Proposition 2.7, we obtain

$$U_n(x;a,b) = \frac{1}{2^n\sqrt{(x+a)^2-4b}}\left\{\left(x+a+\sqrt{(x+a)^2-4b}\right)^n - \left(x+a-\sqrt{(x+a)^2-4b}\right)^n\right\},$$

$$V_n(x;a,b) = \frac{1}{2^n}\left\{\left(x+a+\sqrt{(x+a)^2-4b}\right)^n - \left(x+a-\sqrt{(x+a)^2-4b}\right)^n\right\}. \tag{2.32}$$

From the last expression, we also see $V_n(2x;0,1) = 2T_n(x)$

A sequence of the generalized Gegenbauer-Humbert polynomials $\{P_n^{\lambda,y,C}(x)\}_{n\geq 0}$ is defined by the expansion (see, e.g., Gould [13] and He et al. [14]):

$$\Phi(t) \equiv (C - 2xt + yt^2)^{-\lambda} = \sum_{n\geq 0} P_n^{\lambda,y,C}(x)t^n, \tag{2.33}$$

where $\lambda > 0$, y and $C \neq 0$ are real numbers. As special cases of (2.33), we consider $P_n^{\lambda,y,C}(x)$ as follows (see [14]):

$$\begin{aligned}
&P_n^{1,1,1}(x) = U_n(x), \text{ Chebyshev polynomial of the second kind,}\\
&P_n^{1/2,1,1}(x) = \varphi_n(x), \text{ Legendre polynomial,}\\
&P_n^{1,-1,1}(x) = P_{n+1}(x), \text{ Pell polynomial,}\\
&P_n^{1,-1,1}\left(\frac{x}{2}\right) = F_{n+1}(x), \text{ Fibonacci polynomial,}\\
&P_n^{1,2,1}\left(\frac{x}{2}\right) = \Phi_{n+1}(x), \text{ Fermat polynomial of the first kind,}\\
&P_n^{1,2a,2}(x) = D_n(x,a), \text{ Dickson polynomial,}
\end{aligned} \tag{2.34}$$

where a is a real parameter and $F_n = F_n(1)$ is the Fibonacci number

Theorem 2.10. Let $x \neq \pm\sqrt{Cy}$. The generalized Gegenbauer-Humbert polynomials $\{P_n^{1,y,C}(x)\}_{n\geq 0}$ defined by expansion (2.33) can be expressed as

$$P_n^{1,y,C}(x) = C^{-n-2}\left[\frac{x+\sqrt{x^2-Cy}}{2\sqrt{x^2-Cy}}\left(x+\sqrt{x^2-Cy}\right)^n - \frac{x-\sqrt{x^2-Cy}}{2\sqrt{x^2-Cy}}\left(x-\sqrt{x^2-Cy}\right)^n\right]. \tag{3.35}$$

Proof. Taking derivative with respect to x to the two sides of (2.33) yields

$$2\lambda(x - yt)(C - 2xt + yt^2)^{-\lambda-1} = \sum_{n\geq 1} nP_n^{\lambda,y,C}(x)t^{n-1}. \tag{2.36}$$

Then, substituting the expansion of $(C - 2xt - yt^2)^{-\lambda}$ of (2.33) into the left-hand side of (2.36) and comparing the coefficients of term t^n on both sides, we obtain

$$C(n+1)P_{n+1}^{\lambda,y,C}(x) = 2x(\lambda+n)P_n^{\lambda,y,C}(x) - y(2\lambda+n-1)P_{n-1}^{\lambda,y,C}(x). \tag{2.37}$$

By transferring $n+1 \mapsto n$, we have

$$P_n^{\lambda,y,C}(x) = 2x\frac{\lambda+n-1}{Cn}P_{n-1}^{\lambda,y,C}(x) - y\frac{2\lambda+n-2}{Cn}P_{n-2}^{\lambda,y,C}(x) \tag{2.38}$$

for all $n \geq 2$ with

$$\begin{aligned}
&P_0^{\lambda,y,C}(x) = \Phi(0) = C^{-\lambda},\\
&P_1^{\lambda,y,C}(x) = \Phi'(0) = 2\lambda x C^{-\lambda-1}.
\end{aligned} \tag{2.39}$$

Thus, if $\lambda = 1$, $P_n^{1,y,C}(x)$ satisfies linear recurrence relation

$$P_n^{1,y,C}(x) = \frac{2x}{C} P_{n-1}^{1,y,C}(x) - \frac{y}{C} P_{n-2}^{1,y,C}(x), \quad n \geq 2,$$

$$P_0^{1,y,C}(x) = C^{-1}, \qquad P_1^{1,y,C}(x) = 2xC^{-2}. \tag{2.40}$$

Therefore, we solve $t^2 - pt - q = 0$, where $p = 2x/C$ and $q = -y/C$, for t, and obtain solutions:

$$\alpha = \frac{1}{C}\left\{x + \sqrt{x^2 - Cy}\right\},$$

$$\beta = \frac{1}{C}\left\{x - \sqrt{x^2 - Cy}\right\}, \tag{2.41}$$

where $x \neq \pm\sqrt{Cy}$. Hence, Proposition 2.7 gives the formula of $P_n^{1,y,C}(x)$ $(n \geq 2)$ as

$$P_n^{1,y,C}(x) = C^{-2}\left[\frac{x + \sqrt{x^2 - Cy}}{2\sqrt{x^2 - Cy}}\alpha^n - \frac{x - \sqrt{x^2 - Cy}}{2\sqrt{x^2 - Cy}}\beta^n\right], \tag{2.42}$$

where α and β are shown as (2.41). This completes the proof.

Remark 2.11. We may use recurrence relation (2.40) to define various polynomials that were defined using different techniques. Comparing recurrence relation (2.40) with the relations of the generalized Fibonacci and Lucas polynomials shown in Example 2.9, with the assumption of $P_0^{1,y,C}=0$ and $P_1^{1,y,C} = 1$, we immediately know that

$$P_n^{1,1,1}(x) = 2xP_{n-1}^{1,1,1}(x) - P_{n-2}^{1,1,1}(x) = U_n(2x; 0, 1) \tag{2.43}$$

defines the Chebyshev polynomials of the second kind,

$$P_n^{1,-1,1}(x) = 2xP_{n-1}^{1,-1,1}(x) - P_{n-2}^{1,-1,1}(x) = U_n(2x; 0, -1) \tag{2.44}$$

defines the Pell polynomials, and

$$P_n^{1,-1,1}\left(\frac{x}{2}\right) = xP_{n-1}^{1,-1,1}\left(\frac{x}{2}\right) + P_{n-2}^{1,-1,1}\left(\frac{x}{2}\right) = U_n(x; 0, -1) \tag{2.45}$$

defines the Fibonacci polynomials.

In addition, in [15], Lidl et al. defined the Dickson polynomials are also the special case of the generalized Gegenbauer-Humbert polynomials, which can be defined uniformly using recurrence relation (2.40), namely,

$$D_n(x; a) = xD_{n-1}(x; a) - aD_{n-2}(x; a) = P_n^{1,2a,2}(x) \tag{2.46}$$

with $D_0(x; a) = 2$ and $D_1(x; a) = x$. Thus, the general terms of all of above polynomials can be expressed using (2.35).

Example 2.12. For $\lambda = y = C = 1$, using (2.35), we obtain the expression of the Chebyshev polynomials of the second kind:

$$U_n(x) = \frac{\left(x + \sqrt{x^2 - 1}\right)^{n+1} - \left(x - \sqrt{x^2 - 1}\right)^{n+1}}{2\sqrt{x^2 - 1}}, \tag{2.47}$$

where $x^2 \neq 1$. Thus, $U_2(x) = 4x^2 - 1$.

For λ =C= 1 and y =−1, formula (2.35) gives the expression of a Pell polynomial of degree n + 1:

$$P_{n+1}(x) = \frac{\left(x+\sqrt{x^2+1}\right)^{n+1} - \left(x-\sqrt{x^2+1}\right)^{n+1}}{2\sqrt{x^2+1}}. \tag{2.48}$$

Thus, $P_2(x) = 2x$.

Similarly, let λ= C =1 and y= −1, the Fibonacci polynomials are

$$F_{n+1}(x) = \frac{\left(x+\sqrt{x^2+4}\right)^{n+1} - \left(x-\sqrt{x^2+4}\right)^{n+1}}{2^{n+1}\sqrt{x^2+4}}, \tag{2.49}$$

and the Fibonacci numbers are

$$F_n = F_n(1) = \frac{1}{\sqrt{5}}\left\{\left(\frac{1+\sqrt{5}}{2}\right)^n - \left(\frac{1-\sqrt{5}}{2}\right)^n\right\}, \tag{2.50}$$

which has been presented in Example 2.5.

Finally, for λ =C= 1 and y =2, we have Fermat polynomials of the first kind:

$$\Phi_{n+1}(x) = \frac{\left(x+\sqrt{x^2-2}\right)^{n+1} - \left(x-\sqrt{x^2-2}\right)^{n+1}}{2\sqrt{x^2-2}}, \tag{2.51}$$

where $x^2 \neq 2$. From the expressions of Chebyshev polynomials of the second kind, Pell polynomials, and Fermat polynomials of the first kind, we may get a class of the generalized Gegenbauer-Humbert polynomials with respect to y defined as follows.

Definition 2.13. The generalized Gegenbauer-Humbert polynomials with respect to y, denoted by $P_n^{(y)}(x)$, are defined by the expansion

$$(1-2xt+yt^2)^{-1} = \sum_{n\geq 0} P_n^{(y)}(x)t^n, \tag{2.52}$$

by

$$P_n^{(y)}(x) = 2xP_{n-1}^{(y)}(x) - yP_{n-2}^{(y)}(x), \tag{2.53}$$

or equivalently, by

$$P_n^{(y)}(x) = \frac{\left(x+\sqrt{x^2-y}\right)^{n+1} - \left(x-\sqrt{x^2-y}\right)^{n+1}}{2\sqrt{x^2-y}} \tag{2.54}$$

with $P_0^{(y)}(x) = 1$ and $P_1^{(y)}(x) = 2x$, where $x^2 \neq y$. In particular, $P_n^{(-1)}(x)$, $P_n^{(1)}(x)$ and $P_n^{(2)}(x)$ are, respectively, Pell polynomials, Chebyshev polynomials of the second kind, and Fermat polynomials of the first kind.

IDENTITIES CONSTRUCTED FROM RECURRENCE RELATIONS

From (2.2) we have the following result.

Proposition 3.1. A sequence $\{a_n\}n\geq 0$ of order 2 satisfies linear recurrence relation (2.1) if and only if it satisfies the nonhomogeneous linear recurrence relation of order 1 with the form

$$a_n = \alpha a_{n-1} + d\beta^{n-1}, \tag{3.1}$$

where d is uniquely determined. In particular, if β =1, then $a_n = \alpha a_{n-1} + d$ is equivalent to an – $(\alpha+1)a_{n-1} - \alpha a_{n-2}$, where $d = a_1 - \alpha a_0$.

Proof. The necessity is clearly from (2.1). We now prove sufficiency. If sequence {an} satisfies the nonhomogeneous recurrence relation of order 1 shown in (3.1), then by substituting n 1 into the above equation, we obtain d $= a_1 - \alpha a_0$. Thus, (3.1) can be written as

$$a_n - \alpha a_{n-1} = (a_1 - \alpha a_0)\beta^{n-1}, \tag{3.2}$$

which implies that $\{a_n\}$ satisfies the linear recurrence relation of order 2: an $pa_{n-1} = qa_{n-2}$ with p α = β and q =–αβ. In particular, if β=1, then p α + 1 and q –α, which yields the special case of the proposition.

An obvious example of the special case of Proposition 3.1 is the Mersenne number an $2^n - 1 = (n \geq 0)$, which satisfies the linear recurrence relation of order 2: $a_n = 3a_{n-1} - 2a_{n-2}$ (with $a_0 = 0$ and $a_1 = 1$) and the nonhomogeneous recurrence relation of order 1: $a_n = 2a_{n-1} + 1$ (with a0+0). It is easy to check that sequence $a_n = (k^n - 1)/(k-1)$ satisfies both the homogeneous recurrence relation of order 2, $a_n = (k+1)a_{n-1} - ka_{n-2}$, and the nonhomogeneous recurrence relation of order 1, $a_n = ka_{n-1} + 1$, where $a_0 = 0$ and $a_1 = 1$.

We now use (3.2) to prove some identities of Fibonacci and Lucas numbers and generalized Gegenbauer-Humbert polynomials. Let $\{F_n\}$ be the Fibonacci sequence0/mm/. From (3.2),

$$F_n - \alpha F_{n-1} = (F_1 - \alpha F_0)\beta^{n-1} = \beta^{n-1}, \tag{3.3}$$

where the last step is due to α + β=1. Therefore, we give a simple identity

$$F_n = \alpha F_{n-1} + \beta^{n-1} = \frac{1+\sqrt{5}}{2}F_{n-1} + \left(\frac{1-\sqrt{5}}{2}\right)^{n-1}, \tag{3.4}$$

which is shown in [16, (8.2) page 122] by Koshy. Similarly, we have

$$\beta F_n = \beta\alpha F_{n-1} + \beta^n = \beta^n - F_{n-1}, \tag{3.5}$$

where the last step is due to $\alpha\beta=-1$. The above identity can be written as

$$\left(\frac{1-\sqrt{5}}{2}\right)^n=\frac{1-\sqrt{5}}{2}F_n+F_{n-1}. \tag{3.6}$$

The same argument yields $\alpha^n=\alpha F_n+F_{n-1}$, or equivalently

$$\left(\frac{1+\sqrt{5}}{2}\right)^n=\frac{1+\sqrt{5}}{2}F_n+F_{n-1}. \tag{3.7}$$

Identities (3.6) and (3.7) were proved by using different method in (16, page 78).

Let $\{L_n\}$ be the Lucas number sequence with $L_0=2$ and $L1=1$, which satisfies recurrence relation (2.2) with the same α and β for the Fibonacci number sequence. Then, using the same argument, we have

$$L_n-\alpha L_{n-1}=(L_1-\alpha L_0)\beta^{n-1}=(1-2\alpha)\beta^{n-1}=-\sqrt{5}\beta^{n-1}. \tag{3.8}$$

Thus

$$L_n-\left(\frac{1+\sqrt{5}}{2}\right)L_{n-1}=-\sqrt{5}\left(\frac{1-\sqrt{5}}{2}\right)^{n-1}, \tag{3.9}$$

or equivalently,

$$-L_{n+1}+\left(\frac{1+\sqrt{5}}{2}\right)L_n=\sqrt{5}\left(\frac{1-\sqrt{5}}{2}\right)^n \tag{3.10}$$

(see [16, page 129]).

We now extend the above results regarding Fibonacci and Lucas numbers to more general sequences presented by Niven et al. in [17]. Let $\{G_n\}_{n\geq 0}$ and $\{H_n\}_{n\geq 0}$ be two sequences defined, respectively, by the linear recurrence relations of order 2:

$$G_n=pG_{n-1}+qG_{n-2}, \qquad H_n=pH_{n-1}+qH_{n-2} \tag{3.11}$$

with initial conditions $G_0=_0$ and $G_1=1$ and $H_0=2$ and $H_1=p$, respectively. Clearly, if p= q =1, then G_n and H_n are, respectively, Fibonacci and Lucas numbers. From (3.2), we immediately have

$$G_n-\frac{p+\sqrt{p^2+4q}}{2}G_{n-1}=\left(\frac{p-\sqrt{p^2+4q}}{2}\right)^{n-1}. \tag{3.12}$$

Multiplying $(p-\sqrt{p^2+4q})/2$ to both sides of the above equation yields

$$\frac{p-\sqrt{p^2+4q}}{2}G_n + qG_{n-1} = \left(\frac{p-\sqrt{p^2+4q}}{2}\right)^n. \tag{3.13}$$

Similarly, we obtain

$$\frac{p+\sqrt{p^2+4q}}{2}G_n + qG_{n-1} = \left(\frac{p+\sqrt{p^2+4q}}{2}\right)^n. \tag{3.14}$$

When p = q = 1, the last two identities are (3.6) and (3.7), respectively. Using (3.2) we can also obtain the identity

$$\frac{p+\sqrt{p^2+4q}}{2}H_n - H_{n+1} = \sqrt{p^2+4q}\left(\frac{p-\sqrt{p^2+4q}}{2}\right)^n, \tag{3.15}$$

which implies (3.10) when p =q= 1.

Aharonov et al. (see [18]) have proved that the solution of any sequence of numbers that satisfies a recurrence relation of order 2 with constant coefficients and initial conditions a_0 =0 and a_1= 1 can be expressed in terms of Chebyshev polynomials. For instance, the authors show $F_n = i^{-n}U_n(i/2)$ and $L_n = 2i^{-n}T_n(i/2)$. Thus, we have identities

$$\left(\frac{1+\sqrt{5}}{2}\right)^n = \frac{1+\sqrt{5}}{2i^n}U_n\left(\frac{i}{2}\right) + i^{-(n-1)}U_n\left(\frac{i}{2}\right),$$

$$\sqrt{5}\left(\frac{1-\sqrt{5}}{2}\right)^n = 2i^{-n}T_n\left(\frac{i}{2}\right) - \left(\frac{1+\sqrt{5}}{i^{n-1}}\right)T_{n-1}\left(\frac{i}{2}\right). \tag{3.16}$$

In [19], Chen and Louck obtained $F_{n+1} = i^n U_n(-i/2)$. Thus we have identity

$$\left(\frac{1+\sqrt{5}}{2}\right)^n = i^{n-1}\frac{1+\sqrt{5}}{2}U_{n-1}\left(\frac{-i}{2}\right) + i^{n-2}U_{n-2}\left(\frac{-i}{2}\right). \tag{3.17}$$

Identities (3.6) and (3.7) can be used to prove the following radical identity given by Sofo in [20]:

$$(\alpha F_n + F_{n-1})^{1/n} + (-1)^{n+1}(F_{n+1} - \alpha F_n)^{1/n} = 1. \tag{3.18}$$

Identity (3.7) shows that the first term on the left-hand side of (3.18) is simply α. Assume the sum in the third parenthesis on the left-hand side of (3.18) is c, then

$$\beta c = \beta F_{n+1} - \alpha\beta F_n = F_n + \beta F_{n+1} = \beta^{n+1}, \tag{3.19}$$

where the last step is from (3.6) with transform $n \mapsto n+1$. Thus, we have

$c = \beta^n$. If n is odd, the left-hand side of (3.18) is $\alpha + \beta = 1$. If n is even, the left-hand side of (3.18) becomes $\alpha - |\beta| = \alpha + \beta = 1$, which completes the proof of Sofo's identity.

In general, let {an} be a sequence of order 2 satisfying linear recurrence relation (1.1) or equivalently (2.1). Then we sum up our results as follows.

Theorem 3.2. Let $\{an\}_{n\geq 0}$ be a sequence of numbers or polynomials defined by the linear recurrence relation an $a_n = pa_{n-1} + qa_{n-2}\ (n \geq 0)$ with initial conditions a_0 and a_1, and let p =α + β and q= −αβ.

Then we have identity

$$a_n = \alpha a_{n-1} + (a_1 - \alpha a_0)\beta^{n-1}. \tag{3.20}$$

In particularly, if $\{a_n = P_n^{\lambda,y,C}(x)\}$, the sequence of the generalized Gegenbauer-Humbert polynomial is defined by (2.33), then we obtain the polynomial identity:

$$P_n^{1,y,C}(x) = \alpha P_{n-1}^{1,y,C}(x) + C^{-2}(2x - \alpha C)\beta^{n-1}, \tag{3.21}$$

where α and β are shown in (2.41). For C 1 (i.e., the generalized Gegenbauer-Humbert polynomials with respect to y), we denote $P_n^{(y)}(x) \equiv P_n^{1,y,1}(x)$ and have

$$P_n^{(y)}(x) = \left(x + \sqrt{x^2 - y}\right)P_{n-1}^{(y)}(x) + \left(x - \sqrt{x^2 - y}\right)^n. \tag{3.22}$$

Actually, (3.22) can also be proved directly. Similarly, for the Chebyshev polynomials of the first kind $T_n(x)$, we have the identity

$$\left(x + \sqrt{x^2 - 1}\right)T_{n-1}(x) - T_n(x) = \sqrt{x^2 - 1}\left(x - \sqrt{x^2 - 1}\right)^{n-1}. \tag{3.23}$$

Let x =cos θ, the above identity becomes

$$e^{i\theta}\cos((n-1)\theta) - \cos(n\theta) = i\sin\theta e^{-i(n-1)\theta}, \tag{3.24}$$

which is equivalent to cos nθ cos(n – 1)θ cos θ – sin(n – 1)θ sin θ

Another example is from the sequence $\{a_n\}$ shown in Corollary 2.2 with a_0= 1 and a1 p. Then (3.20) gives the identity $a_n = \alpha a_{n-1} + \beta^n$, where $\alpha = (p + \sqrt{p^2 + 4})/2$ and $\beta = -1/\alpha$. Similar to (3.6) and (3.7), for those sequences $\{a_n\}$ with a_0 =1 and a1= p, we obtain identities

$$\left(\frac{p-\sqrt{p^2+4}}{2}\right)^{n+1} = \frac{p-\sqrt{p^2+4}}{2}a_n + a_{n-1},$$

$$\left(\frac{p+\sqrt{p^2+4}}{2}\right)^{n+1} = \frac{p+\sqrt{p^2+4}}{2}a_n + a_{n-1}. \quad (3.25)$$

When p=1, the above identities become (3.6) and (3.7), respectively. Similarly, we can prove

$a_{n+k} = \alpha^k a_n + \beta^n a_k$, where $\alpha = (p+\sqrt{p^2+4})/2$ and $\beta = -1/\alpha$.

It is clear that if $1/a_n$ is bounded and $|\beta| < 1$, from (3.20) we have

$$\lim_{n\to\infty}\frac{a_n}{a_{n-1}} = \alpha. \quad (3.26)$$

Therefore,

$$\lim_{n\to\infty}\frac{F_n}{F_{n-1}} = \lim_{n\to\infty}\frac{L_n}{L_{n-1}} = \frac{1+\sqrt{5}}{2}. \quad (3.27)$$

The method presented in this paper cannot be extended to the higher-order setting. However, we may use the idea and a similar argument to derive some identities of sequences of order greater than 2. For instance, for a sequence $\{a_n\}$ of numbers or polynomials that satisfies the linear recurrence relation of order 3:

$$a_n = pa_{n-1} + qa_{n-2} + ra_{n-3}, \quad n \geq 3, \quad (3.28)$$

we set the equation

$$t^3 - pt^2 - qt - r = 0. \quad (3.29)$$

Using transform t= s + p/3, we can change the equation to the standard form $s^3 + as + b = 0$, which can be solved by Vieta's substitution $s = u - a/(3u)$. The formulas for the three roots, denoted by α, β, γ, are sometimes known as Cardano's formula. Thus, we have

$$\alpha+\beta+\gamma = p, \quad \alpha\beta+\beta\gamma+\gamma\alpha = -q, \quad \alpha\beta\gamma = r. \quad (3.30)$$

Denote $y_n = a_n - \alpha a_{n-1}$. Then (3.28) can be written as

$$y_n = (\beta+\gamma)y_{n-1} - \beta\gamma y_{n-2}. \quad (3.31)$$

From Propositions 2.1 or 2.7, one may obtain

$$y_n = \begin{cases} \left(\frac{y_1 - \beta y_0}{\gamma - \beta}\right)\gamma^n - \left(\frac{y_1 - \gamma y_0}{\gamma - \beta}\right)\beta^n, & \text{if } \gamma \neq \beta; \\ n y_1 \gamma^{n-1} - (n-1) y_0 \gamma^n, & \text{if } \gamma = \beta. \end{cases}$$

(3.32)

Therefore, from the identity $y_n = \gamma y_{n-1} + \beta^{n-1}(y_1 - \gamma y_0)$, we obtain identity in terms of a_n:

$$\beta^{n-1}(y_1 - \gamma y_0) = a_n - \alpha a_{n-1} - \gamma(a_{n-1} - \alpha a_{n-2})$$
$$= a_n - (\alpha + \gamma)a_{n-1} + \alpha\gamma a_{n-2},$$

(3.33)

or equivalently,

$$a_n = (\alpha+\gamma)a_{n-1} - \alpha\gamma a_{n-2} + \beta^{n-1}(a_1 - (\alpha+\gamma)a_0 + \alpha\gamma a_{-1}),$$

(3.34)

where a_{-1} can be found uniquely from $y_2 = (\beta+\gamma)y_1 - \beta\gamma y_0$ and $y_0 = a_0 - \alpha a_{-1}$, , that is

$$a_2 - \alpha a_1 = (\beta+\gamma)(a_1 - \alpha a_0) - \beta\gamma(a_0 - \alpha a_{-1}),$$

(3.35)

or equivalently,

$$a_{-1} = \frac{a_2 - p a_1 - q a_0}{r}, \quad r \neq 0.$$

(3.36)

We have seen the equivalence between the homogeneous recurrence relation of order 3, in (3.28), and the nonhomogeneous recurrence relation of order 2, in (3.34).

Remark 3.3. Similar to the particular case shown in Proposition 3.1, we may find the equivalence between the nonhomogeneous recurrence relation of order 2, $a_n = pa_{n-1} + qa_{n-2} +$+k, and the homogeneous recurrence relation of order 3, $a_n = (p+1)a_{n-1} + (q-p)a_{n-2} + qa_{n-3}$, where $k = a_2 - pa_1 - qa_0$.

Example 3.4. As an example, we consider the tribonacci number sequence generated by $a_n = a_{n-1} + a_{n-2} + a_{n-3}\ (n \geq 3)$. Solving $t^3 - t^2 - t - 1 = 0$, we obtain

$$\beta = \frac{1}{3} + \frac{1}{3}(19 - 3\sqrt{33})^{1/3} + \frac{1}{3}(19 + 3\sqrt{33})^{1/3},$$

$$\alpha = \frac{1}{3} - \frac{1}{6}\left(1 + i\sqrt{3}\right)(19 - 3\sqrt{33})^{1/3} - \frac{1}{6}\left(1 - i\sqrt{3}\right)(19 + 3\sqrt{33})^{1/3},$$

$$\gamma = \frac{1}{3} - \frac{1}{6}\left(1 - i\sqrt{3}\right)(19 - 3\sqrt{33})^{1/3} - \frac{1}{6}\left(1 + i\sqrt{3}\right)(19 + 3\sqrt{33})^{1/3}.$$

(3.37)

Substituting α, β, γ, and y_0 =0 (with the assumption a_{-1}= 0) and y1 1 into (3.33), we obtain an identity regarding the tribonacci number sequence {a_n 0, 1, 1, 2, 4, 7, 13, 24,...}:

$$a_n - \frac{1}{3}(2-a-b)a_{n-1} + \frac{1}{9}\left(-3-a-b+a^2+b^2\right)a_{n-2} = \frac{1}{3^{n-1}}(1+a+b)^{n-1}, \tag{3.38}$$

where $a = (19-3\sqrt{33})^{1/3}$ and $b = (19+3\sqrt{33})^{1/3}$.
For the sequence defined by

$$a_n = 6a_{n-1} - 11a_{n-2} + 6a_{n-3}, \quad n \geq 3, \tag{3.39}$$

with initial conditions $a_0 = a_1 = 1$ and $a_2 = 2$. Then, the first few numbers of the sequence are {1, 1, 2, 7, 26, 91,...}. The three roots of $t^3 - 6t^2 + 11t - 6 = 0$ are $\alpha = 1, \beta = 2$, and $\gamma = 3$. Therefore, by assuming $a_{-1} = 7/6$, we obtain the corresponding $y_0 = -1/6$ and $y_1 = 0$ and the following identity for the above-defined sequence:

$$a_n - 4a_{n-1} + 3a_{n-2} = 2^{n-2} \tag{3.40}$$

for all $n \geq 2$. From [21] by Have, $a_n = \left\{ {n+1 \atop 3} \right\} + 1\ (n \geq 0)$ where $\left\{ {n \atop k} \right\}$ are Stirling numbers of the second kind. Hence, we obtain an identity of the Stirling numbers of the second kind:

$$\left\{ {n+1 \atop 3} \right\} - 4\left\{ {n \atop 3} \right\} + 3\left\{ {n-1 \atop 3} \right\} = 2^{n-2}. \tag{3.41}$$

The idea to reduce a linear recurrence relation of order 3 to order 2 can be extended to the higher order cases. In general, if we have a sequence $\{a_n\}$ satisfying the linear recurrence relation of order r:

$$a_n = \sum_{k=1}^{r} p_k a_{n-k}. \tag{3.42}$$

Assume the equation $t^r - \sum_{k=1}^{r} p_k t^{r-k} = 0$ has solutions $\alpha_k\ (1 \leq k \leq r)$. Denote $y_n = a_n - \alpha_1 a_{n-1}$. Then the above recurrence relation can be reduced to

$$y_n = y_{n-1}\sum_{k=2}^{r}\alpha_k - y_{n-2}\sum_{2\leq i\neq j\leq r}\alpha_i\alpha_j + y_{n-3}\sum_{2\leq i\neq j\neq k\leq r}\alpha_i\alpha_j\alpha_k - \cdots + (-1)^r y_{n-r+1}\prod_{k=2}^{r}\alpha_k, \tag{3.43}$$

a linear recurrence relation of order $r-1$ for sequence $\{y_n\}$. Using this process, we may obtain the explicit formula of an and/or identities in terms of an if we know the solution of the last equation and/or the identities in terms of sequence $\{y_n\}$.
The process shown in Proposition 3.1 can be applied conversely to elevate a nonhomogenous recurrence relation of order n to a homogeneous recurrence relation of order n + 1.

SOLUTIONS OF ALGEBRAIC EQUATIONS AND DIFFERENTIAL EQUATIONS

The results presented in Sections 2 and 3 have more applications. In this section, we will discuss the applications in the solutions of algebraic equations and initial value problems of second-order ordinary differential equations.

First, we consider roots of polynomials $p(x) = x^n - xF_n - F_{n-1}$ or the solution of $p(x) = 0$, where $\{Fn\}_{n\geq 0}$ is the Fibonacci sequence0/mm/. Using the identity $\alpha^n = \alpha F_n + F_{n-1}$, we immediately know that the largest root of p(x) is

$$x = \alpha = \frac{1+\sqrt{5}}{2}. \tag{4.1}$$

Indeed, p(x) only changes its coefficient signs once, which implies that it has only one positive root α and all of its other roots must be negative, for example, $\beta = (1-\sqrt{5})/2$. In [22], Wall proved the largest root of p(x) is α using a more complicated manner.

We may write the identity $\alpha^n = \alpha F_n + F_{n-1}$ as

$$\left(\frac{1+\sqrt{5}}{2}\right)^n = \left(\frac{1+\sqrt{5}}{2}\right)F_n + F_{n-1} = \frac{L_n + \sqrt{5}F_n}{2}, \tag{4.2}$$

where L_n is the nth Lucas number. Similarly, we have

$$\left(\frac{1-\sqrt{5}}{2}\right)^n = \left(\frac{1-\sqrt{5}}{2}\right)F_n + F_{n-1} = \frac{L_n - \sqrt{5}F_n}{2}. \tag{4.3}$$

Multiplying the last two equations side by side yields $(-1)^n = (L_n^2 - 5F_n^2)/4$, or equivalently, $L_n^2 = F_n^2 + 4(-1)^n$. The last expression means $5F_n^2 + 4(-1)^n$ is a perfect square, or equivalently, if n is a Fibonacci number, then $5n^2 \pm 4$ is a perfect square. This result is a part of Gessel's results in [23], but the method we used seems simpler. In addition, the above result also shows that the Pell's equation $x^2 - 5y^2 \pm 4$ has a solution $(x, y) = (L_n, F_n)$.

The above results on the Fibonacci sequence0/mm/ can be extended to the sequence $\{Gn\}_{n>0}$ shown in [17] and **Section 3.** Consider polynomial $p(x) = x^n - xG_n - qG_{n-1}$ with $q \geq 1$. We can see the largest root of $p(x)$ is $(p + \sqrt{p^2 + 4q})/2$, which implies Wall's result when p =q= 1. In addition, because of

$$\left(\frac{p+\sqrt{p^2+4q}}{2}\right)^n = \frac{p+\sqrt{p^2+4q}}{2}G_n + qG_{n-1}$$

$$= \frac{1}{2}\left(\sqrt{p^2+4q}G_n + H_n\right),$$

$$\left(\frac{p-\sqrt{p^2+4q}}{2}\right)^n = \frac{p-\sqrt{p^2+4q}}{2}G_n + qG_{n-1}$$

$$= \frac{1}{2}\left(-\sqrt{p^2+4q}G_n + H_n\right), \tag{4.4}$$

we have

$$(-q)^n = \frac{1}{4}\left(H_n^2 - \left(p^2+4q\right)G_n^2\right), \tag{4.5}$$

or equivalently,

$$H_n^2 = \left(p^2+4q\right)G_n^2 + 4(-q)^n. \tag{4.6}$$

Hence, $(p^2+4q)G_n^2+4(-q)^n$ is a perfect square, which implies the special case for the Fibonacci sequence0/mm/ that has been presented. Therefore, Pell's equation $x^2-(p^2+4q)y^2 = \pm 4q^n \ (p^2+4q>0)$ has a solution $(x,y) = (H_n, G_n)$
We now use the method presented in Section 2 to reduce an initial problem of a second order ordinary differential equation:

$$y'' - py' - qy = 0, \qquad y'(0) = A, \qquad y(0) = B, \tag{4.7}$$

of second-order ordinary differential equation with constant coefficients to the problem of linear equations. Let $p, q \neq 0$, and let α and β be solution(s) of $x^2 - px - q = 0$. Denote

$$v(x) := y'(x) - \alpha y(x). \tag{4.8}$$

Then $v'(x) = y''(x) - \alpha y'(x)$ and the original initial problem of the second order is split into two problems of first order by using the method shown in (2.2) for n 2:

$$v'(x) = \beta v(x), \qquad v(0) = A - \alpha B;$$

$$y'(x) - \alpha y(x) = v(x), \qquad y(0) = B. \tag{4.9}$$

Thus, we obtain the solutions

$$v(x) = (A - \alpha B)e^{\beta x},$$

$$y(x) = \begin{cases} \frac{1}{\alpha - \beta}((A - \beta B)e^{\alpha x} - (A - \alpha B)e^{\beta x}), & \text{if } \alpha \neq \beta; \\ e^{\alpha x}((A - \alpha B)x + B), & \text{if } \alpha = \beta. \end{cases} \quad (4.10)$$

The above technique can be extended to the initial problems of higher-order ordinary differential equations. In this paper, we presented an elementary method for construction of the explicit formula of the sequence defined by the linear recurrence relation of order 2 and the related identities. Some other applications in solutions of algebraic and differential equations and some extensions to the higher dimensional setting are also discussed. However, besides those applications, more applications in combinatorics and the combinatorial explanations of our given formulas still remain much to be investigated.

ACKNOWLEDGMENTS

Dedicated to Professor L. C. Hsu on occasion of his 90th birthday. The authors wish to thank the referees for their helpful comments and suggestions.

REFERENCES

1. T. Mansour, "A formula for the generating functions of powers of Horadam's sequence," The Australasian Journal of Combinatorics, vol. 30, pp. 207–212, 2004.
2. A. F. Horadam, "Basic properties of a certain generalized sequence of numbers," The Fibonacci Quarterly, vol. 3, pp. 161–176, 1965.
3. L. Comtet, Advanced Combinatorics: The Art of Finite and Infinite Expansions, D. Reidel, Dordrecht, The Netherlands, 1974.
4. L. C. Hsu, Computational Combinatorics, Shanghai Scientific & Techincal, Shanghai, China, 1st edition, 1983.
5. G. Strang, Linear Algebra and Its Applications, Academic Press, New York, NY, USA, 2nd edition, 1980.
6. H. S. Wilf, Generatingfunctionology, Academic Press, Boston, Mass, USA, 1990.
7. A. T. Benjamin and J. J. Quinn, Proofs that Really Count: The Art of Combinatorial Proof, vol. 27 of The Dolciani Mathematical Expositions, Mathematical Association of America, Washington, DC, USA, 2003.
8. C. Falbo, "The golden ratio—a contrary viewpoint," The College Mathematics Journal, vol. 36, no. 2, pp. 123–134, 2005.
9. G. E. Bergum, L. Bennett, A. F. Horadam, and S. D. Moore, "Jacobsthal polynomials and a conjecture concerning Fibonacci-like matrices," The Fibonacci Quarterly, vol. 23, pp. 240–248, 1985.
10. J. C. Mason and D. C. Handscomb, Chebyshev Polynomials, Chapman & Hall/CRC, Boca Raton, Fla, USA, 2003.
11. T. J. Rivlin, Chebyshev Polynomials: From Approximation Theory to Algebra and Number Theory, Pure and Applied Mathematics (New York), John Wiley & Sons, New York, NY, USA, 2nd edition, 1990.
12. R. André-Jeannin, "Differential properties of a general class of polynomials," The Fibonacci Quarterly, vol. 33, no. 5, pp. 453–458, 1995.
13. H. W. Gould, "Inverse series relations and other expansions involving Humbert polynomials," Duke Mathematical Journal, vol. 32, pp. 697–711, 1965.
14. T.-X. He, L. C. Hsu, and P. J.-S. Shiue, "A symbolic operator approach to several summation formulas for power series. II," Discrete

Mathematics, vol. 308, no. 16, pp. 3427–3440, 2008.

15. R. Lidl, G. L. Mullen, and G. Turnwald, Dickson Polynomials, vol. 65 of Pitman Monographs and Surveys in Pure and Applied Mathematics, Longman Scientific & Technical, Harlow, UK; John Wiley & Sons, New York, NY, USA, 1993.
16. T. Koshy, Fibonacci and Lucas Numbers with Applications, Pure and Applied Mathematics (New York), Wiley-Interscience, New York, NY, USA, 2001.
17. I. Niven, H. S. Zuckerman, and H. L. Montgomery, An Introduction to the Theory of Numbers, John Wiley & Sons, New York, NY, USA, 5th edition, 1991.
18. D. Aharonov, A. Beardon, and K. Driver, "Fibonacci, Chebyshev, and orthogonal polynomials," The American Mathematical Monthly, vol. 112, no. 7, pp. 612–630, 2005.
19. W. Y. C. Chen and J. D. Louck, "The combinatorial power of the companion matrix," Linear Algebra and Its Applications, vol. 232, pp. 261–278, 1996.
20. A. Sofo, "Generalization of radical identity," The Mathematical Gazette, vol. 83, pp. 274–276, 1999.
21. R. L. Haye, "Binary relations on the power set of an n-element set," Journal of Integer Sequences, vol. 12, no. 2, article 09.2.6, 2009.
22. C. R. Wall, "Problem 32," The Fibonacci Quarterly, vol. 2, no. 1, p. 72, 1964.
23. I. Gessel, "Problem H-187," The Fibonacci Quarterly, vol. 10, no. 4, pp. 417–419, 1972.

12

On The Partial Finite Sums of the Reciprocals of the Fibonacci Numbers

Andrew YZ Wang and Peibo Wen

School of Mathematical Sciences, University of Electronic Science and Technology of China, Chengdu, 611731, P.R. China

ABSTRACT

In this article, we obtain two interesting families of partial finite sums of the reciprocals of the Fibonacci numbers, which substantially improve two recent results involving the reciprocal Fibonacci numbers. In addition, we present an alternative and elementary proof of a result of Wu and Wang.

MSC: 11B39

Keywords: Fibonacci numbers; partial sums; reciprocal

INTRODUCTION

The Fibonacci sequence [1], Sequence A000045 is defined by the linear recurrence relation

Citation: Andrew YZ Wang and Peibo Wen "On the partial finite sums of the reciprocals of the Fibonacci numbers" Journal of Inequalities and Applications 2015 2015:73. https://doi.org/10.1186/s13660-015-0595-6

$$F_n = F_{n-1} + F_{n-2} \quad \text{for } n \geq 2,$$

where F_n is the nth Fibonacci number with $F_0 = 0$ and $F_1 = 1$. There exists a simple and non-obvious formula for the Fibonacci numbers,

$$F_n = \frac{1}{\sqrt{5}}\left(\frac{1+\sqrt{5}}{2}\right)^n - \frac{1}{\sqrt{5}}\left(\frac{1-\sqrt{5}}{2}\right)^n.$$

The Fibonacci sequence plays ĄL an important role in the theory and applications of mathematics, and its various properties have been investigated by many authors; see [2, 3, 4, 5].

In recent years, there has been an increasing interest in studying the reciprocal sums of the Fibonacci numbers. For example, Elsner *et al.* [6, 7, 8, 9] investigated the algebraic relations for reciprocal sums of the Fibonacci numbers. In [10], the partial infinite sums of the reciprocal Fibonacci numbers were studied by Ohtsuka and Nakamura. They established the following results, where $\lfloor\cdot\rfloor$ denotes the floor function.

Theorem 1.1

For all $n \geq 2$,

$$\left\lfloor \left(\sum_{k=n}^{\infty} \frac{1}{F_k}\right)^{-1} \right\rfloor = \begin{cases} F_{n-2}, & \text{if } n \text{ is even;} \\ F_{n-2} - 1, & \text{if } n \text{ is odd.} \end{cases} \quad (1.1)$$

Theorem 1.2

For each $n \geq 1$,

$$\left\lfloor \left(\sum_{k=n}^{\infty} \frac{1}{F_k^2}\right)^{-1} \right\rfloor = \begin{cases} F_n F_{n-1} - 1, & \text{if } n \text{ is even;} \\ F_n F_{n-1}, & \text{if } n \text{ is odd.} \end{cases} \quad (1.2)$$

Further, Wu and Zhang [11, 12] generalized these identities to the Fibonacci polynomials and Lucas polynomials and various properties of such polynomials were obtained.

Recently, Holliday and Komatsu [13] considered the generalized Fibonacci numbers which are defined by

$$G_{n+2} = aG_{n+1} + G_n, \quad n \geq 0$$

with $G_0=0$ and $G_1=1$, and a is a positive integer. They showed that

$$\left\lfloor \left(\sum_{k=n}^{\infty} \frac{1}{G_k}\right)^{-1} \right\rfloor = \begin{cases} G_n - G_{n-1}, & \text{if } n \text{ is even and } n \geq 2; \\ G_n - G_{n-1} - 1, & \text{if } n \text{ is odd and } n \geq 1 \end{cases} \quad (1.3)$$

And

$$\left\lfloor \left(\sum_{k=n}^{\infty} \frac{1}{G_k^2} \right)^{-1} \right\rfloor = \begin{cases} aG_nG_{n-1} - 1, & \text{if } n \text{ is even and } n \geq 2; \\ aG_nG_{n-1}, & \text{if } n \text{ is odd and } n \geq 1. \end{cases} \quad (1.4)$$

Morc rccently, Wu and Wang [14] studied the partial finite sum of the reciprocal Fibonacci numbers and deduced that, for all n≥4,

$$\left\lfloor \left(\sum_{k=n}^{2n} \frac{1}{F_k} \right)^{-1} \right\rfloor = F_{n-2}. \quad (1.5)$$

Inspired by Wu and Wang's work, we obtain two families of partial finite sums of the reciprocal Fibonacci numbers in this paper, which significantly improve Ohtsuka and Nakamura's results, Theorems 1.1 and 1.2. In addition, we present an alternative proof of (1.5).

RECIPROCAL SUM OF THE FIBONACCI NUMBERS

We first present several well-known results on Fibonacci numbers, which will be used throughout the article. The detailed proofs can be found in [5].

Lemma 2.1

Let n≥1, *we have*

$$F_n^2 - F_{n-1}F_{n+1} = (-1)^{n-1} \quad (2.1)$$

and

$$F_aF_b + F_{a+1}F_{b+1} = F_{a+b+1} \quad (2.2)$$

if a and b are positive integers.

As a consequence of (2.2), we have the following result.

Corollary 2.2

For all n≥1, *we have*

$$F_{2n} = F_{n-1}F_n + F_nF_{n+1}, \quad (2.3)$$

$$F_{2n+1} = F_n^2 + F_{n+1}^2, \quad (2.4)$$

$$F_{2n+1} = F_{n-1}F_{n+1} + F_nF_{n+2}. \quad (2.5)$$

It is easy to derive the following lemma and we leave the proof as a simple exercise.

Lemma 2.3

For each n≥1, *we have*

$$F_{n+1}F_{n+2} - F_{n-1}F_n = F_{2n+1}. \qquad (2.6)$$

We now establish two inequalities on Fibonacci numbers which will be used later.

Lemma 2.4

If n≥6, *then*

$$F_{n-2}F_{n-1} > F_{n+1}. \qquad (2.7)$$

Proof

$$\begin{aligned} F_{n-2}F_{n-1} - F_{n+1} &= F_{n-2}F_{n-1} - (F_{n-1} + F_n) \\ &= F_{n-2}F_{n-1} - F_{n-1} - (F_{n-2} + F_{n-1}) \\ &= (F_{n-2} - 2)F_{n-1} - F_{n-2}. \end{aligned}$$

Since $n \geq 6, F_{n-2} - 2 > 1$. So

$$F_{n-2}F_{n-1} - F_{n+1} > F_{n-1} - F_{n-2} > 0,$$

which completes the proof.

Lemma 2.5

For each n≥3, *we have*

$$F_{3n-1}(F_n + F_{n-3}) > F_{n-2}F_{n-1}F_nF_{n+1}. \qquad (2.8)$$

Proof

Applying (2.2), we get

$$F_{3n-1} = F_{n-1}F_{2n-1} + F_nF_{2n}.$$

Thus

$$F_{3n-1}(F_n + F_{n-3}) \geq (F_{n-1}F_{2n-1} + F_nF_{2n})F_n > F_n^2F_{2n} > F_{n-1}F_nF_{2n}.$$

Employing (2.3), we have

$$F_{2n} > F_nF_{n+1} > F_{n-2}F_{n+1},$$

which yields the desired equation (2.8).

The following are some inequalities on the sum of reciprocal Fibonacci numbers.

Proposition 2.6

For all n≥2, *we have*

$$\sum_{k=n}^{2n} \frac{1}{F_k} > \frac{1}{F_{n-2}+1}. \qquad (2.9)$$

Proof

For all k≥2,

$$\begin{aligned}\frac{1}{F_{k-2}+1} - \frac{1}{F_k} - \frac{1}{F_{k-1}+1} &= \frac{F_k - F_{k-2} - 1}{F_k(F_{k-2}+1)} - \frac{1}{F_{k-1}+1} \\ &= \frac{(F_{k-1}-1)(F_{k-1}+1) - F_kF_{k-2} - F_k}{(F_{k-2}+1)(F_{k-1}+1)F_k} \\ &= \frac{F_{k-1}^2 - F_kF_{k-2} - 1 - F_k}{(F_{k-2}+1)(F_{k-1}+1)F_k}.\end{aligned}$$

Invoking (2.1), we obtain $F_{k-1}^2 - F_kF_{k-2} = (-1)^k$. Therefore,

$$\frac{1}{F_{k-2}+1} - \frac{1}{F_k} - \frac{1}{F_{k-1}+1} = \frac{(-1)^k - 1 - F_k}{(F_{k-2}+1)(F_{k-1}+1)F_k}.$$

Now we have

$$\begin{aligned}\sum_{k=n}^{2n} \frac{1}{F_k} &= \frac{1}{F_{n-2}+1} - \frac{1}{F_{2n-1}+1} + \sum_{k=n}^{2n} \frac{(-1)^{k-1}+1+F_k}{(F_{k-2}+1)(F_{k-1}+1)F_k} \\ &> \frac{1}{F_{n-2}+1} - \frac{1}{F_{2n-1}+1} + \sum_{k=n}^{2n} \frac{1}{(F_{k-2}+1)(F_{k-1}+1)} \\ &> \frac{1}{F_{n-2}+1} + \frac{1}{(F_{n-2}+1)(F_{n-1}+1)} - \frac{1}{F_{2n-1}+1}.\end{aligned}$$

Because of (2.5), we have

$$\begin{aligned}F_{2n-1}+1-(F_{n-2}+1)(F_{n-1}+1) &= F_{2n-1} - F_{n-2}F_{n-1} - F_{n-2} - F_{n-1} \\ &= F_{n-2}^2 + F_{n-1}F_{n+1} - F_n \\ &> 0.\end{aligned}$$

Thus, we arrive at

$$\sum_{k=n}^{2n} \frac{1}{F_k} > \frac{1}{F_{n-2}+1}.$$

This completes the proof.

Proposition 2.7

Assume that m ≥ 2. Then, for all even integers n ≥ 4, we have

$$\sum_{k=n}^{mn} \frac{1}{F_k} < \frac{1}{F_{n-2}}.$$

Proof

By elementary manipulations and (2.1), we deduce that

$$\frac{1}{F_{k-2}} - \frac{1}{F_k} - \frac{1}{F_{k-1}} = \frac{(-1)^k}{F_{k-2}F_{k-1}F_k}, \quad k \geq 3.$$

Hence, for n ≥ 3, we have

$$\sum_{k=n}^{mn} \frac{1}{F_k} = \frac{1}{F_{n-2}} - \frac{1}{F_{mn-1}} + \sum_{k=n}^{mn} \frac{(-1)^{k-1}}{F_{k-2}F_{k-1}F_k}. \tag{2.11}$$

Since n is even,

$$\sum_{k=n}^{mn} \frac{(-1)^{k-1}}{F_{k-2}F_{k-1}F_k} < 0,$$

from which we conclude that

$$\sum_{k=n}^{mn} \frac{1}{F_k} < \frac{1}{F_{n-2}}.$$

The proof is complete.

Proposition 2.8

If n ≥ 5 is odd, then

$$\sum_{k=n}^{2n} \frac{1}{F_k} < \frac{1}{F_{n-2}}. \tag{2.12}$$

Proof

It is straightforward to check that the statement is true when n=5. Now we assume that n≥7. Since n is odd, we have

$$\sum_{k=n+1}^{2n} \frac{(-1)^{k-1}}{F_{k-2}F_{k-1}F_k} < 0.$$

Applying (2.7) and (2.6) yields

$$\begin{aligned}\frac{1}{F_{n-2}F_{n-1}F_n} - \frac{1}{F_{2n-1}} &< \frac{1}{F_nF_{n+1}} - \frac{1}{F_{2n-1}} \\ &= \frac{F_{2n-1} - F_nF_{n+1}}{F_nF_{n+1}F_{2n-1}} \\ &= -\frac{F_{n-2}F_{n-1}}{F_nF_{n+1}F_{2n-1}} \\ &< 0.\end{aligned}$$

Employing (2.11) and the above two inequalities, (2.12) follows immediately.

Proposition 2.9

Let m≥3 *be given. If* n≥3 *is odd, we have*

$$\sum_{k=n}^{mn} \frac{1}{F_k} > \frac{1}{F_{n-2}}. \tag{2.13}$$

Proof

It is easy to see that

$$\sum_{k=3}^{3m} \frac{1}{F_k} > \frac{1}{F_1},$$

thus (2.13) holds for n=3. Now we assume that n≥5.

Based on (2.11) and using the fact *n* is odd, we have

$$\sum_{k=n}^{mn} \frac{1}{F_k} = \frac{1}{F_{n-2}} + \frac{1}{F_{n-2}F_{n-1}F_n} - \frac{1}{F_{n-1}F_nF_{n+1}} - \frac{1}{F_{mn-1}} + \sum_{k=n+2}^{mn} \frac{(-1)^{k-1}}{F_{k-2}F_{k-1}F_k}.$$

It is clear that

$$\sum_{k=n+2}^{mn} \frac{(-1)^{k-1}}{F_{k-2}F_{k-1}F_k} > 0.$$

Since m≥3 and invoking (2.8), we obtain

$$F_{mn-1}(F_{n-3} + F_n) \geq F_{3n-1}(F_{n-3} + F_n) > F_{n-2}F_{n-1}F_nF_{n+1},$$

which implies

$$\frac{1}{F_{n-2}F_{n-1}F_n} - \frac{1}{F_{n-1}F_nF_{n+1}} - \frac{1}{F_{mn-1}} = \frac{F_{n+1}-F_{n-2}}{F_{n-2}F_{n-1}F_nF_{n+1}} - \frac{1}{F_{mn-1}}$$
$$= \frac{F_{n-3}+F_n}{F_{n-2}F_{n-1}F_nF_{n+1}} - \frac{F_{n-3}+F_n}{F_{mn-1}(F_{n-3}+F_n)}$$
$$> 0.$$

Therefore, (2.13) also holds for n≥5.

Now we state our main results on the sum of reciprocal Fibonacci numbers.

Theorem 2.10

For all n≥4, *we have*

$$\left\lfloor \left(\sum_{k=n}^{2n}\frac{1}{F_k}\right)^{-1} \right\rfloor = F_{n-2}. \qquad (2.14)$$

Proof

Combining (2.9), (2.10), and (2.12), we conclude that, for all n≥4,

$$\frac{1}{F_{n-2}+1} < \sum_{k=n}^{2n}\frac{1}{F_k} < \frac{1}{F_{n-2}},$$

from which (2.14) follows immediately.

Remark

Identity (2.14) was first conjectured by Professor Ohtsuka, the first author of [10]. Based on the formula of F_n and using analytic methods, Wu and Wang [14] presented a proof of (2.14). In contrast to Wu and Wang's work, the techniques we use here are more elementary.

Theorem 2.11

If m≥3 *and* n≥2, *then*

$$\left\lfloor \left(\sum_{k=n}^{mn}\frac{1}{F_k}\right)^{-1} \right\rfloor = \begin{cases} F_{n-2}, & \text{if } n \text{ is even;} \\ F_{n-2}-1, & \text{if } n \text{ is odd.} \end{cases} \qquad (2.15)$$

Proof

It is clear that

$$\left\lfloor \left(\sum_{k=2}^{2m} \frac{1}{F_k} \right)^{-1} \right\rfloor = F_0. \tag{2.16}$$

Combining (2.9) and (2.10), we find that, for all even integers n≥4,

$$\frac{1}{F_{n-2}+1} < \sum_{k=n}^{mn} \frac{1}{F_k} < \frac{1}{F_{n-2}}. \tag{2.17}$$

Thus (2.16) and (2.17) show that, for all m≥3,

$$\left\lfloor \left(\sum_{k=n}^{mn} \frac{1}{F_k} \right)^{-1} \right\rfloor = F_{n-2},$$

provided that n≥2 is even.

Next we aim to prove that, for m≥3m≥3 and all odd integers n≥3,

$$\left\lfloor \left(\sum_{k=n}^{mn} \frac{1}{F_k} \right)^{-1} \right\rfloor = F_{n-2} - 1.$$

If n = 3, we can readily see that

$$\sum_{k=3}^{3m} \frac{1}{F_k} > 1,$$

thus (2.18) holds for n=3. So in the rest of the proof we assume that n≥5n≥5.

It is not hard to derive that, for all k≥5,

$$\frac{1}{F_{k-2}-1} - \frac{1}{F_k} - \frac{1}{F_{k-1}-1} = \frac{(-1)^k - 1 + F_k}{F_k(F_{k-2}-1)(F_{k-1}-1)} > 0.$$

Hence, we get

$$\sum_{k=n}^{mn} \frac{1}{F_k} < \frac{1}{F_{n-2}-1} - \frac{1}{F_{mn-1}-1} < \frac{1}{F_{n-2}-1}.$$

Finally, combining (2.19) with (2.13) yields (2.18). □

Remark

As m→∞, (2.15) becomes (1.1). Hence our result, Theorem 2.11, substantially improves Theorem 1.1.

RECIPROCAL SQUARE SUM OF THE FIBONACCI NUMBERS

We first give several preliminary results which will be used in our later proofs.

Lemma 3.1

For all n≥1,

$$F_nF_{n+1} - F_{n-1}F_{n+2} = (-1)^{n-1}. \tag{3.1}$$

Proof

It is easy to show that

$$\begin{aligned} F_nF_{n+1} - F_{n-1}F_{n+2} &= F_nF_{n+1} - F_{n-1}(F_n + F_{n+1}) \\ &= F_nF_{n+1} - F_nF_{n-1} - F_{n-1}F_{n+1} \\ &= F_n(F_{n+1} - F_{n-1}) - F_{n-1}F_{n+1} \\ &= F_n^2 - F_{n-1}F_{n+1}. \end{aligned}$$

Employing (2.1), the desired result follows.

Proposition 3.2

Given an integer m≥2 *and let* n≥3 *be odd, we have*

$$\sum_{k=n}^{mn} \frac{1}{F_k^2} < \frac{1}{F_{n-1}F_n}. \tag{3.2}$$

Proof

It is straightforward to check that, for each k ≥ 2,

$$\frac{1}{F_{k-1}F_k}-\frac{1}{F_k^2}-\frac{1}{F_{k+1}^2}-\frac{1}{F_{k+1}F_{k+2}}$$
$$=\frac{F_kF_{k+1}^2F_{k+2}-F_{k-1}F_{k+1}^2F_{k+2}-F_{k-1}F_k^2F_{k+2}-F_{k-1}F_k^2F_{k+1}}{F_{k-1}F_k^2F_{k+1}^2F_{k+2}}$$
$$=\frac{F_kF_{k+1}(F_{k+1}F_{k+2}-F_{k-1}F_k)-F_{k-1}F_{k+2}(\Gamma_{k+1}^2+\Gamma_k^2)}{F_{k-1}F_k^2F_{k+1}^2F_{k+2}}$$
$$=\frac{(F_kF_{k+1}-F_{k-1}F_{k+2})F_{2k+1}}{F_{k-1}F_k^2F_{k+1}^2F_{k+2}}$$
$$=\frac{(-1)^{k-1}F_{2k+1}}{F_{k-1}F_k^2F_{k+1}^2F_{k+2}},$$

where the last equality follows from (3.1).

Since n is odd, we have

$$\frac{1}{F_{n-1}F_n}-\frac{1}{F_n^2}-\frac{1}{F_{n+1}^2}-\frac{1}{F_{n+1}F_{n+2}}>0.$$

If m is even, then

$$\sum_{k=n}^{mn}\frac{1}{F_k^2}<\frac{1}{F_{n-1}F_n}-\frac{1}{F_{mn}F_{mn+1}}<\frac{1}{F_{n-1}F_n}.$$

If m is odd, then

$$\sum_{k=n}^{mn}\frac{1}{F_k^2}<\frac{1}{F_{n-1}F_n}-\frac{1}{F_{mn-1}F_{mn}}+\frac{1}{F_{mn}^2}<\frac{1}{F_{n-1}F_n}.$$

Thus, (3.2) always holds.

Proposition 3.3

Let n be odd, then we have

$$\sum_{k=n}^{2n}\frac{1}{F_k^2}>\frac{1}{F_{n-1}F_n+1}. \tag{3.3}$$

Proof

Invoking (2.1), we can readily derive that

$$\frac{1}{F_{k-1}F_k+1}-\frac{1}{F_k^2}-\frac{1}{F_kF_{k+1}+1}=\begin{cases}-\frac{2F_{k-1}F_k+1}{F_k^2(F_{k-1}F_k+1)(F_kF_{k+1}+1)}, & \text{if } k \text{ is odd;}\\ -\frac{2F_kF_{k+1}+1}{F_k^2(F_{k-1}F_k+1)(F_kF_{k+1}+1)}, & \text{if } k \text{ is even.}\end{cases}$$

Now we have

$$\begin{aligned}\sum_{k=n}^{2n}\frac{1}{F_k^2} &= \frac{1}{F_{n-1}F_n+1}+\left(\frac{2F_{n-1}F_n+1}{F_n^2(F_{n-1}F_n+1)(F_nF_{n+1}+1)}\right.\\ &\quad+\frac{2F_{n+1}F_{n+2}+1}{F_{n+1}^2(F_nF_{n+1}+1)(F_{n+1}F_{n+2}+1)}+\cdots\\ &\quad\left.+\frac{2F_{2n}F_{2n+1}+1}{F_{2n}^2(F_{2n-1}F_{2n}+1)(F_{2n}F_{2n+1}+1)}\right)-\frac{1}{F_{2n}F_{2n+1}+1}\\ &> \frac{1}{F_{n-1}F_n+1}+\frac{2F_{n-1}F_n+1}{F_n^2(F_{n-1}F_n+1)(F_nF_{n+1}+1)}-\frac{1}{F_{2n}F_{2n+1}+1}.\end{aligned}$$

It is obvious that $2F_{k-1}F_k \geq F_{k-1}F_{k+1}$. From (2.1) and the fact that n is odd, we obtain

$$\frac{2F_{n-1}F_n+1}{F_n^2}\geq\frac{F_{n-1}F_{n+1}+1}{F_n^2}=\frac{F_n^2}{F_n^2}=1,$$

which implies that

$$\sum_{k=n}^{2n}\frac{1}{F_k^2}>\frac{1}{F_{n-1}F_n+1}+\frac{1}{(F_{n-1}F_n+1)(F_nF_{n+1}+1)}-\frac{1}{F_{2n}F_{2n+1}+1}.$$

By (2.3) and (2.4), we have

$$\begin{aligned}F_{2n}F_{2n+1}+1 &> (F_{n-1}F_n+F_nF_{n+1})\left(F_n^2+F_{n+1}^2\right)\\ &> (F_{n-1}F_n+1)(F_nF_{n+1}+1),\end{aligned}$$

from which we conclude that

$$\sum_{k=n}^{2n}\frac{1}{F_k^2}>\frac{1}{F_{n-1}F_n+1}.$$

The proof is complete.

Proposition 3.4

Suppose that m≥2 *and* n>0 *is even. Then*

$$\sum_{k=n}^{mn}\frac{1}{F_k^2}<\frac{1}{F_{n-1}F_n-1}. \tag{3.4}$$

Proof

Applying (2.1), we can rewrite F_k^4 k as

$$F_k^4 = F_{k-1}F_k^2F_{k+1} + (-1)^{k-1}F_k^2. \quad (3.5)$$

In addition,

$$\begin{aligned}(F_{k-1}F_k - 1)(F_kF_{k+1} - 1) &= F_{k-1}F_k^2F_{k+1} - F_{k-1}F_k - F_kF_{k+1} + 1 \\ &= F_{k-1}F_k^2F_{k+1} - 2F_{k-1}F_k - F_k^2 + 1. \end{aligned} \quad (3.6)$$

Combining (3.5) and (3.6) yields

$$\begin{aligned}\frac{1}{F_{k-1}F_k - 1} - \frac{1}{F_k^2} - \frac{1}{F_kF_{k+1} - 1} &= \frac{F_k^2}{(F_{k-1}F_k - 1)(F_kF_{k+1} - 1)} - \frac{1}{F_k^2} \\ &= \frac{2F_{k-1}F_k - 1 + F_k^2 + (-1)^{k-1}F_k^2}{(F_{k-1}F_k - 1)(F_kF_{k+1} - 1)F_k^2} \\ &\geq \frac{2F_{k-1}F_k - 1}{(F_{k-1}F_k - 1)(F_kF_{k+1} - 1)F_k^2} \\ &> 0.\end{aligned}$$

Therefore,

$$\sum_{k=n}^{mn} \frac{1}{F_k^2} < \frac{1}{F_{n-1}F_n - 1} - \frac{1}{F_{mn}F_{mn+1} - 1} < \frac{1}{F_{n-1}F_n - 1},$$

which completes the proof.

Proposition 3.5

If n>0 *is even, then*

$$\sum_{k=n}^{2n} \frac{1}{F_k^2} > \frac{1}{F_{n-1}F_n}. \quad (3.7)$$

Proof

Employing (2.1), we can deduce that

$$\frac{1}{F_{k-1}F_k} - \frac{1}{F_k^2} - \frac{1}{F_kF_{k+1}} = \frac{(-1)^{k-1}}{F_{k-1}F_k^2F_{k+1}}.$$

Hence, since n is even, we have

$$\sum_{k=n}^{2n}\frac{1}{F_k^2}=\frac{1}{F_{n-1}F_n}+\sum_{k=n}^{2n}\frac{(-1)^k}{F_{k-1}F_k^2F_{k+1}}-\frac{1}{F_{2n}F_{2n+1}}$$
$$=\frac{1}{F_{n-1}F_n}+\left(\frac{1}{F_{n-1}F_n^2F_{n+1}}-\frac{1}{F_nF_{n+1}^2F_{n+2}}-\frac{1}{F_{2n}F_{2n+1}}\right)$$
$$+\sum_{k=n+2}^{2n-1}\frac{(-1)^k}{F_{k-1}F_k^2F_{k+1}}+\frac{1}{F_{2n-1}F_{2n}^2F_{2n+1}}.$$

It is easy to see that

$$\sum_{k=n+2}^{2n-1}\frac{(-1)^{k-1}}{F_{k-1}F_k^2F_{k+1}}>0,$$

thus

$$\sum_{k=n}^{2n}\frac{1}{F_k^2}>\frac{1}{F_{n-1}F_n}+\left(\frac{1}{F_{n-1}F_n^2F_{n+1}}-\frac{1}{F_nF_{n+1}^2F_{n+2}}-\frac{1}{F_{2n}F_{2n+1}}\right).$$

We claim that

$$\frac{1}{F_{n-1}F_n^2F_{n+1}}-\frac{1}{F_nF_{n+1}^2F_{n+2}}-\frac{1}{F_{2n}F_{2n+1}}>0.$$

First, by (2.6), we have

$$\frac{1}{F_{n-1}F_n^2F_{n+1}}-\frac{1}{F_nF_{n+1}^2F_{n+2}}-\frac{1}{F_{2n}F_{2n+1}}=\frac{F_{2n+1}}{F_{n-1}F_n^2F_{n+1}^2F_{n+2}}-\frac{F_{2n+1}}{F_{2n}F_{2n+1}^2}.$$

It follows from (2.3), (2.4), and (2.5) that

$$F_{2n}>F_{n-1}F_n,$$

$$F_{2n+1}>F_{n+1}^2,$$

$$F_{2n+1}>F_nF_{n+2},$$

which implies that

$$F_{2n}F_{2n+1}^2>F_{n-1}F_n^2F_{n+1}^2F_{n+2}.$$

Thus we obtain

$$\frac{1}{F_{n-1}F_n^2F_{n+1}}-\frac{1}{F_nF_{n+1}^2F_{n+2}}-\frac{1}{F_{2n}F_{2n+1}}>0,$$

which yields the desired (3.7).

Now we introduce our main result on the square sum of reciprocal Fibonacci numbers.

Theorem 3.6

For all n≥1 *and* m≥2, *we have*

$$\left\lfloor \left(\sum_{k=n}^{mn} \frac{1}{F_k^2} \right)^{-1} \right\rfloor = \begin{cases} F_n F_{n-1}, & \text{if } n \text{ is odd;} \\ F_n F_{n-1} - 1, & \text{if } n \text{ is even.} \end{cases} \quad (2.8)$$

Proof

We first consider the case when n is odd. If n=1, the result is clearly true. So we assume that n≥3.

It follows from (3.3) that

$$\sum_{k=n}^{mn} \frac{1}{F_k^2} \geq \sum_{k=n}^{2n} \frac{1}{F_k^2} > \frac{1}{F_{n-1}F_n + 1}. \quad (3.9)$$

Employing (3.2) and (3.9) yields

$$\frac{1}{F_{n-1}F_n + 1} < \sum_{k=n}^{mn} \frac{1}{F_k^2} < \frac{1}{F_{n-1}F_n},$$

which implies that, if n>0n>0 is odd, we have

$$\left\lfloor \left(\sum_{k=n}^{mn} \frac{1}{F_k^2} \right)^{-1} \right\rfloor = F_n F_{n-1}.$$

We now consider the case where n>0 is even. It follows from (3.7) that

$$\sum_{k=n}^{mn} \frac{1}{F_k^2} \geq \sum_{k=n}^{2n} \frac{1}{F_k^2} > \frac{1}{F_{n-1}F_n}. \quad (3.10)$$

Combining (3.4) and (3.10), we arrive at

$$\frac{1}{F_{n-1}F_n} < \sum_{k=n}^{mn} \frac{1}{F_k^2} < \frac{1}{F_{n-1}F_n - 1},$$

from which we find that, if n>0n>0 is even,

$$\left\lfloor \left(\sum_{k=n}^{mn} \frac{1}{F_k^2} \right)^{-1} \right\rfloor = F_n F_{n-1} - 1.$$

This completes the proof.

Remark

Theorem 1.2 can be regarded as the limiting case as m→∞ in (3.8).

Competing interests

The authors declare that they have no competing interests.

Authors' contributions

All authors contributed equally to deriving all the results of this article, and read and approved the final manuscript.

ACKNOWLEDGEMENTS

This work was supported by the National Natural Science Foundation of China (No. 11401080) and the Research Project of Education Teaching Reform of University of Electronic Science and Technology of China (No. 2013XJYEL032). The authors would like to thank the referees for helpful comments leading to an improvement of an earlier version.

REFERENCES

1. Sloane, NJA: The On-Line Encyclopedia of Integer Sequences. https://oeis.org (1991). Accessed 30 Apr 1991
2. Duncan, RL: Applications of uniform distribution to the Fibonacci numbers. Fibonacci Q. 5, 137-140 (1967)
3. Karaduman, E: An application of Fibonacci numbers in matrices. Appl. Math. Comput. 147, 903-908 (2004)
4. Ma, R, Zhang, WP: Several identities involving the Fibonacci numbers and Lucas numbers. Fibonacci Q. 45, 164-170 (2007)
5. Vorobiev, NN: Fibonacci Numbers. Springer, Basel (2002)
6. Elsner, C, Shimomura, S, Shiokawa, I: Algebraic relations for reciprocal sums of Fibonacci numbers. Acta Arith. 130, 37-60 (2007)
7. Elsner, C, Shimomura, S, Shiokawa, I: Algebraic relations for reciprocal sums of odd terms in Fibonacci numbers. Ramanujan J. 17, 429-446 (2008)
8. Elsner, C, Shimomura, S, Shiokawa, I: Algebraic independence results for reciprocal sums of Fibonacci numbers. Acta Arith. 148, 205-223 (2011)
9. Elsner, C, Shimomura, S, Shiokawa, I: Algebraic relations for reciprocal sums of even terms in Fibonacci numbers. J. Math. Sci. 180, 650-671 (2012)
10. Ohtsuka, H, Nakamura, S: On the sum of reciprocal Fibonacci numbers. Fibonacci Q. 46/47, 153-159 (2008/2009)
11. Wu, ZG, Zhang, WP: The sums of the reciprocals of Fibonacci polynomials and Lucas polynomials. J. Inequal. Appl. 2012, Article ID 134 (2012)
12. Wu, ZG, Zhang, WP: Several identities involving the Fibonacci polynomials and Lucas polynomials. J. Inequal. Appl. 2013, Article ID 205 (2013)
13. Holliday, SH, Komatsu, T: On the sum of reciprocal generalized Fibonacci numbers. Integers 11A, Article ID 11 (2011)
14. Wu, ZG, Wang, TT: The finite sum of reciprocal of the Fibonacci numbers. J. Inn. Mong. Norm. Univ. Nat. Sci. 40, 126-129 (2011)

SECTION 6: GRAPH ALGORITHMS

13

Shortest Augmenting Paths for Online Matchings on Trees

Bartłomiej Bosek[1], Dariusz Leniowski[2], Piotr Sankowski[2] and Anna Zych-Pawlewicz[2]

[1] Theoretical Computer Science Department, Faculty of Mathematics and Computer Science, Jagiellonian University, Krakow, Poland ´

[2] Institute of Computer Science, University of Warsaw, Warsaw, Poland

ABSTRACT

The shortest augmenting path (SAP) algorithm is one of the most classical approaches to the maximum matching and maximum flow problems, e.g., using it Edmonds and Karp (J. ACM 19(2), 248–264 1972) have shown the first strongly polynomial time algorithm for the maximum flow problem. Quite astonishingly, although it has been studied for many years already, this approach is far from being fully understood. This is exemplified by the online bipartite matching problem. In this problem a bipartite graph $G = (W \uplus B, E)$ is being revealed online, i.e., in each round one vertex from

Citation: Bosek, B., Leniowski, D., Sankowski, P. et al. "Shortest Augmenting Paths for Online Matchings on Trees" Theory Comput Syst (2018) 62: 337. https://doi.org/10.1007/s00224-017-9838-x

B with its incident edges arrives. After arrival of this vertex we augment the current matching by using shortest augmenting path. It was conjectured by Chaudhuri et al. (INFOCOM'09) that the total length of all augmenting paths found by SAP is $O(n \log n)$. However, no better bound than $O(n^2)$ is known even for trees. In this paper we prove an $O(n \log^2 n)$ upper bound for the total length of augmenting paths for trees.

Keywords: Online matchings, Bipartite matchings, Approximate matchings, Shortest augmenting paths, Dynamic graph algorithms

INTRODUCTION

The shortest augmenting path (SAP) algorithm is one of the most classical approaches to the maximum matching and maximum flow problems. Using this idea Edmonds and Karp in 1972 have shown the first strongly polynomial time algorithm for the maximum flow problem [5]. Quite astonishingly, although this idea is one of the most basic algorithmic techniques, it is far from being fully understood. It is easier to talk about it by introducing the online bipartite matching problem. In this problem a bipartite graph $G = (W \uplus B, E)$ is being revealed online, i.e., in each round one vertex from B with its incident edges arrives. After arrival of this vertex we augment this matching by using shortest augmenting path. It was conjectured by Chaudhuri et al. [4] that the total length of augmenting paths found by SAP is $O(n \log n)$. However, no better bound than $O(n^2)$ is known even for trees. Proving this conjecture would have quite striking consequences even for maximum flow problem, as it would show that the total length of augmenting paths in unit capacity networks in Edmonds-Karp algorithm is $O(m \log n)$. This consequence is obtained via the bipartite line graph construction that is used to reduce the max-flow problem to maximum matching problem [10]. The obtained bipartite line graph has 2m vertices.

Our paper contributes to the study of SAP algorithm by showing that in the case of trees the total length of all augmenting paths is bounded by $O(n \log^2 n)$. This result is obtained via the application of the heavy-light decomposition of trees [16] combined with charging technique that carefully assigns shortest augmenting paths to the structure of the tree. Although, this result seems to be restricted only to trees we be believe that it constitutes the first nontrivial progress towards resolving the above conjecture. Moreover, we actually conjecture here that trees are the worst-case examples for this problem. It seems that adding more edges can only help the SAP algorithm. In addition to that we explain why SAP is harder to analyze than other augmenting path algorithms, even though it seems way more natural.

RELATED WORK

The online bipartite matching problem with augmentations has recently received increasing research attention [3, 4, 6, 7]. There are several reasons to study this problem. First of all, it provides a simple solution to the online bipartite matching algorithms used in many modern applications such as online advertising (e.g., Google Ads) [12] or client-server assignment [4]. Secondly, they could give rise to new effective offline bipartite matching algorithms as in [3]. Those new algorithms provide new insights to the old problem that was studied for decades.

In this paper we concentrate on bounding the total length of augmenting paths and not on the running time. With this respect, it was shown that if the vertices of B appear in a random order, the expected total paths' length for SAP is O(n log n) [4]. The worst-case total length of paths remains an open question even for trees. In the class of trees the authors of [4] proposed a different augmenting path algorithm that achieves total paths' length of O(n log n). On the other hand, for general bipartite graphs greedy ranking algorithm [3] guarantees $O(n\sqrt{n})$ total length of paths.

Another point of view is given by the dynamic matching algorithms. Most papers in this area consider edge updates in a general fully-dynamic model which allows for both insertions and deletions intermixed with each other. We note, however, that the exact results in this model [9, 15] do not imply any bound on the number of changes to the matching. Much faster update times can be achieved by constant approximate algorithms, for example [1, 14], which achieve polylogarithmic and logarithmic update times. Yet, the 2-approximation can be obtained in our setting by trivial greedy algorithm that preforms no changes at all.

Better approximation factor of 3 2 was achieved by [13] in $O(\sqrt{m})$ update time, and then improved by Gupta and Peng to $(1 + \varepsilon)$ in $O(\sqrt{m}\varepsilon^{-2}$) [8]. The $O(\sqrt{m})$ barrier was broken by Bernstein and Stein who gave a $(\frac{3}{2} + \varepsilon)$-approximation algorithm that achieves $O(m^{1/4}\varepsilon^{-2.5})$ update time [2]. The same paper proposes an $(1 + \varepsilon)$- approximation algorithm in very fast $O(\alpha(\alpha + \log n) + \varepsilon^{-4}(\alpha + \log n) + \varepsilon^{-6})$ update time for the special case of bipartite graphs with constant arboricity. However, when allowing approximation in our model a much better results are possible. An $(1 + \varepsilon)$ approximation in $O(m\varepsilon-1)$ total time and with $O(n\varepsilon^{-1})$ total length of paths was shown in [3].

PRELIMINARIES

We consider the following matching problem. Let W and B be two sets of vertices over which the bipartite graph will be formed. The set W (called white vertices) is given up front to the algorithm, whereas the vertices in B (black vertices) arrive online. We denote by $G_t = \langle W \uplus B_t, E_t \rangle$ the bipartite graph after the t'th black vertex has arrived. The graph Gt is constructed online in the following manner. We start with $G_0 = \langle W \uplus B_0, E_0 \rangle = \langle W \uplus \emptyset, \emptyset \rangle$. In turn t a new vertex $b_t \in B$ together with all its incident edges $E(b_t)$ is revealed and Gt is defined as:

$$\begin{cases} E_t = E_{t-1} \cup E(b_t), \\ B_t = B_{t-1} \cup \{b_t\}. \end{cases}$$

The goal of our algorithm is to compute for each G_t the maximum size matching M_t . For simplicity we assume that we add in total |W| black vertices. The final graph $G_{|W|}$ which is obtained in this process will be denoted by $G = (\uplus W\ B,E)$. We denote $n = |W| = |B|$ and $m = |E|$.

For every $t \in [n]$, we add orientation to edges of the graph G_t. This orientation is induced by matching M_t: the matched edges are oriented towards black vertices, while the unmatched edges are oriented towards white vertices. When a new vertex b_t arrives, we get an intermediate orientation $G^{int}t = (E^{int}t, B_t)$, where the edges of b^t are oriented towards its neighbors, and the rest of the edges is oriented according to M_{t-1}. Note that $G^{int}t$ and G_{t-1} differ only by one vertex b_t. Any simple directed path in $G^{int}t$ from b_t to some unmatched white vertex is an augmenting path. In turn t, if b_t can be matched, the edges of $G^{int}t$ are reoriented along augmenting path π_t chosen by the algorithm, and the resulting orientation is G_t. The unmatched white vertices are called seeds. We denote the set of seeds after turn t as:

$$S_t = \{w \in W : wb \notin M_t \text{ for any } b \in B\}.$$

So in turn t the augmenting paths in G^{int}_t are the directed paths from b_t to some $s \in S_{t-1}$. We refer to the seed of the path π_t from turn t as s_t , where st $\in S_{t-1}$. We represent a path as a graph consisting of path vertices and path edges. We use the notation $v \overset{\pi}{\rightarrow} v'$ to denote that a (directed) path π starts in v and ends in v , and v $-\!\!\rightarrow$ v' to denote a connection via a directed edge. We use the notation $v \in \pi$ and $\rho \subseteq \pi$ to state that a vertex $v \in V(\pi)$ and that a path ρ is a subgraph of π, respectively. We also denote the length of a path π as |π|. Throughout the paper, when we write "at time t", what we formally mean is "in G^{int}_t".

The next thing we define is a set of vertices D_t called dead at time t. The set D_t is defined as the set of vertices in G^{int}_t that cannot reach S_{t-1} via a directed path in G^{int}_t. Observe, that if at some point there is no directed path from a vertex to a seed, never again there will be such a path. If a vertex is dead, all vertices reachable from it are dead as well. Hence, no alternating path can enter such a dead region and reorient its edges to make some vertices alive. In other words, $Dt \subseteq D_{t+1}$ for every time moment t. Detailed matching-independent proof of this fact can be found in Section 1.2.2 of [11]. The vertices of D_t are called dead, while the remaining vertices are called alive.

We now define the effective degree of a black vertex b in turn t as the number of it's non-dead out-neighbors:

$$\text{degeff}_t(b) = |\Gamma_t(b) \setminus \mathcal{D}_t|$$

where $\Gamma_t(b)$ is the set of vertices v such that b $\longrightarrow$ v in G^{int}_t , referred to sometimes as out-neighbours of b. In particular $\text{degeff}_t(b_t)$ is the number of all non-dead neighbors (in the undirected sense) of b_t , as all the edges adjacent to b_t are directed towards its neighbors.

Since we consider in this paper the special case when G_t is a tree at any time t, from now on we will refer to G as T, and to G_t as T_t .

SHORTEST PATHS ON TREES

In this section we study the shortest augmenting path (SAP) algorithm, which in each turn chooses the shortest among all available augmenting paths. We start by giving an easy argument, that the total length of augmenting paths for SAP is O(n log n) if all vertices b_t satisfy $\text{degeff}_t(b_t) > 1$. This shows that the difficult case is to deal with vertices of effective degree 1.

Lemma 1 If for each t ∈ [n] it holds that $\text{degeff}_t(b_t) > 1$, then the total length of all augmenting paths applied by SAP is bounded by O(n log n).

Proof Due to the definition of effective degree, every vertex b_t connects at least two trees T_1 and T_2 that contain a directed path connecting b_t with a seed. Let T_1 be a smaller of the two trees. The length of the shortest path πt from b_t to a seed is at most the size of T_1. We charge the cost of π_t to $|\pi_t|$ arbitrary vertices of T_1. During the course of the SAP algorithm, every vertex can be charged at most O(log n) times, as each time it is charged, the size of its tree doubles. The total charge is hence O(n log n). The main result of this paper and the subject of the remainder of this section is the bound for the general case, stated in the following theorem.

Theorem 1 The total length of augmenting paths applied by SAP is $O(n \log^2 n)$.

In order to prove Theorem 1 we introduce a few definitions and observations. The core of our proof is the concept of a dispatching vertex.

Definition 1 A black vertex $b \in B$ is called dispatching at time t if b is the closest vertex to b_t on the path πt that satisfies $degeff_t(b) > 1$. If there is no such vertex at time t we define s_t to be the dispatching vertex. We denote a dispatching vertex in time t as $dis(\pi_t)$. Moreover, for every dispatching black vertex b we define $tlas_t(b)$ as the moment when b is dispatching for the last time.

So every path π_t applied by SAP has a uniquely defined dispatching vertex $dis(\pi_t)$ assigned to it. The first observation we make is that we only have to care about suffixes of π_t's starting with $dis(\pi_t)$.

Definition 2 We split path πt into two segments $\pi t = \mu_t \rho_t$, where ρ_t is the suffix of πt such that $dis(\pi_t)\ \rho_t \xrightarrow{\rho_t} st$. Path $\mu_t = \pi_t \setminus \rho_t$ is the remaining part of π_t (a possibly empty prefix that ends in a vertex preceding $dis(\pi_t)$). We refer to the above defined suffixes as dispatching paths.

Lemma 2 The total length of paths μ_t is linear in the size of the tree T , i.e.,

$$\sum_{t\in[n]} |\mu_t| \in \mathcal{O}(n).$$

Proof The lemma holds due to Observation 2, proven below, which states that vertices of μ_t die at the time t when π_t is applied. With this observation it is clear that the time μ_t passes through a vertex is the last time SAP visits that vertex. So every vertex in the tree is visited by μ_t for any t at most once.

Observation 2 Vertices of μ_t die at the time t when π_t is applied.

Proof At the time when π_t is applied, all vertices on μ_t have effective degree equal to 1, i.e., they have only one alive directed out-neighbour – their successor on μ_t . If we reverse the edges, the only chance for the vertices of μ_t to be alive is the last vertex b_t . This vertex, however, becomes dead because its only alive out-neighbour is removed. As a consequence the whole path dies.

To bound the total length of augmenting paths π_t , it remains to bound the total length of dispatching paths: $\sum_{t\in[n]} |\rho_t|$. Consider the case when $dis(\pi t)$

= st . Then the path ρ_t consists of a single vertex st . The total sum of paths ρ_t satisfying this case is thus O(n). It remains to consider the sum over all dispatching paths ρ_t that start in a black dispatching vertex. These non-trivial dispatching paths will be, from now on, the focus of our attention. In other words, our goal is to bound the following sum.

Lemma 3 The total length of non-trivial dispatching paths is O(n log² n):

$$\sum_{\substack{t\in[n]:\\ \mathrm{dis}(\pi_t)\in B}} |\rho_t| \in \mathcal{O}(n\log^2 n).$$

For the sake of clarity, we split the proof of Lemma 3 into two steps, presented as Lemmas 5 and 6. More precisely, we partition the dispatching paths depending on whether t < tlast(dis(π_t)) or t = tlast(dis(π_t)), that is, if the dispatching path in question, is the last for its dispatching vertex (cf. Definition 1). In what follows, paths that satisfy the former condition are called non-final and their total length is bounded in Lemma 5, while final paths start at b ∈ B when b is a dispatching vertex for the last time; their total length is the subject of Lemma 6.

These results will complete the proof of Theorem 1. However, before we jump into their proofs, we first briefly recall the heavy-light decomposition introduced by Sleator and Tarjan in [16] and state a related technical result, Lemma 4.

For a tree T rooted at r, the original technique partitions its edges into heavy and light, depending on whether the size of the subtree is strictly bigger than half of the size of the subtree rooted at parent. More precisely, let v be any vertex of T other than root r and set p_v to be its parent, then an edge {v, p_v} is heavy if and only if |subtree(v)| > $\frac{1}{2}$ |subtree(p_v)|, where subtree(x) is a subtree of T rooted in x ∈ V (T). Non-heavy edges are called light.

Observe, that because of the size requirements, each time we traverse a light edge away from the root r, the size of the current subtree halves. In other words, for any vertex v of T there are at most $\lfloor \log_2 |T| \rfloor$ light edges on the simple path from r to v. Note that each vertex can have at most two heavy incident edges, thus heavy edges form vertex-disjoint paths. Moreover, paths are of much simpler structure than arbitrary trees, hence allow for more efficient handling despite being possibly numerous.

For convenience, in this paper, we use a slightly modified version, that is, each non-leaf node selects exactly one heavy edge – the edge to the child that has the greatest number of descendants (breaking ties arbitrarily). In particular an edge may be considered heavy, even if the subtree is strictly smaller than half of the size of the current tree. Just like in the original technique, the selected edges form the paths of the decomposition (each non-leaf vertex has at least one and at most two heavy edges), which we call heavy paths. By heavy-path(v) we denote the heavy path to which vertex v belongs, while level : V (T) → $\mathbb{N}$ is the number of light edges on the simple path from a vertex to the root. Observe that level $(v) \leq \lfloor \log_2 |T| \rfloor$ for any vertex v of T.

Lemma 4 Let T be any unrooted tree of size n. For any vertex v let $S^v = \langle S_0^v, S_1^v, \ldots \rangle$ be the sequence of subtrees of v (i.e., the connected components of T \ {v}) ordered descending by their size, that is, $|V(S_i^v)| \geq |V(S_{i+1}^v)|$. Then for:

$$\Psi(v) = \sum_{i=2}^{|S^v|-1} |V(S_i^v)|,$$

we have $\sum_{v \in V(T)} \Psi(v) \in \mathcal{O}(n \log n)$.

Proof Let r be a centroid point of T, that is, a vertex such that $|V(S_0^r)| \leq \frac{1}{2}|V(T)|$. We root T at r, and perform the heavy-light decomposition of T . Observe that for all vertices v ≠ r we have that S_0^vcontains r (it corresponds to the parent of v) and S_1^vcorresponds to the biggest child of v. In other words, at most $S_0^v$and S_1^vcan be connected by heavy edges, all the other subtrees $S_2^v, S_3^v, \cdots$are connected by light edges.

Now we take an arbitrary vertex w and calculate how many times it can appear in $\sum_{v \in V(T)} \Psi(v)$. Suppose v is a vertex that counts w in Ψ (v), then the first edge on the path from v to w has to be light. Moreover, $S_0^v$is not counted in Ψ (v), so that path cannot pass through the parent of v. Because of that v has to be an ancestor of w. However, there are at most O(log n) light edges on any path from w to the root r for any w. In other words, there can be at most O(log n) vertices that count w in its sum of. Summing that for all vertices of T we get the desired bound of O(n log n).

With the help of Lemma 4 we can tackle the first part of Lemma 3, that is, the sum of the lengths of non-final dispatching paths.

Lemma 5 The total length of non-final dispatching paths is O(n log n):

$$\sum_{\substack{t \in [n]: \\ b = \mathrm{dis}(\pi_t) \in B \\ t < \mathrm{tlast}(b)}} |\rho_t| \in \mathcal{O}(n \log n)$$

Proof Recall that the path π_t starts in the newly added vertex b_t . So in turn t either b_t is dispatching, or it dies. At any later time t' > t at which b_t is dispatching π_t' does not begin with b_t and hence one of b_t's neighbours dies based on Observation 2.

Consider a fixed vertex b and let $W_b^{\mathcal{D}} \subseteq W$ be the set of neighbors of b that die in turns when b is dispatching. The first time b is dispatching no neighbour of b dies, so $\mathcal{D}_t \cap W_b^{\mathcal{D}} = \emptyset$The second time b is dispatching, it has at least one dead neighbour and set $\mathcal{D}_t \cap W_b^{\mathcal{D}}$ has exactly one element, namely the white vertex that preceded b on μ_t. More generally, the k-th time b is dispatching, $\mathcal{D}_t \cap W_b^{\mathcal{D}}$has k − 1 elements.

Suppose that the total number of times b is dispatching equalsl, in particular we know that at some point of time $\mathcal{D}_t \cap W_b^{\mathcal{D}}$ will have l − 1 elements. When b is dispatching for the k-th time set $\mathcal{D}_t \cap W_b^{\mathcal{D}}$ has only k − 1 members. In other words, b has l − k neighbors which are at that turn not yet in $\mathcal{D}_t \cap W_b^{\mathcal{D}}$, and thus alive. Furthermore b has at least two white neighbors that do not belong to $W_b^{\mathcal{D}}$ and are alive at the time when b is dispatching for the last time. Therefore, in total b has at least l −k +2 alive white out-neighbours.

We say that a subtree hangs from the neighbour w of b, if it is obtained by the removal of b from T and it contains w. Suppose that we discard two neighbors of b with the heaviest trees hanging from them, i.e., two heaviest neighbours. Then for k = l − 1 we have at least one alive neighbor, for k = l − 2 we have at least two alive neighbors, that is, at least one alive neighbor other than the neighbor used at k = l − 1, and so on. In other words, for any k<l we can find a distinct, not already assigned, alive neighbor w different than the two heaviest neighbors of b. However, the size of the subtree hanging from that neighbour bounds the length of the shortest augmenting path starting at b. Therefore, we can bound the total length of non-final paths dispatching at b by the total size of all subtrees of b except the two heaviest. Summing that up over the whole tree gives us a O(n log n) upper bound, as shown by the previous lemma, Lemma 4.

We now move on to the second part of Lemma 3, namely we bound the total length of final dispatching paths.

Lemma 6 The sum of lengths of ρ_t such that t = tlast(dis(π_t)) is bounded from above by O(n $\log^2$ n):

$$\sum_{\substack{t \in [n]: \\ b=\mathrm{dis}(\pi_t) \in B \\ t=\mathrm{tlast}(b)}} |\rho_t| \in \mathcal{O}(n \log^2 n).$$

To prove the above statement we will need a more fine-grained analysis than before. The problem with the shortest path approach is that its structure and the structure of matchings are very different. To close this gap we introduce yet another family of augmenting paths that relies much more on the structure of the tree. Obviously, because the shortest paths are shorter than any other path, any upper bound on the total length of the aforementioned new family of augmenting paths is an upper bound for the shortest paths as well.

Proof Similarly to Lemma 4, we root T at a centroid point and preform the heavy light decomposition of the tree. That is, each vertex selects an edge to its largest subtree, which we call heavy, while all other edges are considered light.

We define λ_t as one of the M_{t-1}-augmenting paths that connect the newly added vertex b_t to an unmatched white vertex. To be more precise, for each path λ connecting b_t with a free white vertex in turn t let the tuple

$$\langle \mathrm{level}(v_0),\ \mathrm{level}(v_1),\ \mathrm{level}(v_2),\ \ldots \rangle$$

represent λ, where v_0, v_1, v_2,... is the sequence of vertices of λ. We define λ_t as the path represented by the lexicographically last tuple. Observe that λ_t leaves any heavy path as soon as possible.

Now note that all M_{t-1}-augmenting paths starting in b_t are the same up to $\mathrm{dis}(\pi_t)$ and let $\mathrm{dis}(\lambda_t) = \mathrm{dis}(\pi_t)$. To bound the length of λ_t , we split it into three parts as follows:

$$\begin{aligned} \lambda_t' &= \mu_t, \\ \lambda_t'' &= (\lambda_t \setminus \lambda_t') \cap \text{heavy-path}(\mathrm{dis}(\lambda_t)), \\ \lambda_t''' &= (\lambda_t \setminus \lambda_t') \setminus \text{heavy-path}(\mathrm{dis}(\lambda_t)). \end{aligned}$$

Then λ_t'' follows heavy-path $(\mathrm{dis}(\lambda_t))$ at most up to the closest vertex b such that $\mathrm{tlast}(b) > t$, namely:

$$|\lambda_t''| \leq \min\{\mathrm{dist}(b, \mathrm{dis}(\lambda_t)) \mid b \in B\,(\text{heavy-path}(\mathrm{dis}(\lambda_t)))\,,\ \mathrm{tlast}(b) > t\}. \tag{1}$$

The reason for this is that any vertex b with $\mathrm{tlast}(b) > t$ has at least three alive neighbors, with at least two of them reachable from b in $G^{\mathrm{int}}t$, and at least one not on heavy-path $(\mathrm{dis}(\lambda_t))$. Any such b is a vertex at which λ_t'' can leave heavy-path$(\mathrm{dis}(\lambda_t))$.

Consider an arbitrary heavy path H and the set D of vertices on H that are ever dispatching in the entire run of the algorithm. Using $D_0 = D$ we split H into at least $d_0 = |D_0|$ non-empty fragments $h_0, h_1,\ldots,h_{d0-1}$. Each such part $h \in \{h_0, h_1,\ldots,h_{d0-1}\}$ has at least one of its endpoints in D_0, and usually both, unless h is the first or the last fragment. Thus, we can assign h to one of its ending vertices with preference for earlier turn tlast if there are two available. Formally $f_0 : \{h_0, h_1,\ldots,h_{d0-1}\} \rightarrow D_0$, where:

$$f_0(h) = \underset{b \in V(h) \cap D_0}{\arg\min} \ \text{tlast}(b).$$

Due to Inequality (1) in the previous paragraph we have:

$$\left|\lambda''_{\text{tlast}(f_0(h_i))}\right| \leq |h_i| \text{ for any } 0 \leq i < d_0.$$

This means that the length of H bounds the total length of λ''s related to the dispatching vertices in the image of f_0. However, as at most two h_i's can be assigned to the same vertex of D_0, the image of f_0 constitutes at least a half of D_0.

To take care of the rest of D_0, we iterate this reasoning. We construct a sequence of sets $D_0 \supseteq D_1 \supseteq \ldots$, each step halving the size of D_i. More precisely, we set:

$$D_i = D_{i-1} \setminus \left\{ f_{i-1}\left(h_j^{i-1}\right) \mid 0 \leq j < d_{i-1} \right\},$$

where $h_0^i, h_1^i, \ldots, h_{d_i-1}^i$ are the parts of H after the split by D_i and functions $f_i : \{h_0^i, h_1^i, \ldots, h_{d_i-1}^i\} \rightarrow D_i$ are defined as:

$$f_i(h^i) = \underset{b \in V(h^i) \cap D_i}{\arg\min} \ \text{tlast}(b).$$

In other words, at most log n copies of H cover all λ'' paths related to H. Summing this up over all heavy paths gives us:

$$\sum_{\substack{t \in [n]: \\ b = \text{dis}(\pi_t) \in B \\ t = \text{tlast}(b)}} \left|\lambda''_t\right| \in \mathcal{O}(n \log n).$$

Furthermore, it also means that for any v ∈ V (H) at most log n of λ''' paths may start in v. That is, log n copies of all non-heavy subtrees of v ∈ V (H) cover all λ''' paths starting in v, which, by Lemma 4, implies

$$\sum_{\substack{t \in [n]: \\ b = \text{dis}(\pi_t) \in B \\ t = \text{tlast}(b)}} \left|\lambda'''_t\right| \in \mathcal{O}(n \log^2 n).$$

From the last two bounds we infer the statement of the Lemma 6. This also completes the proof of Theorem 1.

PLAYING AGAINST AN ADVERSARY

In the last section of this paper we discuss a quite surprising characteristic of Theorem 1 and its implications. Namely, nowhere in the proofs of Lemmas 3, 4, 5 and 6 we rely on the shape of any particular matching at any given turn, or even on the fact that these matchings are related to each other. To be more specific, we depend only on the structure of the tree, the properties of the dead and alive vertices, and the cardinality of the matchings. This leads us to a generalization of the setting in question and a respective counterpart of Theorem 1.

We define the adversarial dynamic augmenting path setting as a setup similar to the one from Section 3 in which, as before, each turn we are given a single black vertex with all its edges. However, the matching we use to calculate the shortest augmenting path is not the one produced by the algorithm in the previous turn, but some arbitrary matching of the same cardinality provided by the adversary. In particular, the edges might be oriented, wherever possible, away from the newly added vertex, thus making the augmenting paths the longest possible. Nonetheless, because we do not depend on the structure of the matching, the total length of all such augmenting paths is still small.

Corollary 1 If the graph in the above setting is a tree, then the total length of all the shortest augmenting paths is $O(n \log^2 n)$.

It seems that this is true also in general bipartite graphs, and thus we form the following conjecture.

Conjecture 1 The total length of all the shortest augmenting paths in the setting above, that is, with the matching changing arbitrarily each turn, is still $O(n \log n)$ worst case for any bipartite graph.

The ramifications of that conjecture are twofold. First, it suggests a new perspective and a new research angle in which we are allowed to change the matching to fit into some schema. That could possibly lengthen the paths in the process, but it might make the problem a bit more predictable and less dynamic, hence, in some aspects, easier. Second, it might allow for better algorithms. A matching procedure based on the above idea could alter the calculated matching during some turns in a random way, thus perhaps making its worst case less bad. As the reasons behind this phenomenon are far from clear, in authors' opinion Conjecture 1 is an interesting open problem.

REFERENCES

1. Baswana, S., Gupta, M., Sandeep, S.: Fully dynamic maximal matching in O(N) update time. In: Proceedings of the 2011 IEEE 52nd Annual Symposium on Foundations of Computer Science, FOCS '11, pp. 383–392. IEEE Computer Society, Washington, DC (2011)
2. Bernstein, A., Stein, C.: Fully dynamic matching in bipartite graphs. 2015 to appear at ICALP (2015)
3. Bosek, B., Leniowski, D., Sankowski, P., Zych, A.: Online Bipartite Matching in Offline Time. In: 55th IEEE Annual Symposium on Foundations of Computer Science, FOCS 2014, pp. 384–393. IEEE Computer Society, Philadelphia (2014)
4. Chaudhuri, K., Daskalakis, C., Kleinberg, R.D., Lin, H.: Online bipartite perfect matching with augmentations. In: 28th IEEE International Conference on Computer Communications, Joint Conference of the IEEE Computer and Communications Societies INFOCOM 2009, pp. 1044–1052. IEEE, Rio De Janeiro (2009)
5. Edmonds, J., Karp, R.M.: Theoretical improvements in algorithmic efficiency for network flow problems. J. ACM 19(2), 248–264 (1972)
6. Grove, E.F., Kao, M.-Y., Krishnan, P., Vitter, J.S.: Online perfect matching and mobile computing. In: Akl, S.G., Dehne, F., Sack, J.-R., Santoro, N. (eds.) Algorithms and Data Structures, volume 955 of Lecture Notes in Computer Science, pp. 194–205. Springer, Berlin (1995)
7. Gupta, A., Kumar, A., Stein, C.: Maintaining assignments online: matching, scheduling, and flows. In: Chekuri, C. (ed.) Proceedings of the Twenty-Fifth Annual ACM-SIAM Symposium on Discrete Algorithms, SODA 2014, pp. 468–479. SIAM, Portland (2014)
8. Gupta, M., Peng, R.: Fully dynamic (1 + e)-approximate matchings. In: 2013 IEEE 54th Annual Symposium on Foundations of Computer Science, vol. 0, pp. 548–557 (2013)
9. Ivkovic, Z., Lloyd, E.L.: Fully dynamic maintenance of vertex cover. In: Leeuwen, J. (ed.) Graph- ´ Theoretic Concepts in Computer Science, volume 790 of Lecture Notes in Computer Science, pp. 99–111. Springer, Berlin (1994)
10. Karp, R.M., Upfal, E., Wigderson, A.: Constructing a perfect matching is in random nc. Combinatorica 6(1), 35–48 (1986)
11. Leniowski, D.: On Maintaining Online Bipartite Matchings with

Augmentations. PhD thesis, University of Warsaw (2015)

12. Mehta, A., Saberi, A., Vazirani, U.V., Vazirani, V.V.: Adwords and Generalized On-Line Matching. In: 46Th Annual IEEE Symposium on Foundations of Computer Science FOCS 2005, pp. 264–273 (2005)
13. Neiman, O., Solomon, S.: Simple deterministic algorithms for fully dynamic maximal matching. In: Proceedings of the Forty-fifth Annual ACM Symposium on Theory of Computing, STOC '13, pp. 745–754. ACM, New York (2013)
14. Onak, K., Rubinfeld, R.: Property Testing. Chapter Dynamic Approximate Vertex Cover and Maximum Matching, pp. 341–345. Springer, Berlin (2010)
15. Sankowski, P.: Faster dynamic matchings and vertex connectivity. In: Proceedings of the Eighteenth Annual ACM-SIAM Symposium on Discrete Algorithms, SODA 07, pp. 118-126. Society for Industrial and Applied Mathematics, Philadelphia (2007)
16. Sleator, D.D., Tarjan, R.E.: A data structure for dynamic trees. J. Comput. Syst. Sci. 26(3), 362–391 (1983)

14

Subgraph-augmented Path Embedding for Semantic User Search on Heterogeneous Social Network

Zemin Liu [1], Vincent W. Zheng[2], Zhou Zhao[3], Hongxia Yang[4], Kevin Chen-Chuan Chang[5], Minghui Wu[6] and Jing Ying[7]

[1]Zhejiang University, China

[2]Advanced Digital Sciences Center, Singapore,

[3]Zhejiang University, China

[4]Alibaba Group, China

[5]University of Illinois at Urbana-Champaign, USA

[6]Zhejiang University City College, China

[7]Zhejiang University, China

ABSTRACT

Semantic user search is an important task on heterogeneous social networks. Its core problem is to measure the proximity between two user objects in the network w.r.t. certain semantic user relation. State-of-the-art solutions often take a path-based approach, which uses the sequences of objects connecting a

Citation: Zemin Liu, Vincent W. Zheng, Zhou Zhao, Hongxia Yang, Kevin Chen-Chuan Chang, Minghui Wu, and Jing Ying. 2018. "Subgraph-augmented Path Embedding for Semantic User Search on Heterogeneous Social Network". InProceedings of The 2018 Web Conference (WWW 2018).ACM, New York, NY, USA, 10 pages. https://doi.org/10.1145/3178876.3186073

query user and a target user to measure their proximity. Despite their success, we assert that path as a low-order structure is insufficient to capture the rich semantics between two users. Therefore, in this paper we introduce a new concept of subgraph-augmented path for semantic user search. Specifically, we consider sampling a set of object paths from a query user to a target user; then in each object path, we replace the linear object sequence between its every two neighboring users with their shared subgraph instances. Such subgraph-augmented paths are expected to leverage both path's distance awareness and subgraph's high-order structure. As it is non-trivial to model such subgraph-augmented paths, we develop a Subgraph-augmented Path Embedding (SPE) framework to accomplish the task. We evaluate our solution on six semantic user relations in three real-world public data sets, and show that it outperforms the baselines.

INTRODUCTION

Heterogeneous social networks are prevalent nowadays [**37**]. Because social networks are human-centric, it is common to observe that users interact with many other types of objects. For example, as shown in Fig. **1**, on a social network, user interact with not only other users, but also college, location and employer. These different types of interactions suggest different semantics of the user-user relationships. For example, Alice and Bob both attend UCLA, thus they are schoolmates; whereas Chris and Donna both work for Facebook, thus they are colleagues. Therefore, it gives us a unique opportunity to do semantic user search. In general, semantic user search is a task that given a query user (e.g., Alice) on a heterogeneous social network and a semantic relation (e.g., schoolmates), we want to find the other users (e.g., Bob) that meet that relation with the query user. Such semantic user search is very useful [**19, 22**]. For example, we can use it to find colleagues, schoolmates and families on social networks such as Facebook and LinkedIn, or find advisors and advisees on academic networks such as DBLP.

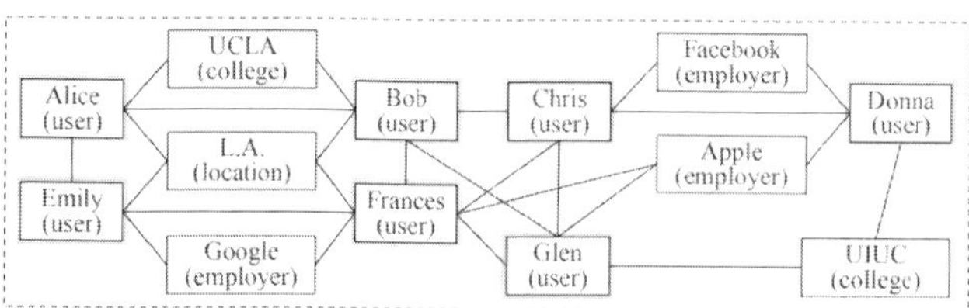

Figure 1: Semantic user search on a heterogeneous social network with rich user interactions with different objects.

Traditionally, path-based approach is used for solving semantic user search. This is because in semantic user search, the target user is often not immediately linked to the query user. A plausible choice is then to consider paths from the query user to the target user, and see whether they match the desired semantic relation. For example, Meta-Path Proximity (MPP) [**30**] first relies on domain experts to specify a few path patterns (i.e., metapaths) that indicate the desired semantic relation, and then enumerates the number of metapath instances between the query user and a target user. Path Ranking Algorithm (PRA) [**18**] first enumerates bounded-length (relation) path patterns; then it recursively defines a score for each path pattern; finally, for a target node, its proximity to the query node is computed as a linear combination of its corresponding path instances. Recently deep learning starts to exploit learning representations for the paths between a query node and a target node, and then using them for proximity estimation. For example, ProxEmbed [**20**] first samples a number of paths from the query node to the target node, and then uses a recurrent neural network to embed each path as a vector; finally it aggregates multiple path embedding vectors for proximity estimation.

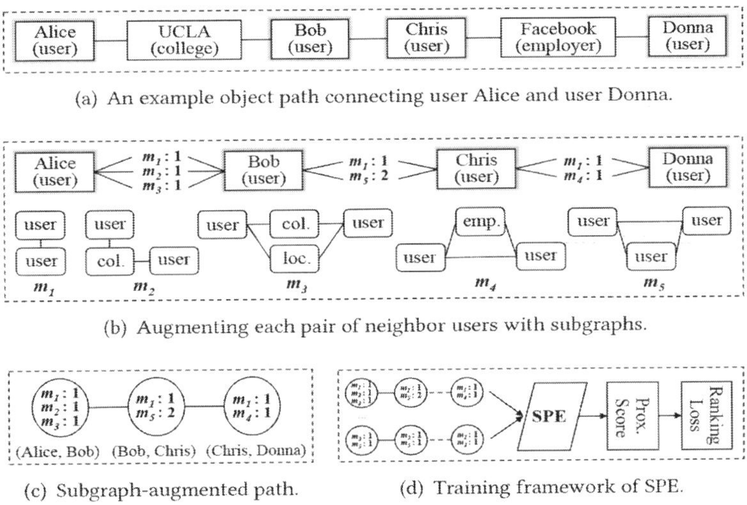

(a) An example object path connecting user Alice and user Donna.

(b) Augmenting each pair of neighbor users with subgraphs.

(c) Subgraph-augmented path.

(d) Training framework of SPE.

Figure 2: Combining path›s distance awareness and subgraph's higher-order structure for semantic user search.

Despite the success of such a path-based approach, we assert that path as a low-order structure is insufficient to capture the rich semantics between two users. Consider an object path connecting a query user Alice and a target user Donna in Fig. **2**(a) . In fact, Alice and Bob not only attended the same college UCLA, but also live in the same city L.A.. Such information is missing in the path, but it is possible to be captured by some higher-order subgraph structure. We are inspired by the state-of-the-art work on exploting subgraph patterns to organize complex networks [**3**]. Suppose we already have some offline mined subgraph patterns in Fig. **2**(b), such as user-user (m_1), user-college-user (m_2), user-college & location-user (m_3) and so on. Then we can replace the linear object sequence between Alice and Bob with richer subgraph instances for m_1, m_2 and m_3. In this way, we have a more complete picture of the semantic relation between Alice and Bob. We envision that, once we better understand the semantic relation between every two neighboring users in a path, we can better estimate the proximity between the query user and the target user. Note that, we focus on augmenting the neighoring user objects only. There are two reasons of avoiding augmenting any two neighboring objects regardless of their types. Firstly, in semantic user search, we wish to directly model the semantic relation between users. Secondly, by constraining the subgraph patterns to involve two users, we can significantly reduce the number of subgraph patterns and thus greatly improve the efficiency in offline subgraph indexing, as suggested in [**11**].

In this paper, we introduce a new concept of subgraph-augmented path for semantic user search. Specifically, we consider sampling a set of object paths from a query user to a target user; then in each object path, we replace the linear object sequence between its every two neighboring users with their shared subgraph instances. Such subgraph-augmented paths are expected to leverage both path's distance awareness (i.e., able to model multi-hop connections between a query user and a target user) and subgraph's high-order structure (i.e., able to use more complex structures than linear sequences). Given these subgraph-augmented paths as new inputs, we aim to embed them into low-dimensional vectors and then aggregate them for proximity estimation. In this work, we assume the subgraph patterns and subgraph instances are given as inputs. Such an assumption is mild in practice, because frequent subgraphs are useful, and often offline mined as basic graph indexing to support many useful applications [**10**]. For example, frequent subgraphs are used for fraud detection in Alibaba[1], and user/content recommendation in Twitter [**13**]. There also exist efficient algorithms to mine frequent subgraph patterns and match subgraph instances [**9**, **31**].

However, embedding subgraph-augmented path (or, s-path for abbreviation) is not trivial. A straightforward approach is to apply ProxEmbed

[20]. For each s-path, we represent each of its node as a key-value pair (Def. 3.5), where the key is the end user pair, and the value denotes the number of each subgraph instances shared by these two users. Then we apply a recurrent neural network to encode each node in the s-path, and finally we pool all the output vectors of the nodes as one. For multiple s-paths, we use distance discounted pooling to de-emphasize those long paths. Yet, such a straightforward approach overlooks two challenges. First of all, subgraphs are structural and noisy. To represent a node in an s-path, we have to take into account the structure of each subgraph, as well as the fact that not all the subgraphs are useful for a particular semantic user relation (e.g., m_5 is less indicative than m_2 for schoolmates). Secondly, s-paths are noisy in and among themselves. In each s-path, its nodes are not equally useful for a semantic relation; e.g., if Alice and Donna are truly schoolmates, then node (Alice, Bob) in the s-path, which implies a schoolmates relation, is more important than the other nodes in the same s-path. Similarly, not all the s-paths are equally useful either; e.g., an s-path constructed from Alice–Emily–Frances–Donna in Fig. 1 is less indicative than the one in Fig. 2(c) for the schoolmates relation, since it has no clear signal for that relation.

To model s-paths for semantic user search, we develop a novel Subgraph-augmented Path Embedding (SPE) framework. In SPE, we first represent an object in an s-path with an aggregation of its subgraphs' embedding vectors. Specifically, we construct a structural similarity matrix among the subgraphs, and based on this matrix we learn an embedding vector for each subgraph to preserve the structural similarity. Then, we introduce the attention mechanism [40] to automatically weigh the subgraphs in aggregation to represent each s-path's node. To deal with the noise in and among s-paths, we also introduce the attention mechanism to automatically weigh each node in an s-path and each s-path between two users. In all, we have a three-layer attention architecture on subgraphs, s-path's nodes and s-paths. Finally, we embed all the s-paths together into a vector, based on which we compute the proximity score and thus the ranking loss for model training.

We summarize our contributions as follows.

- We introduce a new concept of subgraph-augmented path, which for the first time systematically combine path's distance awareness and subgraph's high-order structure to solve semantic user search.
- We develop a novel SPE framework to embed these subgraph-

augmented paths for user proximity estimation.

- We evaluate SPE on six semantic user relations in three public data sets, and show it outperforms the state-of-the-art baselines.

RELATED WORK

Earlier graph semantic search work such as Personalized PageRank [**17**] and SimRank [**16**] often consider homogeneous networks as input, and they do not differentiate semantic classes. Recent work starts to consider the rich network structure in heterogeneous networks. For example, Supervised Random Walk (SRW) [**1**] tries to bias a random walk on the network, so as to ensure the resulting ranking result on the network to be consistent with the ground truth. MPP [**30**] and PRA [**18**] try to match the paths between a query node and a target node with some supervision (i.e., either metapath patterns or ground truth labels) to see whether certain semantic relation holds. Meta-Graph Proximity (MGP) [11] considers more general subgraph patterns than metapaths. It first identifies a few frequent subgraph patterns as metagraphs; then it leverages the supervision to automatically learn which metagraph is indicative for a desired semantic relation; finally it counts the number of indicative metagraphs between two users to measure their proximity. Thanks to the higher-order structure of subgraph, MGP improves the path-based methods such as SRW and MPP. But MGP lacks distance awareness; i.e., if a query user and a target user are multi-hop away and no metagraph is shared by them, then their proximity becomes (close to) zero. Both MPP and MGP can be seen as exploiting explicit graph features for proximity estimation.

With the development of neural networks, some recent studies start to consider learning "implicit" graph features for proximity estimation. For example, in graph embedding, DeepWalk [**27**], LINE [**32**], node2vec [**12**], and many more [**23**, **25**, **26**] all try to learn an embedding vector for each node in the graph, which can preserve the graph structure. In particular, metapath2vec [**8**] extends DeepWalk by using metapath to guide the random walk for better node embedding. Struc2vec [**28**] exploits additional structural equivalence of two nodes for node embedding. A comprehensive survey of graph embedding is recently available [**5**]. One possible approach to make use of such a node-level embedding for semantic search is to aggregate two nodes' embedding vectors (e.g., first applying a Hadamard product and then multiplying it with a parameter vector) for estimating their proximity. However, such an approach is considered as "indirect", as suggested by

ProxEmbed [**20**], since it does not directly encode the network structure between two possibly distant nodes. In contrast, ProxEmbed expresses the network structure between two objects by a set of paths connecting them, and directly encodes these paths into a proximity embedding vector. However, since ProxEmbed takes object paths as input, it is unable to leverage the readily available subgraphs' high-order structure. Similarly, although D2AGE [**21**] manages to model multiple object paths as one directed acyclic graph for proximity embedding, it is also unable to leverage the offline mined subgraphs.

In the line of graph embedding, there exist several related, yet different concepts. First of all, some recent work exploits the concept of "high-order proximity" in graph embedding [**6**, **38**]. They use higher-order reachability to construct adjacency matrix of a graph, and then run node embedding on this adjacency matrix. There are two major differences with our method: 1) they do not exploit the high-order structure of subgraph patterns; 2) they consider node embedding instead of path embedding. Secondly, some other work models high-order structure by graph convolution [**14**, **24**]. These methods are powerful to capture local graph patterns, but it is not clear how to incorporate the path's distance awareness for the task of semantic user search. Besides, they cannot leverage the readily available subgraph patterns. Thirdly, graph kernel methods [**7**], especially Weisfeiler-Lehman graph kernels [**29**, **41**], try to measure the similarity between two (small) graphs w.r.t. their structures. They often exploit some predefined subgraph structures, such as edges, subtrees and shortest paths. But their goal of measuring similarity between graphs is very different from ours of measuring proximity between nodes. Besides, it is also not clear how to adapt their methods with path distance awareness for our task. Finally, in the field of knowledge base, recent work such as TransE [**4**], TransH [**39**] and TransNet [**34**] has greatly advanced the study of knowledge embedding. However, these methods are not directly applicable to our task, due to the different problem settings. They often require the edges to have explicit descriptions, and aim to generate node/edge embedding instead of path embedding. Besides, it is also not clear how to extend these methods with the subgraph's high-order structure and the path's distance awareness for our task.

PROBLEM FORMULATION

We first introduce terminologies and notations (listed in Table **1**).

Definition 3.1.

A **heterogeneous network** is G = (V, E, C, τ), where V is a set of objects, E is a set of edges between the objects in V, $C = \{c_1, ..., c_K\}$ is a set of distinct object types, and τ: V → C is an object type mapping function.

For example, in Fig. **1**, we have C={user,college,location,employer}C={user,college,location,employer} . For an object of Alice, τ(Alice) = user.
Definition 3.2.
A **subgraph pattern** is $m = (C_m, E_m)$, where C_m is a set of object types, E_m is a set of edges between object types in C_m.
For example, Fig. **2**(b) lists five subgraph patterns $m_1, ..., m_5$. We denote the set of possible subgraph patterns on G as M. As discussed in Sect. **1**, we consider MM as the frequent subgraph patterns offline mined from G, and readily available as the input. Note that unlike C in G, the object types in C_m may be nondistinct. E.g., for subgraph pattern m_1 in Fig. **2**(b), Cm1={user,user} Cm1={user,user} .
Definition 3.3.
An object subgraph $g = (V_g, E_g)$ is a **subgraph instance** of $m = (C_m, E_m)$, if there exists a bijection between the node set of g and m, $\phi: V_g \rightarrow C_m$, such that

- $\forall v \in V_g$, we have $\tau(v) = \phi(v)$;
- $\forall v, u \in V_g$, we have $(v, u) \in E_g$ iff $(\phi(v), \phi(u)) \in E_m$.

For example, Alice-UCLA-Bob is an instance of m_2 in Fig. **2**(b) . We denote the set of possible subgraph instances for MM on G as II . For each m∈Mm∈M , there may be multiple subgraph instances on G.

Table 1: Notations used in this paper.

Notation	**Description**
G, V, E, C	Network G, objects V, edges E, object types C
M	Set of frequent subgraph patterns from G
I	Set of subgraph instances for MM on G
D	Set of training tuples
P	Set of sampled object paths on G
$\hat{\mathcal{P}}$	Set of subgraph-augmented paths constructed from PP
r	A subgraph-augmented node
x	Embedding vector for a subgraph
z	Embedding vector for a subgraph-augmented path

f	Proximity embedding vector between two users
π	Proximity score
S	Structural similarity matrix of subgraph patterns
γ, ℓ	Number of paths per object γ, walk length ℓ
d, d′	Embedding dimensions
ζ	Average number of subgraph instances for a user

Definition 3.4.

An **object path** on G is a sequence of objects $v_1 \rightarrow v_2 \rightarrow \ldots \rightarrow v_t$, where each $v_i \in V$, and t is the path length.

For example, the sequence in Fig. **2**(a) is an object path.

Definition 3.5.

A **subgraph-augmented node** (or "s-node" for abbreviation) r for two user objects u, v ∈ V is a key-value pair, whose key is (u, v) and value is a set of tuples $\{m_1: e_1, \ldots, m_l : e_l\}$. $\forall m_i \in M$, e_i is the number of m_i 's instances between u and v.

For example, in Fig. **2**(b), the s-node for Alice and Bob is defined as r.key = (Alice, Bob), r.value = $\{m_1: 1, m_2: 1, m_3: 1\}$. We skip the subgraphs with zero instance in r.value. As discussed in Sect. **1**, we choose only augmenting two user objects with their shared subgraphs to both focus on user-user semantic relation and improve subgraph indexing efficiency. We leave as future work augmenting any two objects regardless of their types for semantic search.

Definition 3.6.

An **subgraph-augmented path** (or "s-path" for abbreviation) is a sequence of s-nodes $r_1 \rightarrow r_2 \rightarrow \ldots \rightarrow r_t$.

For example, the sequence in Fig. **2**(c) is an s-path.

Problem inputs and outputs. For inputs of our model, we have a heterogeneous network G, a set of readily available frequent subgraph patterns M and their subgraph instances II on G, and finally a set of training tuples D={(qi,vi,ui):i=1,...,n}D={(qi,vi,ui):i=1,...,n} , where for each query user object q_i , user v_i is closer to q_i than user u_i . Besides, we also offline sample some object paths from G as inputs. We take a similar approach as DeepWalk [**27**] for path sampling. Specifically, starting from each object in G, we randomly sample γ object paths, each of length ℓ. As a result, we obtain a set of object paths, denoted as PP . These object paths are indexed to support efficient training and testing. For each query object $q \in \{q_1, \ldots, q_n$

In line 4, for each resulting object path, we get all the neighboring user pairs. In line 6, for each pair of neighboring users, we identify their shared subgraph instances from the pre-indexed subgraph instance set I . After that, we construct an s-node in lines 7 and 8, and further append it to the previous s-node sequence to form an s-path in line 9.

Complexity analysis: as each object path's length is bounded by ℓ, getting the neighboring user pairs in line 4 takes $O(\ell)$. The straightforward approach to get shared subgraph instances between two users has to scan through all the subgraph instances of one user. Denote the average number of subgraph instance for a user on G as ζ. Because in practice the subgraph patterns have limited sizes (e.g., less than six in our experiments), the above straightforward approach to run line 6 takes $O(\zeta)$. This complexity can be further reduced; as suggested by [**11**], if in the subgraph indexing stage only those subgraph patterns that involved at least two users were considered, then both the number of subgraph patterns and the number of subgraph instances can be significantly reduced. By a sophisticated indexing of which subgraph instance matching which two users, we may reduce line 6's complexity to a constant. In all, constructing an s-path from an object path takes $O(\ell + \zeta)$.

S-PATH EMBEDDING

We first introduce how to embed each subgraph-augmented path to a vector $\mathbf{z}(q, v)$, and later aggregate multiple such vectors into a single one $\mathbf{f}(q, v)$. We design a hierarchical neural network for subgraph-augmented path embedding (SPE) as shown in Fig. **3**. As motivated in Sect. **1**, we will take multiple factors into the design. First of all, we embed each subgraph with structural information. Then, we aggregate the subgraph embedding in each subgraph-augmented node (s-node) with attention to obtain an s-node embedding. To model the sequential information of each s-path, we also employ a recurrent neural network architecture to learn an s-path's embedding. Finally, we aggregate all the s-paths' embedding with attention to obtain an overall proximity embedding vector for (q, v). Next we introduce each step of embedding in Fig. **3**.

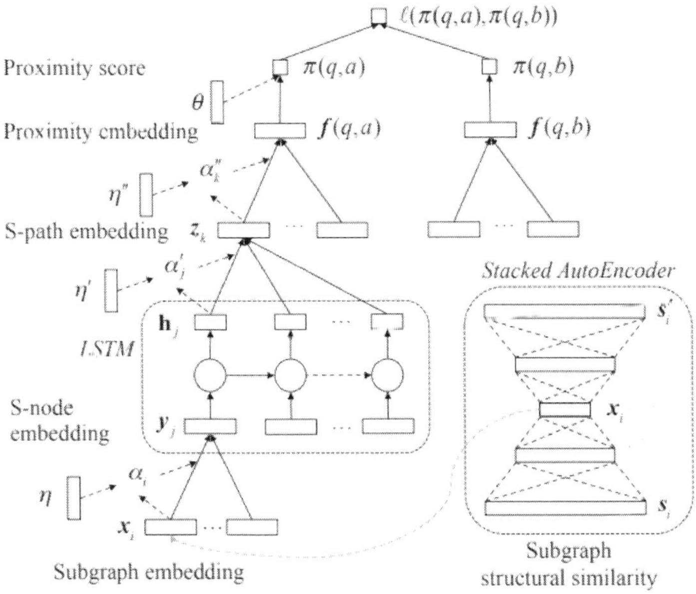

Figure 3: Subgraph-augmented path embedding.

Subgraph Embedding. To take the subgraph structure into account, we are inspired by the structural deep network embedding [**38**] to consider embedding the subgraphs from a subgraph structural similarity matrix. In general, two subgraphs are similar if they share some common structures. Therefore, we adopt the widely used Maximum Common Subgraph (MCS) approach [**35**] to measure the similarity between two subgraphs. Given two subgraphs m_i and m_j, we denote m^* as their MCS. Then the structural similarity between two subgraphs is defined as

$$S(m_i, m_j) = \frac{(|C_{m^*}| + |E_{m^*}|)^2}{(|C_{m_i}| + |E_{m_i}|) \times (|C_{m_j}| + |E_{m_j}|)}. \tag{2}$$

If m^* is bigger, $S(m_i, m_j)$ is bigger. We use stacked AutoEncoder [**2**] to learn an embedding $\mathbf{x}_i \in \mathbb{R}^d$ for each subgraph m_i. Denote $\mathbf{s}_i = [S(m_i, m_1), \ldots, S(m_i, m_{|M|})]^T$. For simplicity, we use a three-layer AutoEncoder to illustrate how we construct $\mathbf{x}_i$ from its $\mathbf{s}_i$. In particular, we define the subgraph embedding vector $\mathbf{x}_i$ for m_i as

$$\hat{\mathbf{s}}_i = \sigma(W^{(b)}\mathbf{x}_i + \mathbf{b}^{(b)}), \tag{3}$$

where W $^{(b)}$ ∈ $\mathbb{R}^{|M|\cdot d}$ and b (b) ∈ $\mathbb{R}^{d}$ are parameters; σ (·) is a sigmoid function. We reconstruct si from xi by

$$\hat{\mathbf{s}}_i = \sigma(W^{(b)}\mathbf{x}_i + \mathbf{b}^{(b)}), \tag{4}$$

where W $^{(b)}$ ∈ $\mathbb{R}^{|M|\cdot d}$ and b (b) ∈ $\mathbb{R}^{|M|}$ are also parameters. Finally, we minimize the reconstruction error:

$$\sum_{i=1}^{|M|} ||\hat{\mathbf{s}}_i - \mathbf{s}_i||_2. \tag{5}$$

Although it is possible to optimize x_i 's together with the proximity embedding later, in this paper we choose to optimize x_i 's from S separately, so as to keep the model simple.

S-Node Embedding. In general, there are multiple subgraphs involved in a subgraph-augmented node (s-node). To differentiate their contributions, we introduce an attention mechanism to automatically learn the weight for each subgraph in s-node embedding. For an s-node r_j .value = {m_1: e_1, ..., m_l : e_l }, we compute the attention score α_i for each subgraph m_i in r_j as

$$\alpha_i = \frac{\exp(\beta_i)}{\sum_{j=1}^{l}\exp(\beta_j)}, \tag{6}$$

$$\beta_i = e_i\,[\boldsymbol{\eta}\cdot\sigma(Q\mathbf{x}_i + \mathbf{b})], \tag{7}$$

where $\boldsymbol{\eta} \in \mathbb{R}^{d'}$, $Q \in \mathbb{R}^{d'\times d}$ and $\mathbf{b} \in \mathbb{R}^{d'}$ are parameters. As a result, we compute the s-node embedding for r_j as

S-Path Embedding. Once having the s-node embeddings in an s-path p_k : $r_1 \rightarrow \ldots \rightarrow r_t$, we learn the s-path embedding by LSTM (Long Short Term Memory) [**15**]. Formally, for each input s-node embedding $\mathbf{y}_j$, we output a vector $\mathbf{h}_j \in \mathbb{R}^{d'}$, by computing a series of neuron activations for an input gate $g_j^{(i)}$, an forget gate $g_j^{(f)}$, a memory cell state $g_j^{(c)}$ and an output gate $g_j^{(o)}$:

$$\mathbf{g}_j^{(i)} = \sigma(W^{(i)}\mathbf{y}_j + U^{(i)}\mathbf{h}_{j-1} + \mathbf{b}^{(i)}), \tag{9}$$

$$\mathbf{g}_j^{(f)} = \sigma(W^{(f)}\mathbf{y}_j + U^{(f)}\mathbf{h}_{j-1} + \mathbf{b}^{(f)}), \tag{10}$$

$$\mathbf{g}_j^{(c)} = \mathbf{g}_j^{(i)} \odot \tilde{\mathbf{g}}_j^{(c)} + \mathbf{g}_j^{(f)} \odot \mathbf{g}_{j-1}^{(c)}, \tag{11}$$

$$\mathbf{g}_j^{(o)} = \sigma(W^{(o)}\mathbf{y}_j + U^{(o)}\mathbf{h}_{j-1} + \mathbf{b}^{(o)}), \tag{12}$$

$$\mathbf{h}_j = \mathbf{g}_j^{(o)} \odot tanh(\mathbf{g}_j^{(c)}), \tag{13}$$

where $\tilde{\mathbf{g}}_{t,j}^{(c)} = tanh(W^{(c)}\mathbf{y}_j + U^{(c)}\mathbf{h}_{j-1} + \mathbf{b}^{(c)})$ and $\odot$ is an element-wise product. We denote the set of LSTM parameters as $\Theta^{(lstm)} = \{W^{(i)} \in \mathbb{R}^{d' \times d}, U^{(i)} \in \mathbb{R}^{d' \times d'}, \mathbf{b}^{(i)} \in \mathbb{R}^{d'}, W^{(f)} \in \mathbb{R}^{d' \times d}, U^{(f)} \in \mathbb{R}^{d' \times d'}, \mathbf{b}^{(f)} \in \mathbb{R}^{d'}, W^{(c)} \in \mathbb{R}^{d' \times d}, U^{(c)} \in \mathbb{R}^{d' \times d'}, \mathbf{b}^{(c)} \in \mathbb{R}^{d'}, W^{(o)} \in \mathbb{R}^{d' \times d}, U^{(o)} \in \mathbb{R}^{d' \times d'}, \mathbf{b}^{(o)} \in \mathbb{R}^{d'}\}$.

To differentiate the contributions of s-nodes, we also compute the attention score α'_j for each s-node r_j as

$$\alpha'_j = \frac{\exp(\beta'_j)}{\sum_{k=1}^{t} \exp(\beta'_k)}, \tag{14}$$

$$\beta'_j = \boldsymbol{\eta}' \cdot \sigma(Q'\mathbf{h}_j + \mathbf{b}'), \tag{15}$$

where $\eta' \in \mathbb{R}^{d'}, Q' \in \mathbb{R}^{d' \times d'}$ and $\mathbf{b}' \in \mathbb{R}^{d'}$, are parameters. As a result, we compute the s-path embedding for p_k as

$$\mathbf{z}_k = \sum_{j=1}^{t} \alpha'_j \mathbf{h}_j. \tag{16}$$

Proximity Embedding. Once having the s-path embedding for each of the s-paths betweenq andv, we can compute their proximity embedding. For a unified notation, we introduce $\hat{\mathcal{P}}$ (q,v) as the spath set between q and v. For asymmetric relations, we define $\hat{\mathcal{P}}$ (q,v) as the s-paths from q to v. For symmetric relations, we define $\hat{\mathcal{P}}$ (q,v) as the s-paths both from q to v and from v to q. To differentiate the contributions of s-paths, we compute the attention score α''_k for each s-path p_t as

$$\alpha''_k = \frac{\exp(\beta''_k)}{\sum_{p_i \in \hat{\mathcal{P}}(q,v)} \exp(\beta''_i)}, \tag{17}$$

$$\beta''_k = \boldsymbol{\eta}'' \cdot \sigma(Q'' \cdot \mathbf{z}_k + \mathbf{b}''), \tag{18}$$

whereη $\eta'' \in \mathbb{R}^{d'}, Q'' \in \mathbb{R}^{d' \times d'}$ and $\mathbf{b}'' \in \mathbb{R}^{d'}$ are parameters. Finally, we compute the proximity embedding for (q,v) as

$$\mathbf{f}(q,v) = \sum_{p_k \in \hat{\mathcal{P}}(q,v)} \alpha''_k \mathbf{z}_k. \tag{19}$$

This f(q,v) encodes the information of all the subgraph-augmented paths between q andv. It will be later used to estimate the proximity score of q and v by Eq. 1.

END-TO-END TRAINING

In training, for each tuple (q_i , v_i , u_i), ∀i = 1, ..., n, we define a ranking loss based on the proximity scores π(q_i , v_i) and π(q_i , u_i). We define the ranking loss function as

$$\ell(\pi(q_i,v_i),\pi(q_i,u_i)) = -\log \sigma_\lambda(\pi(q_i,v_i) - \pi(q_i,u_i)), \tag{20}$$

where $\sigma_\lambda(x) = 1/(1 + e^{-\lambda x})$ and λ > 0 is a parameter. We denote the parameter set of our three-layer attentions for subgraphs, s-nodes and s-paths as $\Theta^{(att)} = \{\eta, Q, \mathbf{b}, \eta', Q', \mathbf{b}', \eta'', Q'', \mathbf{b}''\}$. In total, our model parameters are $\Theta = \{\Theta^{(lstm)}, \Theta^{(att)}, \theta\}$. Our ultimate goal in training is to minimize

$$L(\Theta) = \sum_{i=1}^{n} \ell(\pi(q_i,v_i),\pi(q_i,u_i)) + \mu\, \Omega(\Theta), \tag{21}$$

where μ > 0 is a trade-off parameter, Ω(·) is a regularization function (e.g., the sum of l_2-norm for each parameter in Θ).

Algorithm 2 SPETrain

Require: heterogeneous network G, a set of subgraph patterns $\mathcal{M}$ and its instances $\mathcal{I}$, training tuples $\mathcal{D}$, number of paths per object γ, walk length ℓ, embedding dimensions d and d', trade-off parameters $\{\lambda, \mu\}$.
Ensure: SPE model parameters Θ.

```
Initialize an object path set P = ∅ and an s-path set P̂ = ∅;
for all v ∈ V do
    for i = 1 : γ do
        P ← P ∪ SamplePath(G, v, ℓ);
for all each (q, v, u) ∈ D do
    P̂(q, v) ← SPathConstruct(P, q, v, M, I);
    P̂(q, u) ← SPathConstruct(P, q, u, M, I);
B ← GenerateBatches(D);
for all batch b ∈ B do
    Initialize loss for batch b as L_b = 0;
    for all each (q, v, u) ∈ b do
        Compute f(q, v) with P̂(q, v), d and d′ by Eq. 19;
        Compute f(q, u) with P̂(q, u), d and d′ by Eq. 19;
        L_b = L_b + ℓ(π(q, v), π(q, u)), according to Eq. 20;
    L_b = L_b + μΩ(Θ);
    Update Θ based on L_b by stochastic gradient descent.
return Θ.
```

Training algorithm: we summarize the SPE training algorithm in Alg. 2 . In lines 2–4, we sample object paths on G. In lines 6 and 7, we construct subgraph-augmented paths based on the object paths for each (q, v) and (q, u) in the training tuples. In line 8, we split the training tuples into batches, and then do batch stochastic gradient descent. In lines 12 and 13, we compute the proximity embedding vectors **f**(q, v) and **f**(q, u). In line 14, we compute the ranking loss $\ell(\pi(q, v), \pi(q, u))$. In lines 15 and 16, we accumulate the loss for batch b and do stochastic gradient descent.

Complexity analysis: we analyze the time complexity for Alg. 2 . In lines 2–4, we sample γ object paths of length ℓ starting from each object in G, hence it takes $O(|V|\gamma\ell)$. In lines 5–7, we in total construct $|\hat{\mathcal{P}}|$ s-paths. According to Alg. 1, contructing an s-path takes $O(\ell + \zeta)$. Thus the complexity of lines 7-10 is $O(|\hat{\mathcal{P}}|(\ell+\zeta))$. In line 8, generating the batches takes O(n). In line 12, computing proximity embedding **f**(q, v) requires three steps: 1) embedding an s-node, which takes $O(|M|(d'd+d'))$ given at most |M| subgraphs; 2) embedding an s-path, which takes $O(\ell(d'd + d'^2 + d'))$ given an s-path's length of at most ℓ; 3) proximity embedding, which takes $O(|\hat{\mathcal{P}}(q,v)|(d'2+d'))$Ogiven $|\hat{\mathcal{P}}(q,v)|$s-paths between q and v. In lines 9–16, we essentially compute **f**(q, v)'s and **f**(q, u)'s for all the $(q,v,u)\in D$, which in total takes $O(|\hat{\mathcal{P}}|((d'^2 + d') + \ell(d'd + d'^2 + d') + \ell(|\mathcal{M}|(d'd+d')))) = O(|\hat{\mathcal{P}}|(\ell d'^2 + \ell|\mathcal{M}|d'd + \ell d'))$. Computing the loss in line 14 over all the training tuples takes O(nd'). Updating Θ in line 16 for |B| batches takes (|B|(d'd+d'2)) . Note that we compute the subgraph embedding offline once. It takes $O(|M|^2)$ to construct the structural similarity matrix, and O(|M|d+|M|) to learn the subgraph embedding, which in total is O(|M|2+|M|d) . In summary, the total complexity of Alg. 2 is $O(|\hat{\mathcal{P}}|\zeta + |\mathcal{M}|^2 + |\mathcal{M}|d + |\hat{\mathcal{P}}|\ell(d'^2 + |\mathcal{M}|d'd + d'))$. Since $|\hat{\mathcal{P}}| \leq |V|\gamma\ell$the complexity of Alg. 2 becomes $O(|V|\gamma\ell(\zeta + \ell(d'^2 + |M|d'd + d')) + |M|^2)$

EXPERIMENTS

Heterogeneous social networks. We conducted extensive experiments on three real-world data sets collected by previous studies, namely LinkedIn [**19**], Facebook [**22**] and DBLP [**36**]. Each data set contains objects of various types. In particular, LinkedIn includes the types of user, employer, location and college; Facebook includes user, concentration, degree, school, hometown, last-name, location, employer, work-location and work-project (other types are ignored due to their sparsity or irrelevance); DBLP includes the more noises may exist. Therefore, for the data sets with more types, the parameters need to be carefully tuned so that the embedded subgraph, the learned attentions and loss discount help to filter useful information out of the noises. On the other hand, for the data sets with less types like LinkedIn, the performance is relatively robust because the noises introduced by irrelevant object types are limited.

Based on both NDCG and MAP over all the six semantic relations on three data sets, SPE tends to generate the best performance at $d = 16$. Especially, for classmate and family, when d is too small, the resulting embeddings are unable to capture the rich semantics. When d is too big, it may bring in more noises and increase the number of parameters to learn. The attention dimension d' also tends to suffer from the similar performance decrease when d' is too small or too big. $d' = 16$ is the best setting. For the ranking loss discount parameter λ, we see that $\lambda = 0.1$ usually gives the best results, suggesting the need to discount the ranking loss in Eq. **20**.

CONCLUSION

In this paper, we study the problem of semantic user search in heterogeneous social networks. We exploit the opportunity of integrating the path's distance awareness and the subgraph's high-order structure for learning a better representation of the proximity between two users. We propose a novel Subgraph-augmented Path Embedding (SPE) model. It takes object paths as input, and enriches them into subgraph-augmented paths. Then it addresses the challenges of incorporating the subgraph structure, the subgraph noise and the subgraph-augmented path noise. Finally, it embeds the subgraph-augmented paths between two users into a proximity embedding vector. With such a proximity embedding vector, we can easily measure the proximity between two users for semantic user search. We test SPE with six semantic relations in three public data sets and it improves the state of the art by at least 1.8%–13.2% (NDCG) and 1.4%–16.9% (MAP) with 100 training samples.

In the future, we would like to explore the heterogeneous social networks with rich edge features and graph dynamics.

ACKNOWLEDGMENTS

We thank the support from: Zhejiang Science and Technology Plan Project (No. 2015C01027), National Natural Science Foundation of China (No.

Training algorithm: we summarize the SPE training algorithm in Alg. 2 . In lines 2–4, we sample object paths on G. In lines 6 and 7, we construct subgraph-augmented paths based on the object paths for each (q, v) and (q, u) in the training tuples. In line 8, we split the training tuples into batches, and then do batch stochastic gradient descent. In lines 12 and 13, we compute the proximity embedding vectors **f**(q, v) and **f**(q, u). In line 14, we compute the ranking loss $\ell(\pi(q, v), \pi(q, u))$. In lines 15 and 16, we accumulate the loss for batch b and do stochastic gradient descent.

Complexity analysis: we analyze the time complexity for Alg. 2 . In lines 2–4, we sample γ object paths of length ℓ starting from each object in G, hence it takes $O(|V|\gamma\ell)$. In lines 5–7, we in total construct $|\hat{\mathcal{P}}|$ s-paths. According to Alg. 1, contructing an s-path takes $O(\ell + \zeta)$. Thus the complexity of lines 7-10 is $O(|\hat{\mathcal{P}}|(\ell+\zeta))$. In line 8, generating the batches takes O(n). In line 12, computing proximity embedding **f**(q, v) requires three steps: 1) embedding an s-node, which takes $O(|M|(d'd+d'))$ given at most |M| subgraphs; 2) embedding an s-path, which takes $O(\ell(d'd + d'^2 + d'))$ given an s-path's length of at most ℓ; 3) proximity embedding, which takes $O(|\hat{\mathcal{P}}(q,v)|(d'2+d'))$Ogiven $|\hat{\mathcal{P}}(q,v)|$s-paths between q and v. In lines 9–16, we essentially compute **f**(q, v)'s and **f**(q, u)'s for all the (q,v,u)∈D , which in total takes $O(|\hat{\mathcal{P}}|((d'^2 + d') + \ell(d'd + d'^2 + d')+\ell(|\mathcal{M}|(d'd+d')))) = O(|\hat{\mathcal{P}}|(\ell d'^2+\ell|\mathcal{M}|d'd+\ell d'))$. Computing the loss in line 14 over all the training tuples takes O(nd'). Updating Θ in line 16 for |B| batches takes $(|B|(d'd+d'2))$. Note that we compute the subgraph embedding offline once. It takes $O(|M|^2)$ to construct the structural similarity matrix, and $O(|M|d+|M|)$ to learn the subgraph embedding, which in total is $O(|M|2+|M|d)$. In summary, the total complexity of Alg. 2 is $O(|\hat{\mathcal{P}}|\zeta + |\mathcal{M}|^2 + |\mathcal{M}|d + |\hat{\mathcal{P}}|\ell(d'^2 + |\mathcal{M}|d'd + d'))$. Since $|\hat{\mathcal{P}}|\leq|V|\gamma\ell$the complexity of Alg. 2 becomes $O(|V|\gamma\ell(\zeta + \ell(d'^2 + |M|d'd + d')) + |M|^2)$

EXPERIMENTS

Heterogeneous social networks. We conducted extensive experiments on three real-world data sets collected by previous studies, namely LinkedIn [**19**], Facebook [**22**] and DBLP [**36**]. Each data set contains objects of various types. In particular, LinkedIn includes the types of user, employer, location and college; Facebook includes user, concentration, degree, school, hometown, last-name, location, employer, work-location and work-project (other types are ignored due to their sparsity or irrelevance); DBLP includes paper, author, year, conference and keyword. We organized them into heterogeneous social networks, as summarized in Table **2**.

Table 2: Summary of data sets and subgraphs.

Network	Objects	Edges	Types	Average degree	Subgraph patterns	Subgraph instances
LinkedIn	65,925	220,812	4	6.7	173	604,848,383
Facebook	5,025	100,356	10	39.9	981	2,398,306,414
DBLP	165,728	928,513	5	11.2	88	740,682,735

Ground truth. On LinkedIn, the user relationships are already labeled into different semantic classes. We tested two major classes: schoolmate and colleague. On Facebook, the user relationships are defined by [**11**] with two classes: family and classmate. On DBLP, the adivsor and advisee in a co-author pair are identified based on the website of some faculty members as well as the Mathematics Genealogy and AI Genealogy projects [**36**]. All unidentified co-author pairs are assumed to be negative. As summarized in Table **3**, the advisor and adviseeclasses are asymmetric, whereas the others are all symmetric.

Table 3: Summary of semantic user relations and queries.

Network	Semantic relations	Symmetry	Queries	Results per query
LinkedIn	Schoolmate	Yes	172	16.2
	Colleague	Yes	173	12.8
Facebook	Family	Yes	340	4.0
	Classmate	Yes	904	6.5
DBLP	Advisor	No	2,439	1.3
	Advisee	No	1,204	2.6

Training and testing. On each graph, a user q can be used as a query node, if there exists another user v such that q and v have the desired semantic relation in our ground truth. The number of query users for each network and each semantic relation, and the average number of results per query are shown in Table **3**. We randomly split these queries into two subsets: 20% reserved as training and the rest as testing. We repeated such splitting for 10 times, and averaged any result over these 10 splits. In each split, based on the training queries, we further generated training examples (q, v, u) such that q and v belong to the desired semantic relation whereas q and u do not. For testing, we constructed an ideal ranking for each test query user and each desired semantic relation. We compared this ideal ranking against the ranking generated by various semantic user search algorithms. We adopted NDCG and MAP [**11**] to evaluate the quality of the algorithmic rankings at the top 10 results.

Subgraphs. We repeated the subgraph pattern mining and subgraph instance matching algorithms in [**11**] to obtain the set of frequent subgraph patterns MM and its instances II on each network. Specifically, we first applied GRAMI [**9**] on each graph to mine the set of frequent subgraph patterns. Then we filtered out those clearly non-viable subgraphs: 1) since our ground truth is designed for semantic user search, a viable subgraph must have at least two user objects; 2) a subgraph must contain at least two different types for capturing richer semantics; 3) to further constrain the number of subgraphs, we restricted them to have at most five nodes on LinkedIn and Facebook or six nodes on DBLP, which are found to be adequate in expressing the interactions between two users. The resulting number of subgraph patterns are shown in Table **2**.

Table 4: Running time of offline subgraph indexing.

	LinkedIn	**Facebook**	**DBLP**
Time (hour)	1.77	3.34	4.43

As shown in Table **4**, the subgraph indexing (including mining the frequent set of subgraph patterns, and matching each subgraph pattern with its possible instances on the graph) can be done in a reasonable time (i.e., a few hours). Since subgraph indexing is not the focus of this paper, we consider subgraph indexing as already done and the resulting subgraph patterns/instances are readily available for our subgraph-augmented path embedding algorithm.

Table 5: Result comparison under different amounts of labels (i.e., 10, 100, and 1000).

	Methods	**LinkedIn**						**Facebook**						**DBLP**					
		Schoolmate			*Colleague*			*Classmate*			*Family*			*Advisor*			*Advisee*		
		10	**100**	**1000**	**10**	**100**	**1000**	**10**	**100**	**1000**	**10**	**100**	**1000**	**10**	**100**	**1000**	**10**	**100**	**1000**
NDCG@10	ProxEm-bed	0.646	0.652	0.670	0.561	0.606	0.616	0.796	0.851	0.852	0.584	0.743	0.761	0.753	0.765	0.771	0.374	0.405	0.411
	MGP	0.546	0.568	0.574	0.527	0.546	0.552	**0.797**	0.843	0.849	**0.627**	0.732	0.753	0.618	0.718	0.745	0.385	0.403	0.405
	MPP	0.503	0.504	0.504	0.497	0.510	0.508	0.707	0.803	0.796	0.575	0.647	0.640	0.581	0.662	0.674	0.345	0.380	0.390
	SRW	0.520	0.515	0.515	0.513	0.502	0.503	0.396	0.389	0.394	0.396	0.389	0.394	0.689	0.689	0.689	**0.402**	0.402	0.402
	DWR	0.515	0.518	0.530	0.493	0.506	0.504	0.692	0.689	0.699	0.585	0.575	0.583	**0.764**	0.758	0.661	0.401	0.395	0.392
	PES	0.650	0.676	0.687	0.552	0.647	0.668	0.720	0.851	0.862	0.482	0.748	0.752	0.741	0.768	0.773	0.374	0.406	0.416
	SPE-A (ours)	0.541	0.603	0.640	0.545	0.597	0.630	0.587	0.565	0.547	0.424	0.451	0.411	0.436	0.451	0.449	0.325	0.353	0.356
	SPE (ours)	**0.660**	**0.690**	**0.694**	**0.644**	**0.686**	**0.704**	0.764	**0.866**	**0.869**	0.467	**0.761**	**0.774**	0.763	**0.782**	**0.789**	0.373	**0.413**	**0.417**
MAP@10	ProxEm-bed	0.535	0.541	0.562	0.443	0.492	0.503	**0.735**	0.801	0.803	0.498	0.678	0.711	0.672	0.690	0.701	0.251	0.283	0.297
	MGP	0.288	0.305	0.313	0.317	0.333	0.338	0.673	0.729	0.738	**0.520**	0.651	0.681	0.548	0.663	0.696	0.278	0.280	0.295
	MPP	0.260	0.260	0.259	0.294	0.305	0.304	0.576	0.681	0.675	0.467	0.543	0.534	0.499	0.597	0.611	0.245	0.273	0.281
	SRW	0.275	0.272	0.272	0.294	0.289	0.291	0.312	0.324	0.384	0.259	0.255	0.259	0.630	0.630	0.630	0.294	0.294	0.294
	DWR	0.393	0.391	0.398	0.363	0.369	0.368	0.578	0.575	0.587	0.467	0.455	0.463	**0.683**	0.675	0.541	**0.295**	0.281	0.276
	PES	**0.547**	0.567	0.581	0.432	0.535	0.559	0.617	0.797	0.815	0.354	0.685	0.691	0.667	0.693	0.699	0.267	0.282	0.293
	SPE-A (ours)	0.447	0.508	0.545	0.439	0.490	0.525	0.457	0.436	0.418	0.294	0.326	0.277	0.287	0.297	0.305	0.231	0.254	0.243
	SPE (ours)	**0.547**	**0.581**	**0.587**	**0.529**	**0.575**	**0.602**	0.682	**0.822**	**0.826**	0.368	**0.704**	**0.723**	0.679	**0.700**	**0.703**	0.249	**0.298**	**0.299**

Parameters and environment. For the fair comparison, we use the same object path sampling design and parameters as ProxEmbed [**20**]. Specifically, on LinkedIn, for both schoolmate and colleague we set $\gamma = 20$, $\ell = 20$. On Facebook, we set $\gamma = 40$, $\ell = 80$ for classmate and $\gamma = 20$, $\ell = 80$ for family. On DBLP, we set $\gamma = 20$, $\ell = 80$ for advisor and $\gamma = 20$, $\ell = 40$ for advisee. By default, we set the number of dimension $d' = 12$ for the parameters in $\Theta^{(att)}$, and $\mu = 10^{-4}$ in Eq. **21**. We tune different dimensions d of subgraph embedding and λ for different semantic relations. We run these experiments on Linux servers with 32GB memory, and use Theano [**33**] for SPE implementation and Java jdk-1.8 for path sampling and s-path construction.

Baselines. We compare our SPE with the following state-of-the-art semantic search baselines.

- ProxEmbed [**20**]: ProxEmbed uses object paths to describe the relations between two objects and measure their proximity.
- MGP [**11**]: Meta-Graph Proximity uses the number of meta-graph instances between two objects as features to measure proximity.
- MPP [**30**]: Meta-Path Proximity uses the number of meta-path instances between two objects as features to measure proximity.
- SRW [**1**]: Supervised Random Walk learns edge weights to bias a random walk for generating consistent ranking results with the ground truth. We define each edge's feature as a binary vector based on the types of its two objects.
- DWR: DeepWalk Ranking was introduced in [**20**]. It first learns object embedding by DeepWalk [**27**], then outputs a Hadamard product over two objects' embedding as the proximity embedding.
- PES: ProxEmbed with s-paths is a straightforward solution to model s-paths. It directly feeds s-paths into ProxEmbed without subgraph embedding, s-node embedding and s-path embedding.
- SPE-A: SPE without Attention is a baseline for validating the need to modeling attention. It replaces the attention mechanisms for subgraph embedding, s-node embedding and s-path embedding in SPE with mean pooling, max pooling and max pooling respectively.

For MGP and MPP, we use the same parameter setting as [**11**]. For SRW, we set its regularization parameter $\lambda = 10$, random walk teleportation parameter $\alpha = 0.2$ and loss parameter $b = 0.1$. We set the dimension of DWR as 128, the same as [27]. For SPE-A, we use the same parameter values as our SPE. We input the same object paths to ProxEmbed, DWR, PES, SPE-A and SPE.

Data and code availability. All the three data sets are publicly available online from their corresponding references, as mentioned earlier. We have made our code available online[2].

Table 6: Relative improvement of SPE over the best baselines in each relation when using 100 training tuples.

	NDCG	**MAP**
LinkedIn-schoolmate	5.8% ($p < 0.01$)	7.4% ($p < 0.01$)
LinkedIn-colleague	13.2% ($p < 0.01$)	16.9% ($p < 0.01$)
Facebook-family	2.4% ($p < 0.05$)	3.8% ($p < 0.01$)
Facebook-classmate	1.8% ($p < 0.1$)	2.6% ($p < 0.01$)
DBLP-advisor	2.2% ($p < 0.05$)	1.4% ($p < 0.1$)
DBLP-advisee	2.0% ($p < 0.05$)	1.4% ($p < 0.1$)

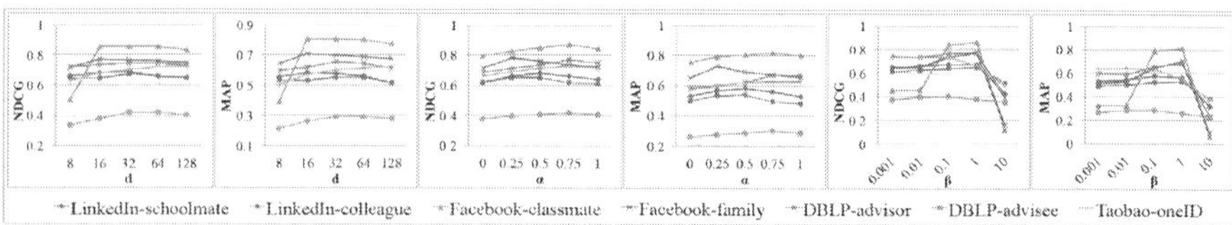

Figure 4: Impact of parameters: subgraph embedding dimension d, attention parameter dimension d′, ranking loss discount λ.

Comparison with Baselines

We compare our proposed SPE with the seven state-of-the-art semantic search baselines introduced above. We test all the methods on the six semantic relations under different amount of training tuples, i.e., 10, 100 and 1000. From the results reported in Table **5**, we make the following observations.

First of all, our SPE generally outperforms the baselines across all the six semantic relations in terms of both NDCG and MAP. The only exception is when training with 10 tuples, SPE does not generate the best performance among all the baselines. This is because SPE has more parameters to learn; when the number of training tuples is small, it does not perform well. As the number of training tuples increases, SPE consistently outperforms others.

Secondly, SPE is better than ProxEmbed since s-paths carry more semantics than simple o-paths used in ProxEmbed. Moreover, SPE is also better than directly applying ProxEmbed on s-paths (i.e., PES). This is because PES is unable to deal with the subgraph structures and noises. Such results validate that, modeling s-paths for proximity embedding is not trivial.

Thirdly, SPE is better than MGP and MPP, showing that feature learning is more effective than feature engineering in proximity learning. Although SPE uses the same subgraph inputs as MGP, SPE further learns subgraph embedding, s-node embedding and s-path embedding from the sampled o-paths. SPE clearly benefits from the constructed subgraph-augmented paths, which leverage both path's distance awareness and subgraph's high-order structure.

Fourthly, SPE outperforms SRW. which uses the biased random walk to guide semantic ranking. SRW seems insensitive to the number of training tuples. Besides, SPE is better than DWR, which uses the Hadamard product over two objects' embedding as the proximity embedding. This observation shows that, the node embedding method, as solving the semantic search in an indirect manner, is less effective for proximity search.

Finally, SPE outperforms SPE-A, which validates the need of modeling attentions. As can be seen in Table **5**, SPE-A consistently generates inferior performance than SPE. This implies that well handling the noise in subgraph and s-paths is importantant.

We summarize the performance improvement of SPE over the best baselines with paired t-test in Table **6**. The largest improvement is observed in LinkedIn-colleague, where SPE improves the best baseline (PES) by relatively 13.2% in terms of NDGG and 16.9% in terms of MAP, with t-test p-values less than 0.01.

Parameter Sensitivity

We also test the parameter sensitivity of SPE using 100 training tuples. We vary the subgraph embedding's dimension d (Eq. **3**), the attention parameters' dimension d′ (Eq. **7**, Eq. **15** and Eq. **18**) and the loss discount λ(Eq. **20**).

As shown in Fig. **4**, Facebook data set is much more sensitive to the parameters setting than the other two data sets (especially LinkedIn). The reason is that the total number of object types in Facebook is much larger than in LinkedIn and DBLP as shown in Table **2**. When searching for users that meet a particular semantic relation type, the irrelevant types may bring in noises to the semantic user search process. The more types there are, the more noises may exist. Therefore, for the data sets with more types, the parameters need to be carefully tuned so that the embedded subgraph, the learned attentions and loss discount help to filter useful information out of the noises. On the other hand, for the data sets with less types like LinkedIn, the performance is relatively robust because the noises introduced by irrelevant object types are limited.

Based on both NDCG and MAP over all the six semantic relations on three data sets, SPE tends to generate the best performance at $d = 16$. Especially, for classmate and family, when d is too small, the resulting embeddings are unable to capture the rich semantics. When d is too big, it may bring in more noises and increase the number of parameters to learn. The attention dimension d' also tends to suffer from the similar performance decrease when d' is too small or too big. $d' = 16$ is the best setting. For the ranking loss discount parameter λ, we see that $\lambda = 0.1$ usually gives the best results, suggesting the need to discount the ranking loss in Eq. **20**.

CONCLUSION

In this paper, we study the problem of semantic user search in heterogeneous social networks. We exploit the opportunity of integrating the path's distance awareness and the subgraph's high-order structure for learning a better representation of the proximity between two users. We propose a novel Subgraph-augmented Path Embedding (SPE) model. It takes object paths as input, and enriches them into subgraph-augmented paths. Then it addresses the challenges of incorporating the subgraph structure, the subgraph noise and the subgraph-augmented path noise. Finally, it embeds the subgraph-augmented paths between two users into a proximity embedding vector. With such a proximity embedding vector, we can easily measure the proximity between two users for semantic user search. We test SPE with six semantic relations in three public data sets and it improves the state of the art by at least 1.8%–13.2% (NDCG) and 1.4%–16.9% (MAP) with 100 training samples.

In the future, we would like to explore the heterogeneous social networks with rich edge features and graph dynamics.

ACKNOWLEDGMENTS

We thank the support from: Zhejiang Science and Technology Plan Project (No. 2015C01027), National Natural Science Foundation of China (No.

61602405), National Research Foundation, Prime Minister's Office, Singapore under its Campus for Research Excellence and Technological Enterprise (CREATE) programme, and National Science Foundation under Grant No. IIS 16-19302. Any opinions, findings, and conclusions or recommendations expressed in this publication are those of the author(s) and do not necessarily reflect the views of the funding agencies.

REFERENCES

1. Lars Backstrom and Jure Leskovec. 2011. Supervised Random Walks: Predicting and Recommending Links in Social Networks. In WSDM. 635–644.
2. Yoshua Bengio. 2009. Learning Deep Architectures for AI. Foundations and Trends in Machine Learning 2, 1 (2009), 1–127.
3. Austin R. Benson, David F. Gleich, and Jure Leskovec. 2016. Higher-order Organization of Complex Networks. Science 353, 6295 (2016), 163–166.
4. Antoine Bordes, Nicolas Usunier, Alberto García-Durán, Jason Weston, and Oksana Yakhnenko. 2013. Translating Embeddings for Modeling Multi-relational Data. In NIPS. 2787–2795.
5. Hongyun Cai, Vincent W. Zheng, and Kevin Chen-Chuan Chang. 2018. A Comprehensive Survey of Graph Embedding: Problems, Techniques and Applications. TKDE (2018).
6. Shaosheng Cao, Wei Lu, and Qiongkai Xu. 2016. Deep Neural Networks for Learning Graph Representations. In AAAI. 1145–1152.
7. Hanjun Dai, Bo Dai, and Le Song. 2016. Discriminative Embeddings of Latent Variable Models for Structured Data. In ICML. 2702–2711.
8. Yuxiao Dong, Nitesh V Chawla, and Ananthram Swami. 2017. metapath2vec: Scalable Representation Learning for Heterogeneous Networks. In KDD. 135–144.
9. Mohammed Elseidy, Ehab Abdelhamid, Spiros Skiadopoulos, and Panos Kalnis. 2014. GRAMI: Frequent Subgraph and Pattern Mining in a Single Large Graph. PVLDB 7, 7 (2014), 517–528.
10. Alessandro Epasto, Silvio Lattanzi, and Mauro Sozio. 2015. Efficient Densest Subgraph Computation in Evolving Graphs. In WWW. 300–310.
11. Yuan Fang, Wenqing Lin, Vincent W. Zheng, Min Wu, Kevin Chen-Chuan Chang, and Xiaoli Li. 2016. Semantic proximity search on graphs with metagraph-based learning. In ICDE. 277–288.
12. Aditya Grover and Jure Leskovec. 2016. node2vec: Scalable Feature Learning for Networks. In KDD.
13. Pankaj Gupta, Venu Satuluri, Ajeet Grewal, Siva Gurumurthy, Volodymyr Zhabiuk, Quannan Li, and Jimmy J. Lin. 2014. Real-Time Twitter Recommendation: Online Motif Detection in Large Dynamic

Graphs. PVLDB 7, 13 (2014), 1379–1380.

14. William L. Hamilton, Rex Ying, and Jure Leskovec. 2017. Inductive Representation Learning on Large Graphs. In NIPS.
15. Sepp Hochreiter and Jürgen Schmidhuber. 1997. Long Short-Term Memory. Neural Comput. 9, 8 (Nov. 1997), 1735–1780.
16. Glen Jeh and Jennifer Widom. 2002. SimRank: A Measure of Structural-context Similarity. In KDD. 538–543.
17. Glen Jeh and Jennifer Widom. 2003. Scaling Personalized Web Search. In WWW. 271–279.
18. Ni Lao and William W. Cohen. 2010. Relational retrieval using a combination of path-constrained random walks. Machine Learning 81, 1 (2010), 53–67.
19. Rui Li, Chi Wang, and Kevin Chen-Chuan Chang. 2014. User profiling in an ego network: co-profiling attributes and relationships. In WWW. 819–830.
20. Zemin Liu, Vincent W. Zheng, Zhou Zhao, Fanwei Zhu, Kevin Chen-Chuan Chang, Minghui Wu, and Jing Ying. 2017. Semantic Proximity Search on Heterogeneous Graph by Proximity Embedding. In AAAI.
21. Zemin Liu, Vincent W. Zheng, Zhou Zhao, Fanwei Zhu, Kevin Chen-Chuan Chang, Minghui Wu, and Jing Ying. 2018. Distance-aware DAG Embedding for Proximity Search on Heterogeneous Graphs. In AAAI.
22. Julian J. McAuley and Jure Leskovec. 2012. Learning to Discover Social Circles in Ego Networks. In NIPS. 548–556.
23. Feiping Nie, Wei Zhu, and Xuelong Li. 2017. Unsupervised Large Graph Embedding. In AAAI.
24. Mathias Niepert, Mohamed Ahmed, and Konstantin Kutzkov. 2016. Learning Convolutional Neural Networks for Graphs. In ICML. 2014–2023.
25. Giannis Nikolentzos, Polykarpos Meladianos, and Michalis Vazirgiannis. 2017. Matching Node Embeddings for Graph Similarity. In AAAI.
26. Mingdong Ou, Peng Cui, Jian Pei, Ziwei Zhang, and Wenwu Zhu. 2016. Asymmetric Transitivity Preserving Graph Embedding. In KDD. 1105–1114.
27. Bryan Perozzi, Rami Al-Rfou, and Steven Skiena. 2014. DeepWalk: Online Learning of Social Representations. In KDD. 701–710.

28. Leonardo F.R. Ribeiro, Pedro H.P. Saverese, and Daniel R. Figueiredo. 2017. Struc2Vec: Learning Node Representations from Structural Identity. In KDD. 385–394.
29. Nino Shervashidze, Pascal Schweitzer, Erik Jan van Leeuwen, Kurt Mehlhorn, and Karsten M. Borgwardt. 2011. Weisfeiler-Lehman Graph Kernels. Journal of Machine Learning Research 12 (2011), 2539–2561.
30. Yizhou Sun, Jiawei Han, Xifeng Yan, Philip S Yu, and Tianyi Wu. 2011. PathSim: Meta Path-Based Top-K Similarity Search in Heterogeneous Information Networks. PVLDB 4, 11 (2011).
31. Zhao Sun, Hongzhi Wang, Haixun Wang, Bin Shao, and Jianzhong Li. 2012. Efficient Subgraph Matching on Billion Node Graphs. PVLDB 5, 9 (2012), 788– 799.
32. Jian Tang, Meng Qu, Mingzhe Wang, Ming Zhang, Jun Yan, and Qiaozhu Mei. 2015. LINE: Large-scale Information Network Embedding. In WWW. 1067–1077.
33. Theano Development Team. 2016. Theano: A Python framework for fast computation of mathematical expressions. CoRR abs/1605.02688 (may 2016).
34. Cunchao Tu, Zhengyan Zhang, Zhiyuan Liu, and Maosong Sun. 2017. TransNet: Translation-Based Network Representation Learning for Social Relation Extraction. In IJCAI. 2864–2870.
35. Rogier J. P. van Berlo, Wynand Winterbach, Marco J. L. de Groot, Andreas Bender, Peter J. T. Verheijen, Marcel J. T. Reinders, and Dick de Ridder. 2013. Efficient calculation of compound similarity based on maximum common subgraphs and its application to prediction of gene transcript levels. IJBRA 9, 4 (2013), 407–432.
36. Chi Wang, Jiawei Han, Yuntao Jia, Jie Tang, Duo Zhang, Yintao Yu, and Jingyi Guo. 2010. Mining advisor-advisee relationships from research publication networks. In KDD. ACM, 203–212.
37. Chi Wang, Rajat Raina, David Fong, Ding Zhou, Jiawei Han, and Greg Badros. 2011. Learning Relevance from Heterogeneous Social Network and Its Application in Online Targeting. In SIGIR. 655–664.
38. Daixin Wang, Peng Cui, and Wenwu Zhu. 2016. Structural Deep Network Embedding. In KDD. 1225–1234.
39. Zhen Wang, Jianwen Zhang, Jianlin Feng, and Zheng Chen. 2014. Knowledge Graph Embedding by Translating on Hyperplanes. In AAAI. 1112–1119.

40. Kelvin Xu, Jimmy Ba, Ryan Kiros, Kyunghyun Cho, Aaron C. Courville, Ruslan Salakhutdinov, Richard S. Zemel, and Yoshua Bengio. 2015. Show, Attend and Tell: Neural Image Caption Generation with Visual Attention. In ICML. 2048–2057.

41. Muhan Zhang and Yixin Chen. 2017. Weisfeiler-Lehman Neural Machine for Link Prediction. In KDD. 575–583.

15

A Hybrid Optimized Weighted Minimum Spanning Tree for the Shortest Intrapath Selection in Wireless Sensor Network

Matheswaran Saravanan[1] and Muthusamy Madheswaran[2]

[1]Department of Computer Science and Engineering, VMKV Engineering College, Salem, Tamil Nadu 636308, India

[2]Department of Electronics and Communication Engineering, Mahendra Engineering College, Namakkal, Tamil Nadu 637503, India

ABSTRACT

Wireless sensor network (WSN) consists of sensor nodes that need energy efficient routing techniques as they have limited battery power, computing, and storage resources. WSN routing protocols should enable reliable multihop communication with energy constraints. Clustering is an effective way to reduce overheads and when this is aided by effective resource allocation, it results in reduced energy consumption. In this work, a novel hybrid evolutionary algorithm called Bee Algorithm-Simulated Annealing

Citation: Matheswaran Saravanan and Muthusamy Madheswaran, "A Hybrid Optimized Weighted Minimum Spanning Tree for the Shortest Intrapath Selection in Wireless Sensor Network," Mathematical Problems in Engineering, vol. 2014, Article ID 713427, 8 pages, 2014. https://doi.org/10.1155/2014/713427

Weighted Minimal Spanning Tree (BASA-WMST) routing is proposed in which randomly deployed sensor nodes are split into the best possible number of independent clusters with cluster head and optimal route. The former gathers data from sensors belonging to the cluster, forwarding them to the sink. The shortest intrapath selection for the cluster is selected using Weighted Minimum Spanning Tree (WMST). The proposed algorithm computes the distance-based Minimum Spanning Tree (MST) of the weighted graph for the multihop network. The weights are dynamically changed based on the energy level of each sensor during route selection and optimized using the proposed bee algorithm simulated annealing algorithm.

INTRODUCTION

Wireless sensor network (WSN) is a cooperative collection of sensor nodes, each having processing capability. Routing in WSN is different from conventional fixed network routing by several ways. WSNs are infrastructureless, have unreliable wireless links, contain sensor nodes that might fail, and its routing protocols face rigorous energy saving requirements [1]. WSN is a distributed real-time system and many routing algorithms have been proposed in literature [2–4]. In the earlier research for distributed systems, it was assumed that wired systems had unlimited power. They had user interfaces, had fixed resources, treated each system node as important, and were location independent. WSNs in contrast are wireless systems with limited power, are constrained in energy consumption, and are real time, with dynamically varying resources [5, 6]. Routing in WSN utilizing minimal energy has been proposed in the literature [7–11]. The power management solutions, at the software level, aim at reducing communications as broadcasting or listening to messages uses up energy. Minimizing message numbers cuts costs and a good MAC protocol ensures reduced collisions and retries. Better routing minimizes the number of messages sent by the use of short paths and congestion avoidance. Factors like efficient neighbor detection, localization, time synchronization, flooding, and query dissemination reduce the number of messages and increase the life of the network. There are varied solutions for scheduling sleep/wake-up patterns [12, 13] with most trying to keep up minimum nodes labeled sentries. The latter provides sensing coverage and allows others to sleep.

The clustering process divides a network into interconnected substructures, called clusters, with each cluster having many Sensor Node (SN) led by a Cluster Head (CH) which is the coordinator in this substructure

[14] as seen in Figure 1. The CH is also a temporary base station which keeps in touch with other CHs. Nodes have four possible states: normal, isolated, cluster head, and gateway. Basically, nodes are in an isolated state with each maintaining the neighbor table where neighbour node information is stored. Electing the CH is the basic step in clustering. Clustering is widely applied in WSN for managing power efficiently [15–17].

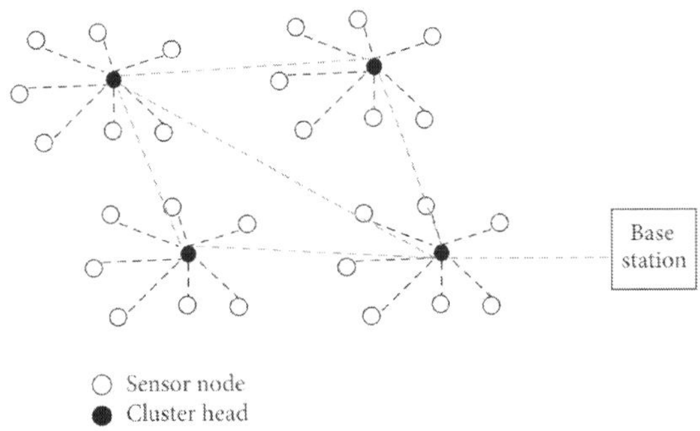

Figure 1: Block diagram of WSN deployment with cluster heads.

A clustering architecture in WSN environment enables features such as network scalability, communication overhead reduction, and fault tolerance. Cluster formation benefits routing as the cluster head and cluster gateways are responsible for the intercluster routing, thus, restricting, creating, and spreading of routing information. Local changes such as nodes changing cluster are updated only in the corresponding clusters and no update is required by the whole network. This significantly reduces information stored by each mobile node.

The major problem in WSNs is experienced by the sensor nodes nearest to the base station that require transmitting more number of packets than that of faraway nodes. Generally clustering algorithms use two methods for extending the lifetime. In the first method, the cluster heads with high residual energy are selected, and in the second method, to distribute the energy utilization between nodes in all clusters, the cluster heads are rotated periodically [18]. To choose the cluster heads, various techniques have been proposed in literature.

Other than clustering technique, tree based routing has been popularly used in WSN due to its energy efficiency [19–21]. Tree based techniques use the concept of selecting a root node before data transmission. A tree-like hierarchical path of nodes is constructed to connect the nodes. The WSN nodes construct a tree which can either be Minimum Spanning Tree (MST) or be optimal tree through which data is transmitted till it reaches the sink. The root node uses tree traversal algorithm to gather data about the children nodes. There are three methods in WSN tree routing protocols [19]. While the first does not create clusters the other two mix the clustering strategy with a tree routing algorithm. The latter strategy reduces low latency, while the tree routing algorithm improves energy efficiency [22, 23]. In cluster based routing strategy, the sensor nodes in the network divide into several clusters. Each cluster then chooses a cluster head randomly or by cluster head election algorithm. The cluster head is responsible for gathering the data from the sensor nodes in its cluster. The aggregated data collected by the cluster head is then transmitted to the sink. Generally, a MST is created to connect the nodes and this reduces energy consumption either by transmitting packets through smaller distances or by reducing number of packets transmitted or by both techniques.

A modified Kruskal's Minimum Spanning Tree (MST) search algorithm based on distributed search by hierarchical clusters was proposed to search the network for small balanced weight routing spanning trees [5]. The proposed technique provided spanning trees with low maximum degree and larger diameter to balance energy consumption in WSN's routing. Based on energy matrix transmission the results proved that this approach extended WSN functional life by more than three times with respect to sensor transmission energy.

An energy efficient spanning tree (EESR) was proposed for multihop routing to increase the lifetime of the network [24]. The EESR provides location of the sensor nodes and base station and produces a sequence of routing paths consisting of suitable number of rounds. The results obtained by simulation reveal that the EESR method outperforms the other existing methods in relation to increasing the lifetime of the network.

A trajectory clustering technique was proposed for the purpose of selecting the cluster heads [25]. In this algorithm, the cluster heads are selected on the basis of traffic and they are periodically rotated. The cluster heads are selected using the trajectory based clustering technique and thus network lifetime is extended.

Guangyan et al. [26] proposed Dynamic Minimal Spanning Tree Routing Protocol (DMSTRP), an innovative cluster-based routing protocol, that enhanced Base Station Controlled Dynamic Clustering Protocol (BCDCP) by means of initiating MSTs rather than clubbing for the purpose of connecting nodes in clusters. When compared to LEACH and BCDCP, the DMSTRP performed well even in large network in terms of network lifetime and delay. A distributed topology control technique was proposed in [27] to enhance energy efficiency and reduce radio interference in WSNs. Each network node makes local decisions about transmission power. These decisions conclude in a network topology preserving global connectivity. The fundamental control technique is the novel Smart Boundary Yao Gabriel Graph (SBYaoGG) and optimization ensures that all network links are symmetric and energy efficient. This technique was effective as compared to other approaches to topology control.

A combined algorithm (COM), a generalization of the MST and SPT, was proposed in [28] which dealt with the issue of executing the operation of Data Aggregation enhanced Convergecast (DAC) in an energy and latency efficient manner. The valuable portion of the total data gathered is approximated by assuming that each and every node in the network consists of a data item and a known application dependent data compression factor.

Multiple Cluster Heads Routing Protocol (MCHRP) [29] was proposed to address cluster head overload. This method improved LEACH by incorporating a decision function which is based on the cluster head's remaining energy, location, and frequency. The decision function selects the main cluster heads and the alternative cluster heads used for data acquisition, data fusion, and data transmission.

Cluster-based and Tree-based Power Efficient Data Collection and Aggregation (CTPEDCDA) protocol by Wang et al. [30] was based on clustering and MST to minimize energy consumption in WSN. MSTs are built by connecting the cluster heads to improve the transmission routing mechanism. Chhabra and Sharma [31] improved the power consumption by improving the first node death. This method combined both the cluster-based and the tree-based protocol to improve evenness of dissipated network energy.

Kumrai et al. [32] proposed evolutionary algorithm heuristically that optimizes the sensing coverage area and the installation cost in WSN by considering the sensor network connectivity as a constraint. The algorithm uses a population of individuals, each of which represents a set of wireless sensor nodes types and positions and evolves them via the proposed genetic operators. The proposed mutation and constraint-domination operators were designed to quickly seek the optimal solutions that meet the WSN installation requirements. Simulation result shows that the sensing coverage and installation cost were improved. Karimi et al. [33] proposed two algorithms such as GP-Leach and HS-Leach. The energy consumption was improved by partitioning the network and using evolutionary algorithms for optimized cluster head selection considering WSN nodes position information and residual energy. The simulation results performed in MATLAB show that the proposed algorithms were more efficient and they increased the lifetime of network.

In this work, a Weighted Minimum Spanning Tree, Bee Algorithm-Simulated Annealing (BASA-WMST) algorithm is proposed. Cluster heads are selected based on the proposed optimization technique and WMST is used to find the shortest intrapath selection within the cluster. The proposed method computes the distance-based minimum spanning tree of the weighted graph for the multihop network. During route selection, the weights are dynamically adjusted based on the mobility, energy level, and distance of each sensor. Section 2presents the problem formulation, Section 3 deals with the proposed methodology in detail, Section 4 shows the experimental results, and Section 5 concludes the paper.

PROBLEM FORMULATION

Tree based routing has the advantage of lower control packet overheads but suffers from approximation error compared to cluster based routing. Cluster based routing provides better energy savings compared to tree based techniques. In this work it is proposed to combine the features of cluster based routing for cluster formation and cluster head selection and use minimum spanning tree for intracluster communication. Ideal clusters are formed when the network parameters like energy spent, lifetime, Packet Delivery Ratio, and end to end delay are optimized. Since most of the network parameters are additive in nature the optimization problem is NP hard. Several metaheuristic techniques including genetic algorithm have been proposed in literature. In this paper, bee algorithm in combination with simulated annealing was chosen due to its faster convergence and its capability to avoid local minima problem.

WSN network can be considered as a connected undirected graph represented by $G = (V, E)$, where V is the vertices made up of $(V_1,..., V_n)$ nodes and E is the edges represented as $(e_{1,2}, e_{1,3},...,e_{i,j},...,e_{n-1,n})$ the connection between nodes [34]. In this work the normalized values of mobility, delay, and remaining energy are represented on the edges. Each edge may then be defined by the attributes represented in positive real numbers and denoted by $w_{i,} = w_{i,j}^1, w_{i,j}^2, \ldots, w_{i,j}^m$

Let $x=x_{1,2}, x_{1,3},\ldots,x_{i,j},\ldots,x_{n-1,n}$ be defined as the connectivity between node i and j:

$$x_{i,j} = \begin{cases} 1 & \text{if } e_{i,j} = 1 \text{ and is selected} \\ 0 & \text{otherwise.} \end{cases} \quad (1)$$

The proposed technique can be formulated as in

$$\min f_1(x) = \sum w_{i,j}^1 x_{i,j}$$

$$\min f_2(x) = \sum w_{i,j}^2 x_{i,j}$$

$$\vdots$$

$$\min f_m(x) = \sum w_{i,j}^m x_{i,j}, \quad (2)$$

where $f_i(x)$ is the objective to be minimized for the problem, $i = 1, \ldots, n-1$; $j = 1, \ldots, n$ subject to $x \in X$. These objectives either can be formulated as a multiobjective function or can be represented as in

$$\min f_i(x) = \alpha \sum w_{i,j}^1 x_{i,j} + \beta \sum w_{i,j}^2 x_{i,j} + \cdots, \quad (3)$$

where $\alpha+\beta+\cdots=1$.

Since node mobility, delay, and remaining energy are used as the edge in the graph, the objective function can be formulated as in

$$\min f_i(x) = \alpha(\min(\text{mobility})) + \beta(\min(\text{delay})) + \gamma(\min(1 - \text{remaining\_energy})). \quad (4)$$

The following assumptions are made for the sensor network.(1)Nodes are dispersed randomly.(2)The energy of sensor nodes is limited and uniform initially.(3)Nodes are location unaware.(4)The transmitting power of the nodes varies depending on the distance to the receiver.(5)Approximate distance is estimated based on the received signal strength.

METHODOLOGY

The node energy model is based on [35]. The energy dissipated to transmit n bit is given in

$$E_{\text{diss\_Tx}} = n\,(\text{Energy\_dissipated\_tranmitter\_electronics} + (\text{Energy\_dissipated\_transmitter\_amplifier} * \text{distance\_squared}))\,. \tag{5}$$

The energy dissipated to receive n bit is given in

$$E_{\text{diss\_Rx}} = n\,(\text{Energy\_dissipated\_receiver\_electronics})\,. \tag{6}$$

Power consumed for a given time period t can be computed by dividing the dissipated energy by time and is given by

$$\frac{E_{\text{diss\_Rx}} + E_{\text{diss\_Tx}}}{t}. \tag{7}$$

The mobility of a node is estimated using the Free Space Path Loss (FSPL) model. The relation between FSPL, frequency of radio signal, and distance between the transmitter and receiver is given by

$$\text{FSPL (dB)} = 20\log(d) + 20\log(f) + k, \tag{8}$$

where d is the distance, f is the frequency, and log is the logarithm to base 10. K is a constant and is equal to 32.44 when frequency is measured in Mhz and distance is measured in Kilometer. Another method to compute the FSPL is using the fade margin and it is given by

$$\text{FSPL} = \text{Energy\_dissipated\_Tx\_electronics} + \text{Energy\_dissipated\_Tx\_amplifier} + \text{Energy\_dissipated\_Rx\_electronics}. \tag{9}$$

Using the two FSPL equations (8) and (9), the distance can be computed by

$$d = \frac{10\,(\text{Free Space Path Loss} - 32.44 - 20\log(f))}{20}. \tag{10}$$

To find the distance travelled by nodes i and j with respect to each other during time n, the distance between the nodes is computed at time t and t=n if high mobility increases the reclustering process and increases the energy consumption. The objective is to form clusters based on low mobility which leads to lower energy consumption and lower delays due to lower link breakages. The mobility of the node can be computed by

$$m = \frac{d_t - d_{t+n}}{D}\left\{\begin{array}{l} > 0.5 \text{ implies high mobility} \\ \leq 0.5 \text{ implies normal mobility} \\ 0 \text{ implies no mobility} \\ < 0 \text{ implies nodes converging} \end{array}\right\}. \quad (11)$$

Each node stores in its neighborhood table the information about its neighbors, as shown in Table 1. Each node broadcasts the Ech Msg, at the beginning of each round which contains residual energies, within radio range r. All nodes within the cluster range of one node are considered as the neighbors of this node. On receiving the Ech Msg nodes update the neighborhood table.

Table 1: Information maintained in the neighborhood table

Parameter	Description
V_i	Node i
V_j	A neighbor node in cluster range of V_i
RE_{V_i}	Residual energy of V_i
Dis_{V_i}	Distance between V_i and V_j
RE_{V_j}	Residual energy of V_j
Ech_Msg	Elect cluster head message
Crt_Msg	Create tree message

The flow chart of the proposed technique is shown in Figure 2. Bee algorithm-simulated annealing algorithm is proposed to avoid the local minima problem faced by bee algorithm and to select the best cluster heads by forming ideal clusters. Clustering is achieved by dividing arbitrarily organized sensors into the best possible number of selfdetermining clusters with cluster head and optimal route to form the initial population. The edge weights between nodes are computed and the objective function is computed. These initial solutions become the initial food source in the proposed BASA algorithm. Once the initial population is found, the bee algorithm is initiated. Each node broadcasts its ID along with its weight W_i to the neighboring nodes and stores the weights W_j of the other nodes within its transmission range.

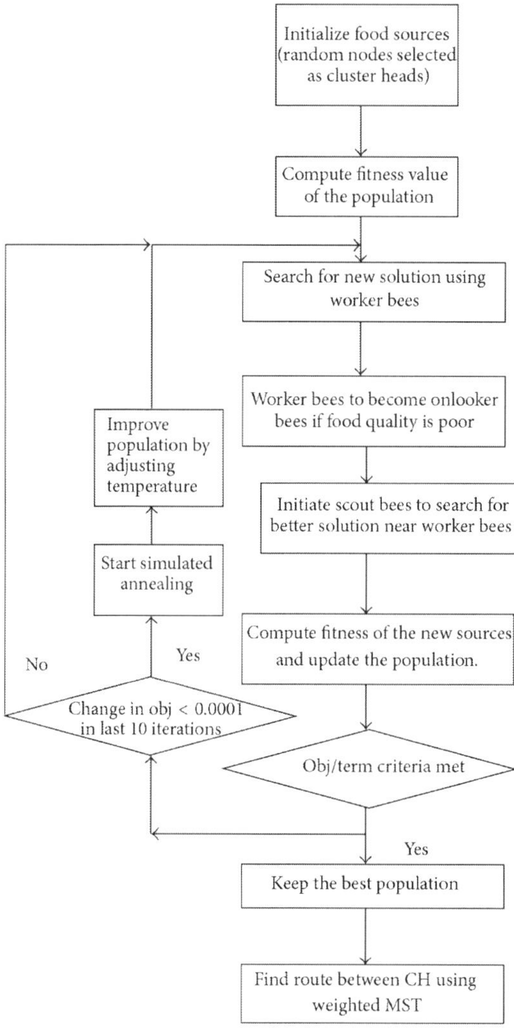

Figure 2: Flowchart of the proposed technique.

Bee's algorithm is a population-based search algorithm inspired by bees foraging behaviour [36]. The algorithm starts with search space being populated by worker bees being placed randomly at the location of the initial food source. The fitness of sites visited by worker bees is evaluated and bees with the best fitness continue to be worker bees. Bees which have visited sites with lower fitness value are delegated to onlooker bees. The location of food source with the best fitness becomes the new search location for better solutions. Effectively a cluster with CH can either add new nodes to increase the cluster or remove some nodes to decrease the cluster.

Similarly, the CH can be rotated within the cluster. This is achieved by searching neighborhoods of selected sites by assigning scout bees to search near the best e sites. Neighborhood searches of the best e sites are made detailed by recruiting more bees other than the selected bees to follow them. This differential recruitment along with scouting is a key bee's algorithm operation.

The probability p_i of selecting a food source i can be determined by using

$$p_i = \frac{\text{fit}_i}{\sum_{n=1}^{\text{SN}} \text{fit}_n}, \tag{12}$$

and the fitness is given as in

$$\text{fitness}(i) = \begin{cases} 1 + \text{abs}(f(i)), & f(i) < 0 \\ \dfrac{1}{(f(i))} + 1, & f(i) \geq 0, \end{cases} \tag{13}$$

where fit_i is the fitness value of ith solution which represents the nectar amount at food source at ith position and SN is the number of employed bees and also a number of food sources.

The process is iterated till the termination criteria are reached or till the improvement in the fitness does not increase by more than 0.001 in the last 10 iterations. If there is no improvement in the solution, the algorithm could have struck in the local minima. Simulated annealing is starting to climb out of the local minima problem.

The simulated annealing (SA) was introduced in 1983 which is based on the ideas formulated in the early 1950s [37]. Simulated annealing is a relatively straight forward algorithm which includes metropolis Monte Carlo method. The metropolis Monte Carlo algorithm is well suited for simulated annealing, since only energetically feasible states can be sampled at any given temperature. Therefore the simulated annealing algorithm starts at a high temperature with simulation of metropolis Monte Carlo algorithm. The temperature is slowly reduced such that the search space becomes smaller for the metropolis simulation, and when the temperature is low then the system has hopefully settled into the most favorable state. Simulated annealing can also be used for searching the optimum solution of the problems by properly determining the initial (high) and final (low) effective temperatures which are used in place of kT (where k is a Boltzmann›s constant) in the acceptance checking and deciding what constitutes a Monte Carlo step [38]. Simulated annealing is a probabilistic method [39] to find global minimum of a cost function that can have several local minima. Simulated annealing emulates the physical process wherein a solid is slowly cooled so that when eventually its structure is frozen, this happens at a minimum energy configuration. Simulated annealing to compute the probability of acceptance:

$$P_a = \begin{cases} e^{-k\Delta E/T} & \Delta E > 0 \\ 1 & \Delta E \le 0, \end{cases} \qquad (14)$$

where ΔE is the difference between the solution error after it has perturbed and the solution error before it was perturbed, T is the current temperature, and k is a suitable constant.

On identifying the potential cluster heads, the MST algorithm is used for tree construction to find the intracluster routes. Suppose that n points are given in different dimensions, then a tree spanning to these points is a set of straight line segments joining pairs of points [40], so that (1) there are no closed loops, (2) a line visits each point at least once, and (3) the tree is connected.

Figure 3 shows an example of a tree of integer segment lengths. If, for example, vertices N3 and N7 are joined, a closed loop is formed and the result would not be a tree. The length of a tree is the sum of its segments lengths. When a set of n points and the lengths of all $(n/2)$ segments are given, a spanning tree of minimum length (MST) is required. The MST is computed using reaching the base station.

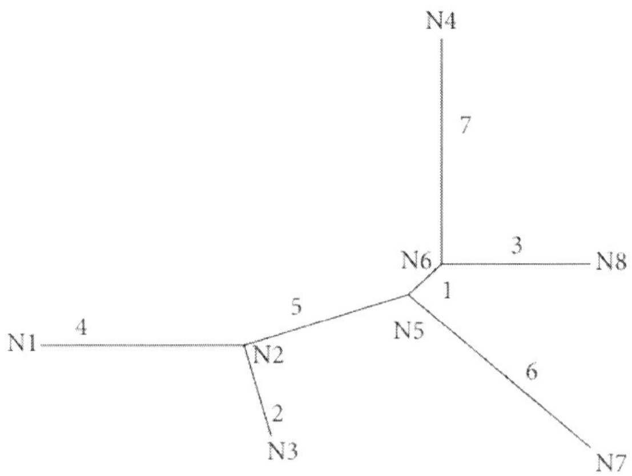

Figure 3: A tree connecting eight vertices.

With traditional MST algorithms, construction cost of a minimum spanning tree is $(m \log n)$, where$m$is the number of graph edges and n is the number of vertices [41].The weight of a tree edge is computed by Euclidean distance between two end points. The average weight w of MST edges based on remaining energy and delay is computed. Any edge with a weight $w>w_{avg}$ is removed leading to a set of disjoint subtrees $t = \{t_1, t_2,...t_i\}$.

The so formed routes are optimal from the spatial perspective since the cluster heads are uniformly distributed over the imperfectly formed wireless sensor network.

EXPERIMENTAL SET-UP AND RESULTS

Experiments were conducted with different number of mobile sensor nodes, spread over an area of 1000 m by 1000 m with the Base Station being stationary at location (500, 500). The simulation parameters and bee algorithm parameters are shown in Table 2.

Table 2: The parameters of the network simulation

Parameters	Values
Initial energy of nodes E_{init}	0.6 J
Amplification coefficient of the free space model E_{fs}	10 pJ-m^2/b
Amplification coefficient of the multipath transmission model E_{amp}	0.0025 pJ-m^2/b
Circuit loss E_{elec}	50 nJ/b
Data packet length	8000 b
Control packet length	120 b
Initial number of worker bees	20
Food source-the initial search space for cluster head (random)	20
Encoding	Binary with
Termination condition	Fitness value <0.0001

Experiments were conducted to simulate the proposed technique and are compared to cluster based routing, GA based cluster formation, and ABC based cluster optimization. Figure 4 shows the number of clusters formed. The proposed BASA WMST technique increases the average number of clusters across by 10.41% compared to cluster based routing for varying number of nodes in the network. ABC produces better clusters compared to GA.

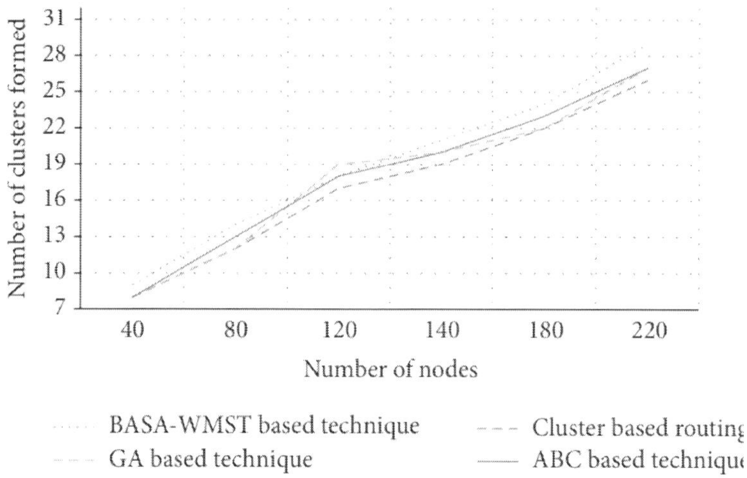

Figure 4: Number of clusters formed.

Figure 5 shows the average end to end delay obtained in the network for different number of nodes. As the number of nodes increases, the performance improvement of the proposed technique is in par with GA based technique showing a small marginal average improvement of 1.27% compared to GA based technique.

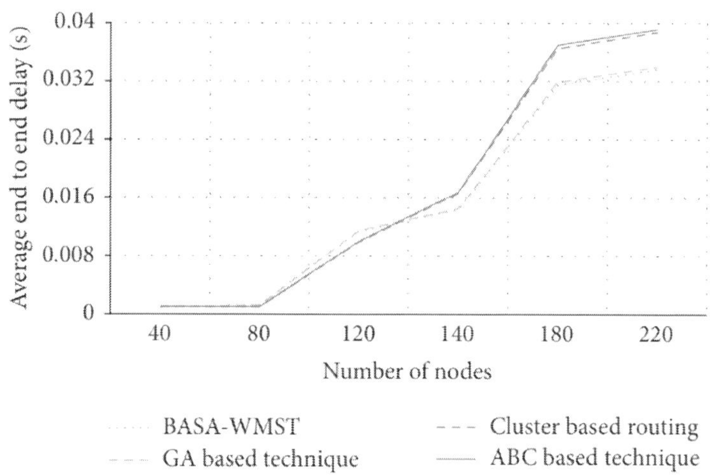

Figure 5: Average end to end delay.

However, both GA and proposed BASA-WMST show significant decrease in end to end delay as the number of nodes is increased. End to end delay decreased over 14% when compared to cluster based technique. This becomes significant for WSN used in streaming applications. Figure 6 shows the Packet Delivery Ratio obtained under different number of nodes.

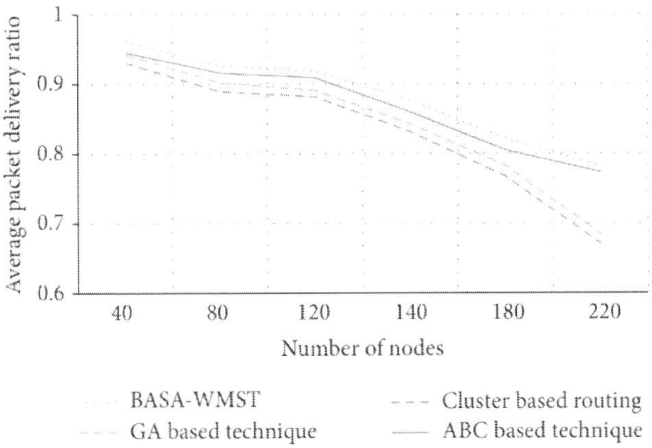

Figure 6: Average Packet Delivery Ratio.

The average PDR in the proposed BASA WMST improved by 6.49% when compared to cluster based routing for various numbers of nodes in the network. Compared to GA the PDR improvement was significant by 4.98% and by 1.47% compared to ABC based technique. Figure 7 shows the life time of the network. ABC and the proposed technique significantly improve the life of the network compared to GA based technique.

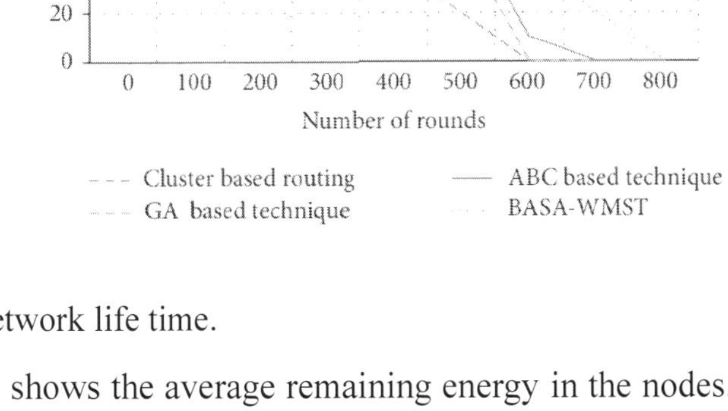

Figure 7: Network life time.

Figure 8 shows the average remaining energy in the nodes. The energy savings are significant in the proposed technique compared to GA based technique. Average energy to the tune of 21.67% is observed compared to GA.

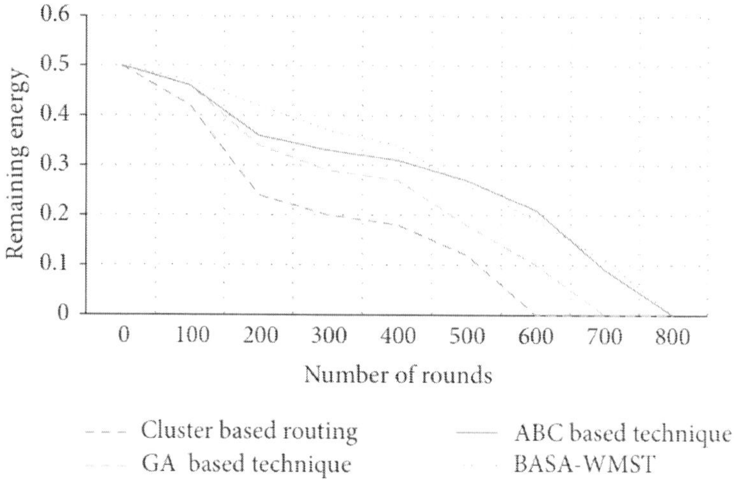

Figure 8: Remaining energy.

CONCLUSION

WSN routing protocols must perform efficiently under mobility and energy constraints. In this paper, clustering is achieved through a hybrid algorithm that divides arbitrarily organized sensors into the best possible number of self-determining clusters with cluster head and optimal route to base station using a novel optimization BASA-WMST. Bee algorithm is incorporated to increase the information exchange among bees and SA is used to escape local optima. Intracluster route is selected from the optimal trees based on weights described. The proposed routing was simulated and compared with conventional cluster based routing and other optimization techniques showing improvements in the QOS.

REFERENCES

1. S. C. Misra and I. Woungang, Guide to Wireless Sensor Networks, vol. 7, Springer, New York, NY, USA, 2009.
2. C. K. Toh, A. N. Le, and Y. Z. Cho, "Load balanced routing protocols for ad hoc mobile wireless networks," IEEE Communications Magazine, vol. 47, no. 8, pp. 78–84, 2009.
3. J. Le, J. C. Lui, and D. M. Chiu, "DCAR: distributed coding-aware routing in wireless networks," IEEE Transactions on Mobile Computing, vol. 9, no. 4, pp. 596–608, 2010.
4. N. S. Moayedian and S. J. Golestani, "Optimal scheduling and routing in wireless networks: a new approach," in Proceedings of the IEEE Wireless Communications and Networking Conference (WCNC '09), pp. 1–6, IEEE, April 2009.
5. A. Gagarin, S. Hussain, and L. T. Yang, "Distributed search for balanced energy consumption spanning trees in wireless sensor networks," in Proceedings of the International Conference on Advanced Information Networking and Applications Workshops (WAINA '09), pp. 1037–1042, Bradford, UK, May 2009.
6. J. N. Al-Karaki and A. E. Kamal, "Routing techniques in wireless sensor networks: a survey," IEEE Wireless Communications, vol. 11, no. 6, pp. 6–28, 2004.
7. T. Watteyne, A. Molinaro, M. G. Richichi, and M. Dohler, "From MANET to IETF ROLL standardization: a paradigm shift in WSN routing protocols," IEEE Communications Surveys and Tutorials, vol. 13, no. 4, pp. 688–707, 2011.
8. B. Priya and S. S. Manohar, "EE-MAC: energy efficient hybrid MAC for WSN," International Journal of Distributed Sensor Networks, vol. 2013, Article ID 526383, 9 pages, 2013.
9. E. Ahvar, S. Ahvar, G. M. Lee, and N. Crespi, "An energy-aware routing protocol for query-based applications in wireless sensor networks," The Scientific World Journal, vol. 2014, Article ID 359897, 9 pages, 2014.
10. X. Zhu, L. Shen, and T. P. Yum, "Hausdorff clustering and minimum energy routing for wireless sensor networks," IEEE Transactions on Vehicular Technology, vol. 58, no. 2, pp. 990–997, 2009.
11. T. W. Kuo and M. J. Tsai, "On the construction of data aggregation

tree with minimum energy cost in wireless sensor networks: NP-completeness and approximation algorithms," in Proceedings of the IEEE Conference on Computer Communications (INFOCOM '12), pp. 2591–2595, IEEE, March 2012.
12. O. Khader, A. Willig, and A. Wolisz, "Distributed wakeup scheduling scheme for supporting periodic traffic in WSNs," in Proceedings of the Wireless Conference, pp. 287–292, IEEE, May 2009.
13. U. Jang, S. Lee, and S. Yoo, "Optimal wake-up scheduling of data gathering trees for wireless sensor networks," Journal of Parallel and Distributed Computing, vol. 72, no. 4, pp. 536–546, 2012.
14. S. Bandyopadhyay and E. J. Coyle, "An energy efficient hierarchical clustering algorithm for wireless sensor networks," in Proceedings of the 22nd Annual Joint Conference on the IEEE Computer and Communications Societies, vol. 3, pp. 1713–1723, IEEE Societies, April 2003.
15. T. Kwon and M. Gerla, "Clustering with power control," in Proceedings of the IEEE Military Communications Conference (MILCOM '99), vol. 2, pp. 1424–1428, November 1999.
16. D. De, A. Sen, and M. D. Gupta, "Cluster based energy efficient lifetime improvement mechanism for WSN with multiple mobile sink and single static sink," in Proceedings of the 3rd International Conference on Computer and Communication Technology (ICCCT '12), pp. 197–199, IEEE, November 2012.
17. M. J. Reddy, P. S. Prakash, and P. C. Reddy, "Homogeneous and heterogeneous energy schemes for hierarchical cluster based routing protocols in WSN: a survey," in Proceedings of the 3rd International Conference on Trends in Information, Telecommunication and Computing, pp. 591–595, Springer, New York, NY, USA, 2013.
18. G. Ma and Z. Tao, "A hybrid energy- and time-driven cluster head rotation strategy for distributed wireless sensor networks," International Journal of Distributed Sensor Networks, vol. 2013, Article ID 109307, 13 pages, 2013.
19. G. Zheng and Z. Hu, "Tree routing protocol with location-based uniformly clustering strategy in WSNs," Journal of Networks, vol. 5, no. 11, pp. 1373–1380, 2010.
20. J. Zhang, Y. Xie, D. Liu, and Z. Zhang, "OCTBR: optimized clustering tree based routing protocol for wireless sensor networks," in Internet of

Things, pp. 192–199, Springer, Berlin, Germany, 2012.

21. X. Wang and H. Qian, "Constructing a 6LoWPAN wireless sensor network based on a cluster tree," IEEE Transactions on Vehicular Technology, vol. 61, no. 3, pp. 1398–1405, 2012.
22. S. S. Satapathy and N. Sarma, "TREEPSI: tree based energy efficient protocol for sensor information," in Proceeding of the IFIP International Conference on Wireless and Optical Communications Networks, Bangalore, India, April 2006.
23. D. Singhal, S. Barjatiya, and G. Ramamurthy, "A novel network architecture for cognitive wireless sensor network," in Proceedings of the International Conference on Signal Processing, Communication, Computing and Networking Technologies (ICSCCN '11), pp. 76–80, IEEE, July 2011.
24. S. Hussain and O. Islam, "An energy efficient spanning tree based multi-hop routing in wireless sensor networks," in Proceedings of the IEEE Wireless Communications and Networking Conference (WCNC '07), pp. 4383–4388, Kowloon, Hong Kong, 2007.
25. M. Hazarath, I. Lucio, and C. Luca, "CAST—a novel trajectory clustering and visualization tool for spatio temporal data," in Proceedings of the 1st International conference on Intelligent Human Computer Interaction (IHCI '09), pp. 169–175, Springer, 2009.
26. H. Guangyan, L. Xiaowei, and H. Jing, "Dynamic minimal spanning tree routing protocol for large wireless sensor networks," in Proceedings of the 1st IEEE Conference on Industrial Electronics and Applications (ICIEA '06), pp. 1–5, Singapore, May 2006.
27. T. M. Chiwewe and G. P. Hancke, "A distributed topology control technique for low interference and energy efficiency in wireless sensor networks," IEEE Transactions on Industrial Informatics, vol. 8, no. 1, pp. 11–19, 2011.
28. S. Upadhyayula and S. K. S. Gupta, "Spanning tree based algorithms for low latency and energy efficient data aggregation enhanced convergecast (DAC) in wireless sensor networks," Ad Hoc Networks, vol. 5, no. 5, pp. 626–648, 2007.
29. D. Tang, X. Liu, Y. Jiao, and Q. Yue, "A load balanced multiple Cluster-heads routing protocol for wireless sensor networks," in Proceedings of the IEEE 13th International Conference on Communication Technology (ICCT '11), pp. 656–660, Jinan, China, September 2011.

30. W. Wang, B. Wang, Z. Liu, L. Guo, and W. Xiong, "A cluster-based and tree-based power efficient data collection and aggregation protocol for wireless sensor networks," Information Technology Journal, vol. 10, no. 3, pp. 557–564, 2011.
31. G. S. Chhabra and D. Sharma, "Cluster-tree based data gathering in wireless sensor network," International Journal of Soft Computing and Engineering, vol. 1, no. 1, pp. 27–31, 2011.
32. T. Kumrai, P. Champrasert, and R. Kuawattanaphan, "Heterogeneous wireless sensor network (WSN) installation using novel genetic operators in a multiobjective optimization evolutionary algorithm," in Proceedings of the 9th International Conference on Natural Computation (ICNC '13), pp. 606–611, IEEE, 2013.
33. M. Karimi, H. R. Naji, and S. Golestani, "Optimizing cluster-head selection in wireless sensor networks using genetic algorithm and harmony search algorithm," in Proceedings of the 20th Iranian Conference on Electrical Engineering (ICEE '12), pp. 706–710, IEEE, Tehran, Iran, May 2012.
34. W. Guo, B. Zhang, G. Chen, X. Wang, and N. Xiong, "A PSO-optimized minimum spanning tree-based topology control scheme for wireless sensor networks," International Journal of Distributed Sensor Networks, vol. 2013, Article ID 985410, 14 pages, 2013.
35. W. R. Heinzelman, A. Chandrakasan, and H. Balakrishnan, "Energy-efficient communication protocol for wireless microsensor networks," in Proceedings of the 33rd Annual Hawaii International Conference on System Siences (HICSS '00), p. 10, January 2000.
36. D. T. Pham, A. Ghanbarzadeh, E. Koc, S. Otri, S. Rahim, and M. Zaidi, "The bees algorithm-a novel tool for complex optimisation problems," in Proceedings of the 2nd Virtual International Conference on Intelligent Production Machines and Systems (IPROMS '06), pp. 454–459, 2006.
37. M. Kumar and D. R. Gupta, "A comparative study using simulated annealing and fast output sampling feedback technique based pss design for single machine infinite bus system modeling," International Journal of Engineering Research and Applications, vol. 2, no. 2, pp. 223–228, 2012.
38. A. Tamilarasi, "An enhanced genetic algorithm with simulated a nnealing for job-shop scheduling," International Journal of Engineering, Science and Technology, vol. 2, no. 1, pp. 144–151, 2010.

39. S. Kirkpatrick, C. D. Gelatt, and M. P. Vecchi, "Optimization by simulated annealing," Science, vol. 220, no. 4598, pp. 671–680, 1983.
40. J. C. Gower and G. J. S. Ross, "Minimum spanning trees and single linkage cluster analysis," Applied Statistics, vol. 18, pp. 54–64, 1969.
41. D. R. Edla and P. K. Jana, "Minimum spanning tree based clustering using partitional approach," in Proceedings of the International Conference on Frontiers of Intelligent Computing: Theory and Applications (FICTA '13), pp. 237–244, Springer, Berlin, Germany, 2013.

SECTION 7:
PERMUTATION GROUPS

16

The Commuting Graph of the Symmetric Group S_n

Timothy Woodcock

Department of Mathematics, Stonehill College Easton, Massachusetts, USA 02357

ABSTRACT

In this paper we analyze the commuting graph of the full symmetric group on n elements, the graph being defined to have the nontrivial group elements as its vertex set, and an edge joining each commuting pair of vertices. We prove that if neither n nor n − 1 is a prime, then the graph has diameter 5; that is, the maximum-length shortest path, over all pairs of vertices, is 5. In the cases where n or n − 1 is a prime, we show that the graph is disconnected. Moreover, the components are completely identified, along with their diameters. This paper reproduces a number of results from the paper of Iranmanesh and Jafarzadeh, [4], but with some generalization and a new approach.

Keywords: Symmetric group, commuting graph, diameter, connected components

Citation: Timothy Woodcock "The commuting graph of the symmetric group S_n" International Journal of Contemporary Mathematical Sciences, Vol. 10, 2015, no. 6, 287-309. http://dx.doi.org/10.12988/ijcms.2015.4553

GROUNDWORK AND MAIN RESULT

For a positive integer n, let S_n denote the symmetric group of all bijective functions on {1, 2, . . . , n}, under composition. Let 1_n denote the identity, or trivial element of S_n. For each subset A of S_n, we define a graph Δ_A, having A \ {1_n} as its vertices, and with an edge joining each pair of elements of A \ {1_n} that commute under the product of S_n. We refer to Δ_A as the commuting graph of A.

For a general graph G = (V, E) with a finite number of vertices, and for v, w ∈ V , the distance from v to w in G is defined to be the minimum number of edges in a path joining these vertices. We denote this distance by $d_G(v, w)$. If no path exists, we let $d_G(v, w) = \infty$. The diameter of G is the maximum value of $d_G(v, w)$ over all v and w in V , where ∞ is understood to be greater than any natural number. For each nonempty subset W of V , we define the induced subgraph G_W of G to have vertex set W, and to include all edges of E that join pairs of vertices in W. If d_{GW} (v, w) is finite for all v, w ∈ W, then G_W is said to be connected. If W is a maximal subset of V such that G_W is connected, then GW is called a component of G. If G has multiple components, it is said to be disconnected.

In a commuting graph Δ_A, a path will be called a commuting path. But it shall be convenient for us to allow consecutive vertices within a such a path to be alike. Of course, this does not affect the distance between any pair of vertices, and thus the diameter of Δ_A is unchanged.

The statement of our main result makes use of the following additional notation. For n ≥ 2 and 2 ≤ m ≤ n, let C_n^m be the collection of all cycles of length m in S_n. Let R_n be the set of all nontrivial elements of S_n, but excluding $C_n^{n-1} \cup C_n^n$. For $\gamma \in S_n$, let $\langle\gamma\rangle$ denote the cyclic subgroup of S_n generated by γ.

Theorem 1.1. Suppose n ≥ 3.

a. If neither n nor n − 1 is a prime, then Δ_{Sn} is connected, of diameter 5.

b. If n − 1 is a prime, then Δ_{Sn} is disconnected, with components $\Delta_{R_n \cup C_n^n}$, and $\Delta_{\langle\gamma\rangle}$ *for* $\gamma \in C_n^{n-1}$. The diameter of $\Delta_{R_n \cup C_n^n}$ *is 1, 3, or 4* according as n is 3, 4, or > 4; the diameter of each $\Delta_{\langle\gamma\rangle}$ is 0 or 1, according as n is 3 or > 3.

c. If n is a prime but n − 1 is not, Δ_{Sn} is disconnected, with components $\Delta_{R_n \cup C_n^{n-1}}$, and $\Delta_{\langle\gamma\rangle}$ *for* $\gamma \in C_n^n$. The diameter of $\Delta_{R_n \cup C_n^{n-1}}$ *is 5*; the diameter of each $\Delta_{\langle\gamma\rangle}$ *is 1*.

Though this result was proved in the paper [4], we take a route to its realization that is substantially different. In particular, we shall prove that 5 is a lower bound for the diameter of Δ_{Sn} for all integers n such that n and n − 1 are composite. In [4, p.132], the minimal case of n = 9 is discussed.

COMMUTING ELEMENTS OF S_N

The results developed in the present section are well known, but organized for completeness.

Definition 2.1. Let π, ρ ∈ S_n. We say that π induces a cycle map on ρ provided that, for each cycle $(a_0 \cdots a_{i-1})$ of ρ, possibly trivial, $(\pi(a_0) \cdots \pi(a_{i-1}))$ remains a cycle of ρ.

Proposition 2.2. Let π and ρ be elements of S_n. Then π and ρ commute if and only if π induces a cycle map on ρ.

Proof. Let $(a_0\, a_1 \cdots a_{i-1})$ be an arbitrary cycle of ρ, and let j be an element of {0, 1, . . . , i − 1}. Assuming that π and ρ commute, we have

$$\rho(\pi(a_j)) = \pi(\rho(a_j)) = \pi\left(a_{(j+1)\bmod i}\right).$$

Hence $(\pi(a_0)\, \pi(a_1) \cdots \pi(a_{i-1}))$ is a cycle of ρ

Conversely if $(\pi(a_0)\, \pi(a_1) \cdots \pi(a_{i-1}))$ is a cycle of ρ, the

$$\rho(\pi(a_j)) = \pi\left(a_{(j+1)\bmod i}\right) = \pi(\rho(a_j)).$$

It follows that π and ρ commute, because a_j may be viewed as a general element of {1, 2, . . . , n

We remark that the statement of 2.2 is symmetric with respect to π and ρ. Thus, π induces a cycle map on ρ if and only if ρ induces a cycle map on

Proposition 2.3. Let π ∈ S_n, and let γ be a cycle of π. Then π and γ commut

Proof. Suppose that $\gamma = (a_0\, a_1 \cdots a_{i-1})$. We observe th

$$(\gamma(a_0)\, \gamma(a_1) \cdots \gamma(a_{i-1})) = (a_1\, a_2 \cdots a_0) = (a_0\, a_1 \cdots a_{i-1}).$$

Also, if $(b_0\, b_1 \cdots b_{k-1})$ is a cycle of π other than γ, then γ fixes each b_j , because γ fixes each element of {1, 2, . . . , n} \ $\{a_0, a_1, \ldots, a_{i-1}\}$. Hen

$$(\gamma(b_0)\, \gamma(b_1) \cdots \gamma(b_{k-1})) = (b_0\, b_1 \cdots b_{k-1}).$$

We conclude that γ induces a cycle map on π, and therefore γ commutes with π, by 2.

Proposition 2.4. Let π, $\rho \in S_n$ be commuting elements, and suppose that $\gamma = (a_0 \cdots a_{i-1})$ is a cycle of π, of unique length over all cycles of π. Then ρ acts on $\{a_0, \ldots, a_{i-1}\}$ as a power of

Proof. Since π and ρ are commuting elements, $(\rho(a_0)\ \rho(a_1) \cdots \rho(a_{i-1}))$ is a cycle of π, by 2.2. Therefore, $(\rho(a_0)\ \rho(a_1) \cdots \rho(a_{i-1})) = \gamma$, because of the uniqueness of the length of γ. Thus we have $\rho(a_0) = a_k$ for some k in $\{0, 1, \ldots, i-1\}$, and furthermore $\rho(a_j) = a_{(k+j)}$ mod i for $0 \leq j \leq i-1$. In other words, ρ acts on $\{a_0, a_1, \ldots, a_{i-1}\}$ in the same manner as γ^k

Definition 2.5. For $\pi \in S_n$, we define the fixed set of π to be the set of all a $\in \{1, 2, \ldots, n\}$ such that $\pi(a) = a$. We denote this set by F ix(π). We define the moved set of π, denoted Move(π), to be $\{1, 2, \ldots, n\} \setminus$ F ix(π).

Proposition 2.6. Suppose that π, $\rho \in S_n$ are commuting elements. Also assume that F ix(π) contains a unique element, a. Then a $\in$ F ix(ρ) as well.

Proof. Since a $\in$ F ix(π), a makes up a trivial cycle of π. Thus by 2.2, $\rho(a)$ constitutes a trivial cycle as well. But π has a unique trivial cycle, because |F ix(π)| = 1. Therefore, $\rho(a) = a$.

Definition 2.7. For π, $\rho \in S_n$, we say that π and ρ are disjoint provided that Move(π) $\cap$ Move(ρ) = $\emptyset$.

Proposition 2.8. If π, $\rho \in S_n$ are disjoint, they commute.

Proof. Since π and ρ are disjoint, π acts as the identity on Move(ρ); thus π permutes F ix(ρ). Likewise, ρ permutes F ix(π).

If a $\in$ F ix(π), $(\rho\pi)(a) = \rho(a)$, of course. And $(\pi\rho)(a) = \rho(a)$, because ρ permutes F ix(π). Hence, $(\rho\pi)(a) = (\pi\rho)(a)$. Similarly, this holds for each a $\in$ F ix(ρ). But, since π and ρ are disjoint, F ix(π) $\cup$ F ix(ρ) = $\{1, 2, \ldots, n\}$. Therefore π and ρ commute

Definition 2.9. Let H be a subgroup of Sn. For a $\in \{1, 2, \ldots, n\}$, we define its orbit under H to be $\{\pi(a) | \pi \in H\}$. We denote this set by $[a]_H$.

Proposition 2.10. For a subgroup H of Sn, the family of all $[a]_H$, $1 \leq a \leq n$, is a partition of $\{1, 2, \ldots, n\}$. Furthermore, a $\in [a]_H$ for each a.

Proof. Since H is a subgroup of S_n, H includes the identity element 1n of S_n. Therefore, for each a $\in \{1, 2, \ldots, n\}$, we have a $\in$ [a]H, because $1_n(a) = a$.

Now suppose that for given elements a, b $\in \{1, 2, \ldots, n\}$, $[a]_H \cap [b]_H$ is nonempty. Then there exist π, $\rho \in$ H such that $\pi(a) = \rho(b)$. Let c $\in$ [a]H, and assume that $\chi(a) = c$, where $\chi \in$ H. Then $(\chi\pi{-}1\rho)$ (b) = c. And, since H is a group, $\chi\pi^{-1}\rho \in {}_H$. Thus c $\in [b]_H$, so [a]H $\subseteq [b]_H$. The reverse containment holds by a parallel argument.

Proposition 2.11. Let H be a subgroup of S_n. If π is in the center of H, and a ∈ F ix(π), then $[a]_H \subseteq$ F ix(π).

Proof. Let b ∈ $[a]_H$, and assume that ρ(a) = b, where ρ ∈ H. Then we have

$$\pi(b) = \pi(\rho(a)) = \rho(\pi(a)) = \rho(a) = b.$$

Therefore, b ∈ F ix(π).

THE COMMUTING GRAPH Δ_{RN}

Theorem 3.1. Suppose n ≥ 4. Then Δ_{Rn} is connected, of diameter 3 or 4, according as n = 4 or n > 4.

The proof of the theorem is realized through a sequence of results. We remark that R_n is empty for n ∈ {1, 2, 3}.

Proposition 3.2. The diameter of Δ_{R4} is 3.

Proof. Let

$$T = \{(1\ 2), (1\ 3), (1\ 4), (2\ 3), (2\ 4), (3\ 4)\},$$
$$D = \{(1\ 2)(3\ 4), (1\ 3)(2\ 4), (1\ 4)(2\ 3)\}.$$

We observe that any element of S4 \ $\{1_4\}$ that is neither a cycle of length 3 nor a cycle of length 4 is a member of T ∪ D; in other words, R_4 = T ∪ D.

Given an arbitrary π ∈ R_4, we claim that π commutes with some ϕ ∈ D. If π ∈ D, we may take ϕ = π, since a group element commutes with itself. If π ∈ T, let us write π = (a b). Then by 2.3, π commutes with ϕ = (a b)(c d) ∈ D, where c and d are the elements of {1, 2, 3, 4} \ {a, b}, in arbitrary order.

Now let ρ ∈ R_4 as well, and assume that ψ ∈ D commutes with ρ. We observe that each individual element of D induces a cycle map on all elements of D; so by 2.2, the elements of D mutually commute. Therefore ϕ and ψ commute, in particular, and so (π, ϕ, ψ, ρ) is a commuting path in Δ_{R4}. Hence $d\Delta_{R4}$ (π, ρ) ≤ 3. We conclude that Δ_{R4} has diameter ≤ 3.

To see that the diameter is ≥ 3, consider the pair σ = (1 2) and τ = (1 3). Suppose that χ ∈ S_4 commutes σ and τ . Then by 2.2, χ induces a cycle map on σ, and on τ . Thus χ({1, 2}) = {1, 2}, and χ({1, 3}) = {1, 3}. Therefore χ(1) = 1, and moreover, {1, 2, 3} ⊆ F ix(χ). This obviously implies that χ = 1^4. Hence there is no commuting path (σ, χ, τ) in ΔS4\{14}; thus there is none in Δ_{R4} . It follows that $d_{\Delta_{R_4}}$ (σ, τ) ≥ 3.

Lemma 3.3. Suppose that n > 4, and let π ∈ R_n. Then π is a product of two disjoint cycles of length n/2 each, or π commutes with a nontrivial cycle of length $< n/2$. Either way, π commutes with a nontrivial cycle of length $\leq n/2$.

Proof. First assume that π is itself a cycle. Then, since R_n includes no cycles of length $n - 1$ or n, by definition, π fixes at least two elements, say a, $b \in \{1, 2, \ldots, n\}$. We observe that π commutes with (a b), by 2.8; and the length of (a b), of course 2, is $< n/2$ because $n > 4$.

Next assume that π is neither a cycle, nor a product of two disjoint cycles of length $n/2$ each. Let γ be a cycle of minimum length over all nontrivial cycles of π. Then the length of γ is $< n/2$, obviously, and γ commutes with π, by 2.3.

Finally, if the cycle decomposition of π consists of two cycles, each of length $n/2$, then in view of 2.3, we may say that π commutes with a nontrivial cycle of length $\leq n/2$.

Proposition 3.4. For $n > 4$, the diameter of Δ_{Rn} is ≤ 4.

Proof. Let $\pi, \rho \in R_n$. First suppose that at least one of π and ρ, say π, commutes with a cycle $\gamma \in R_n$ of length $< n/2$. We observe that ρ commutes with a cycle $\delta \in R_n$ of length $\leq n/2$, by 3.3. If γ and δ are disjoint, on the one hand, then these cycles commute by 2.8. Thus $(\pi, \gamma, \delta, \rho)$ is a commuting path in Δ_{Rn}, and so $d_{\Delta_{R_n}}(\pi, \rho) \leq 3$. On the other hand if Move(γ) $\cap$ Move(δ) $6 \neq \emptyset$, then we have

$$|Move(\gamma) \cup Move(\delta)| \; < \; |Move(\gamma)| + |Move(\delta)| \; < \; 2(n/2).$$

Therefore, $|\text{Move}(\gamma) \cup \text{Move}(\delta)| \leq n - 2$; hence F ix($\gamma$) $\cap$ F ix(δ) contains at least two elements, say a and b. Define $\sigma = (a\ b)$, an element of R_n. We observe that σ commutes with γ and δ by 2.8. Furthermore, $(\pi, \gamma, \sigma, \delta, \rho)$ is a commuting path in Δ_{Rn}. Thus, $d_{\Delta_{R_n}}(\pi, \rho) \leq 4$.

Now assume that neither π nor ρ commutes with a cycle of length $<$ n/2. Then π is a product of two disjoint cycles, each of length n/2, by 3.3. Likewise for ρ. Let us write

$$\pi = (a_1\ a_2\ \cdots\ a_{n/2})\,(b_1\ b_2\ \cdots\ b_{n/2})\,,\ \rho = (c_1\ c_2\ \cdots\ c_{n/2})\,(d_1\ d_2\ \cdots\ d_{n/2})\,.$$

Since Move(π) = Move(ρ) = $\{1, 2, \ldots, n\}$, we may take $a_1 = c_1$. We note that $a_1 \notin \{b_1, b_2, \ldots, b_{n/2}\}$, because the cycles of π are disjoint, and $c_1 \notin \{d_1, d_2, \ldots, d_{n/2}\}$, by the same token. Hence, since $a_1 = c_1$, we realize that

$$\{b_1, b_2, \ldots, b_{n/2}\} \cup \{d_1, d_2, \ldots, d_{n/2}\} \subsetneq \{1, 2, \ldots, n\}.$$

Therefore, $\{b_1, b_2, \ldots, b_{n/2}\} \cap \{d_1, d_2, \ldots, d_{n/2}\}$ 6 $\neq \emptyset$; let us assume that $b_1 = d_1$. Define

$$\varphi = (a_1\ b_1)(a_2\ b_2)\cdots(a_{n/2}\ b_{n/2}),$$
$$\tau = (a_1\ b_1) = (c_1\ d_1),$$
$$\psi = (c_1\ d_1)(c_2\ d_2)\cdots(c_{n/2}\ d_{n/2}).$$

Then $\phi, \tau, \psi \in R_n$, clearly. We claim, furthermore, that $(\pi, \phi, \tau, \psi, \rho)$ is a commuting path in Δ_{Rn}. We observe

$$(\varphi(a_1)\ \varphi(a_2)\ \cdots\ \varphi(a_{n/2})) = (b_1\ b_2\ \cdots\ b_{n/2}),$$
$$(\varphi(b_1)\ \varphi(b_2)\ \cdots\ \varphi(b_{n/2})) = (a_1\ a_2\ \cdots\ a_{n/2}).$$

Thus, ϕ induces a cycle map on π, and hence ϕ and π are commuting elements by 2.2. Similarly, ψ and ρ commute. And τ, being a cycle of both ϕ and ψ, commutes with each of these elements by 2.3. Therefore we have our claim. It follows that $d_{\Delta_{R_n}}(\pi, \rho) \leq 4$,

Proposition 3.5. Suppose $n > 4$. Let m be the odd element of $\{n - 1, n\}$, and define

$$\pi = (1\ 2\ \cdots\ m-2)(m-1\ m) \in R_n, \quad \rho = (2\ 3\ \cdots\ m-1)(1\ m) \in R_n.$$

Then $d_{\Delta_{S_n}}(\pi, \rho) \geq 4.$

Proof. Suppose that $d_{\Delta_{R_n}}(\pi, \rho) \leq 3$. Let (π, ϕ, ψ, ρ) be a commuting path in Δ_{Sn}. We observe that $m \geq 5$; hence the cycles of π, including a trivial one if n is even, have distinct lengths. Therefore by 2.4, there exist positive integers r and s such that

$$\varphi = (1\ 2\ \cdots\ m-2)^r(m-1\ m)^s.$$

Similarly, there exist $t, u \in \mathbb{N}$ such that

$$\psi = (2\ 3\ \cdots\ m-1)^t(1\ m)^u.$$

We also note that ϕ and ψ are nontrivial, being elements of $S_n \backslash \{1_n\}$. Moreover they are commuting elements, by our assumption that they appear consecutively in a commuting path.

Since $m - 2$ is odd, the cycle decomposition of $(1\ 2 \cdots m - 2)r$ does not contain a transposition. Therefore $(m-1\ m)$ is the only potential transposition among the cycles of ϕ. So, if $(m-1\ m)$ is a cycle of ϕ, then $(\psi(m-1)\ \psi(m)) = (m - 1\ m)$ by 2.2. But clearly $\psi(m - 1)$ 6 $\neq$ m. Thus $m - 1$ and m are fixed by ψ. However, this implies that ψ is the identity on all of $\{1, 2, \ldots, n\}$, contradicting that ψ is nontrivial. Hence $(m - 1\ m)$ must not be a cycle of

ϕ, and we therefore conclude that m − 1 and m are fixed by ϕ. By a parallel argument, ψ fixes the elements 1 and m. Thus

$$\varphi = (1\ 2\ \cdots\ m-2)',\quad \psi = (2\ 3\ \cdots\ m-1)'.$$

Now on the one hand, we observe that $\phi(1) \in \{2, 3, \ldots, m-2\}$, because ϕ is nontrivial. On the other hand, since ϕ and ψ commute, and ψ −1 fixes 1, we have $\phi(1) = (\psi\phi\psi^{-1})(1) = \psi(\phi(1))$. Therefore ψ fixes an element of $\{2, 3, \ldots, m-2\}$. But then ψ = id, a contradiction.

We may now give an argument for Theorem 3.1

Proof. We obtain the result by combining 3.2, 3.4, and 3.5. We note that 3.5 implies that π and ρ are at distance ≥ 4 in Δ_{Rn} , a subgraph of Δ_{Sn}.

THE COMMUTING GRAPH $\Delta_{R_n \cup C_n^n}$ FOR EVEN N

We shall prove that if n is even, the result of Theorem 3.1 applies to the larger graph $\Delta_{R_n \cup C_n^n}$

Theorem 4.1. Suppose that n is even, n ≥ 4. Then $\Delta_{R_n \cup C_n^n}$ is connected. Moreover the diameter of the commuting graph is 3 or 4, according as n = 4 or n > 4.

Once again, the theorem is realized through several propositions.

Proposition 4.2. The diameter of $\Delta_{R_4 \cup C_4^4}$ is 3.

Proof. From the proof of 3.2, we recall that D denotes the set of all double transpositions in S_4. In the proof we argued that ΔR4 has diameter ≤ 3 through two observations. In particular, the elements of D mutually commute, and each element of R4 commutes with an element of D. The second of these observations extends to include C_4^4. Indeed, given γ = (a b c d) C_4^4, we observe that γ commutes with γ 2 = (a c)(b d) ∈ D. Thus, as in the proof of 3.2, we conclude that $\Delta_{R_4 \cup C_4^4}$ has diameter ≤ 3.

Now, in demonstrating that the inequality of 3.2 is sharp, we defined σ = (1 2) and τ = (2 3), elements of R4, and argued that $d_{\Delta_{S_4}}(\sigma, \tau) \geq 3$. Thus the distance between σ and τ is ≥ 3 in $\Delta_{R_4 \cup C_4^4}$. We conclude that $\Delta_{R_4 \cup C_4^4}$ has diameter ≥ 3.

Lemma 4.3. Suppose that n ≥ 4. Let $\pi \in R_n$ be a product of i ≥ 2 disjoint nontrivial cycles of a common length. Let $\gamma \in R_n$ be a nontrivial cycle of length j, where j ≤ i. Then there exists an element $\rho \in R_n$ that commutes with π and γ.

Proof. We consider two possibilities. First assume that there exists a nontrivial cycle δ in the decomposition of π that is disjoint from γ. Then π and δ commute, by 2.3, and γ and δ commute as well, by 2.8. We observe that δ is an element of R_n, because its length is at most $n/i \leq n/2 \leq n-2$. Thus δ may serve as ρ.

Now let us assume that for each nontrivial cycle δ in the decomposition of π, Move(γ) ∩ Move(δ) is nonempty. Then the number of nontrivial cycles of π does not exceed the length of γ. In other words $i \leq j$, and hence $i = j$, because the reverse inequality is being assumed. Let m denote the common length of the cycles of π, and suppose that in decomposed form we have

$$\pi = \prod_{k=0}^{i-1} (a_{k,0}\ a_{k,1}\ a_{k,2}\ \cdots\ a_{k,m-1}).$$

Also suppose that $\gamma = (b_0\ b_1\ b_2 \cdots b_{i-1})$, and with no loss in generality, take $a_{k,0} = b_k$ for $0 \leq k < i$. Define

$$\rho = \prod_{l=0}^{m-1} (a_{0,l}\ a_{1,l}\ a_{2,l}\ \cdots\ a_{i-1,l}).$$

We observe that ρ is nontrivial, because $i \geq 2$, and not itself a cycle, because $m \geq 2$. Hence $\rho \in R_n$. Also, ρ commutes with $(a_{0,0}\ a_{1,0}\ a_{2,0} \cdots a_{i-1,0})$, one of its cycles, by 2.3. Thus ρ commutes with γ. Furthermore, for each $k \in \{0, 1, \ldots, i-1\}$, and each $l \in \{0, 1, \ldots, m-1\}$,

$$(\rho\pi)(a_{k,l}) = \rho\left(a_{k,(l+1) \bmod m}\right) = a_{(k+1) \bmod i,\,(l+1) \bmod m},$$
$$(\pi\rho)(a_{k,l}) = \pi\left(a_{(k+1) \bmod i,\,l}\right) = a_{(k+1) \bmod i,\,(l+1) \bmod m}.$$

Hence ρ commutes with π as well, because both of π and ρ fix each element of the s

$$\{1, 2, \ldots, n\} \setminus \{a_{k,l} \,|\, 0 \leq k < i,\ 0 \leq l < m\}.$$

Proposition 4.4. Suppose that n is even, $n > 4$. Let $\pi \in R_n$, and $\gamma \in C_n^n$. Then the distance between π and γ in $\Delta_{R_n \cup C_n^n}$ *is* ≤ 4

Proof. Assume that $\gamma = (a_1\ a_2 \cdots a_n)$. Since n is even, we ha

$$\gamma^{n/2} = (a_1\ a_{n/2+1})(a_2\ a_{n/2+2})\ \cdots\ (a_{n/2}\ a_n) \in R_n.$$

We observe that γ commutes with $\gamma^{n/2}$, because a group element commutes with each of its powers. By 3.3, there exists a nontrivial cycle $\delta \in R_n$ of length $\leq n/2$ that commutes with π, because $\pi \in R_n$. Furthermore by 4.3, there exists

$\rho \in R_n$ commuting with γ n/2 and δ. Hence $(\pi, \delta, \rho, \gamma^{n/2}, \gamma)$ a commuting path in $\Delta_{R_n \cup C_n^n}$. Therefore we have the proposition

Lemma 4.5. Suppose that n is even, and let $\pi \in S_n$ be a product of n/2 disjoint transpositions. Then the order of the centralizer of π in S_n is given

$$(n/2)! \cdot 2^{n/2} = (2)(4)(6)(8)\cdots(n).$$

Proof. In view of 2.2, we must show that the number of elements of S_n that induce a cycle map on π is $(n/2)!\cdot 2^{n/2}$. Suppose that we have the decomposition

$$\pi = (a_1\ b_1)(a_2\ b_2)\cdots(a_{n/2}\ b_{n/2}).$$

Consider a general element of S_n, written in table form as follow

$$\rho = \begin{pmatrix} a_1 & b_1 & a_2 & b_2 & \cdots & a_{n/2} & b_{n/2} \\ x_1 & y_1 & x_2 & y_2 & \cdots & x_{n/2} & y_{n/2} \end{pmatrix}$$

We observe that ρ induces a cycle map on π if and only if

$$\{\{x_1, y_1\}, \{x_2, y_2\}, \ldots \{x_{n/2}, y_{n/2}\}\} = \{\{a_1, b_1\}, \{a_2, b_2\}, \ldots \{a_{n/2}, b_{n/2}\}\}.$$

To produce a table ρ that satisfies this condition, we may first arrange the sets $\{a_i, b_i\}$, $1 \le i \le {}_{n/2}$, into a sequence, then arbitrarily order the pair of elements within each set. The number of ways to complete this process is

$$(n/2)! \cdot 2^{n/2} = (1\cdot 2)(2\cdot 2)(3\cdot 2)\cdots((n/2)\cdot 2) = (2)(4)(6)\cdots(n).$$

Proposition 4.6. Suppose that n is even, n > 4. For $\gamma, \delta \in C_n^n$, the distance between γ and δ in $\Delta_{R_n \cup C_n^n}$ *is* ≤ 4.

Proof. Let $\pi = \gamma^{n/2}$, and $\rho = \delta^{n/2}$. We observe that γ commutes with π, because a group element commutes with each of its powers. We also note that π is a product of n/2 disjoint transpositions, and in particular, $\pi \in R_n$. Likewise, δ commutes with $\rho \in R_n$, a product of n/2 disjoint transpositions. Let H and K be the centralizers of π and ρ in S_n, respectively. By a well-known result of finite group theory, we have

$$|HK| = \frac{|H|\cdot|K|}{|H \cap K|}. \tag{4.1}$$

(Refer to [3, p.39].) It is also well known that $|S_n| = n!$. Therefore $|HK| \le n!$, because $HK \subseteq S_n$. But by 4.5,

$$|H|\cdot|K| = [(2)(4)(6)\cdots(n)]^2 > n!.$$

Hence $|H \cap K| > 1$. Let $\phi \in (H \cap K) \setminus \{id\}$, and note that ϕ commutes with π and ρ. Suppose that $\phi \in C_n^{n-1}$. Then $Fix(\phi)$ contains precisely one element, say a. So by 2.6, $a \in Fix(\pi) \cap Fix(\rho)$. But $Fix(\pi) = Fix(\rho) = \emptyset$, obviously. Thus $\varphi \notin C_n^{n-1}$, and so $\varphi \in R_n \cup C_n^n$. We now realize that $(\gamma, \pi, \phi, \rho, \delta)$ is a commuting path in $\Delta_{R_n \cup C_n^n}$. The proposition follows.

We may now provide an argument for Theorem 4.1

Proof. The assertion for n = 4 is handled in Proposition 4.2. For n > 4 and n even, we combine the results of 3.4, 3.5, 4.4, and 4.6.

We conclude the section by developing a second argument for Proposition 4.6, one which is more enlightening but also more technical. We shall illustrate the construction of a particular element ϕ, based on $\pi = \gamma^{n/2}$ and $\rho = \delta^{n/2}$. Especially noteworthy is that the element ϕ, like π and ρ, will be a product of n/2 disjoint transpositions.

Lemma 4.7. Suppose that n is even, $n \geq 4$. Let $\pi, \rho \in S_n$. Furthermore assume that each of π and ρ is a product of n/2 disjoint transpositions. Then for all nonnegative integers j,

$$Fix\left[\pi(\rho\pi)^j\right] = Fix\left[\rho(\pi\rho)^j\right] = \emptyset. \tag{4.2}$$

Proof. We proceed by induction on j. Clearly we have $Fix(\pi) = Fix(\rho) = \emptyset$, thus (4.2) holds for j = 0. Let j be a nonnegative integer, and inductively assume that (4.2) holds for this particular j. However, suppose that $Fix[\pi(\rho\pi)^{j+1}] 6\neq \emptyset$. Then there exists an element $a \in \{1, 2, \ldots, n\}$ such that $[\pi(\rho\pi)^{j+1}](a) = a$. We note that $\pi -1 = \pi$, because π is a product of disjoint transpositions. Therefore $[\rho(\pi\rho)^j](\pi(a)) = \pi(a)$, and hence $Fix[\rho(\pi\rho)^j]$ is nonempty. This is a contradiction. Thus $Fix[\pi(\rho\pi)^{j+1}] = \emptyset$. And by a parallel argument, $Fix[\rho(\pi\rho)^{j+1}] = \emptyset$. Hence we have the lemma, by induction.

Lemma 4.8. Let n, π, and ρ be as in Lemma 4.7. Let H be the subgroup of S_n generated by π and ρ, and let $A \subseteq \{1, 2, \ldots, n\}$ be an orbit under the natural action of H on $\{1, 2, \ldots, n\}$. Then the elements of A may be arranged into a sequence $(x_0, x_1, \ldots, x_{2k-1})$ such that $\pi(x_{2j}) = x2j+1$ and $\rho(_{x2j+1}) = x_{(2j+2) \bmod 2k}$ for $0 \leq j < k$. In particular, A has even order.

Proof. Let x0 be an arbitrary element of A. For each integer $j \in \mathbb{Z}$, we define

$$x_{2j} = (\rho\pi)^j(x_0), \quad x_{2j+1} = \left[\pi(\rho\pi)^j\right](x_0).$$

Then we have $x_{2j+1} = \pi(x_{2j})$, and $x_{2j+2} = (\rho\pi)(x_{2j}) = \rho_{(x2j+1)}$. We note that each xi is an element of $\{1, 2, \ldots, n\}$. Thus there exists a repeated value in the sequence (x_0, x_1, x_2, . . .). Let m be the minimum positive index such that x_m is an element of {x_0, x_1, . . . , x_{m-1}}. In particular suppose that $x_m = x_l$, where $l \in \{0, 1, \ldots, m-1\}$.

We claim that m and l have the same parity. If l is even, say l = 2i, then for an arbitrary nonnegative integer j,

$$x_{l+2j+1} = \left[\pi(\rho\pi)^{i+j}\right](x_0) = \left[\pi(\rho\pi)^j(\rho\pi)^i\right](x_0) = \left[\pi(\rho\pi)^j\right](x_l).$$

And if l = 2i + 1,

$$x_{l+2j+1} = (\rho\pi)^{i+j+1}(x_0) = \left[\rho(\pi\rho)^j\pi(\rho\pi)^i\right](x_0) = \left[\rho(\pi\rho)^j\right](x_l).$$

But regardless of the parity of l, we see that $x_{l+2j+1} \neq x_l$, by 4.7. Thus we have our claim, because $x_m = x_l$. Assume that m = l + 2k, where k is a positive integer. Since m − 1 and l − 1 have the same parity, $x_m = \pi(x_{m-1})$ and $x_l = \pi(x_{l-1})$, or xm = $\rho(x_{m-1})$ and xl = $\rho(x_{l-1})$. Whichever the case, $x_{m-1} = x_{l-1}$, because π and ρ are injective. It follows that l = 0, because of the minimum condition that we imposed on m. Hence m = 2k.

Now for an arbitrary integer j, we have

$$x_{2j+2k} = (\rho\pi)^{j+k}(x_0) = (\rho\pi)^j(x_{2k}) = (\rho\pi)^j(x_0) = x_{2j}.$$

Therefore,

$$x_{(2j+1)+2k} = x_{2(j+k)+1} = \pi(x_{2j+2k}) = \pi(x_{2j}) = x_{2j+1}.$$

Hence the function $i \mapsto x_i$, defined on $\mathbb{Z}$, has period 2k

We observe that A = {ϕ(x_0)| ϕ ∈ H}. Thus {x_i | i ∈ $\mathbb{Z}$} ⊆ A. We complete the proof by demonstrating the reverse inclusion. Let ϕ be an element of H. Then for some nonnegative integer t, there exist elements $\psi_s \in \{\pi, \pi^{-1}, \rho, \rho^{-1}\}$, $1 \leq s \leq t$, such that $\phi = \psi_1\psi_2\psi_3 \cdots \psi_t$. (See [2, p.62].) However, $\pi^{-1} = \pi$, or equivalently π 2 = id, because π is a product of disjoint transpositions. Likewise, $\rho^{-1} = \rho$. Thus by canceling successive factors of $\psi_1\psi_2\psi_3 \cdots \psi_t$ as long as possible, we obtain

$$\varphi \in \left\{(\rho\pi)^j,\ \pi(\rho\pi)^j,\ (\pi\rho)^j,\ \rho(\pi\rho)^j\right\},$$

for some nonnegative integer j. We have $(\rho\pi)^j(x_0) = x_{2j}$, and $[\pi(\rho\pi)^j](x_0) = x_{2j+1}$. Furthermore, we observe

$$(\pi\rho)^j(x_0) = \left[(\rho^{-1}\pi^{-1})^{-j}\right](x_0) = \left[(\rho\pi)^{-j}\right](x_0) = x_{-2j}.$$

$$\left[\rho(\pi\rho)^j\right](x_0) = \left[\pi(\pi\rho)^{j+1}\right](x_0) = \pi(x_{-2(j+1)}) = x_{-2(j+1)+1}.$$

Therefore $\phi(x_0) \in \{x_i \mid i \in \mathbb{Z}\}$. We conclude that $A \subseteq \{x_i \mid i \in \mathbb{Z}\}$, as desired

Proposition 4.9. Let n, π, and ρ be as in 4.7. Then there exists $\phi \in S_n$, also a product of n/2 disjoint transpositions, commuting with π and ρ

Proof. Let H be the subgroup of S_n generated by π and ρ. Let A be an arbitrary orbit under the natural action of the subgroup H on $\{1, 2, \ldots, n\}$. Then each of the elements π and ρ maps A to itself, bijectively. Let π_A and ρ_A denote the restrictions of π and ρ to A, respectively. Let $(x_0, x_1, x_2, \ldots, x_{2k-1})$ be an arrangement of the elements of A, as in 4.8. For $0 \leq i < k$, define

$$\varphi_A(x_i) = x_{(2k-1)-i}, \quad \varphi_A(x_{(2k-1)-i}) = x_i.$$

Since π is a product of disjoint transpositions, and $\pi(x_{2j}) = x_{2j+1}$ for $0 \leq j < k$, by design, we see that

$$\pi_A = (x_0\ x_1)(x_2\ x_3)(x_4\ x_5)\ \cdots\ (x_{2k-2}\ x_{2k-1}).$$

Applying ϕ_A to the elements x_i within the respective transpositions here, we reverse the sequence of indices and obtain the product

$$(x_{2k-1}\ x_{2k-2})(x_{2k-3}\ x_{2k-4})(x_{2k-5}\ x_{2k-6})\ \cdots\ (x_1\ x_0) = \pi_A.$$

Hence ϕA induces a cycle map on π_A. Furthermore, defining $\varphi = \prod_A \varphi_A$, we realize that ϕ induces a cycle map on $\prod_A \pi_A = \pi$. Therefore ϕ and π are commuting elements, by 2.2.

Now, since ρ is a product of disjoint transpositions, and furthermore, $\rho(_{x2j+1}) = x_{(2j+2) \bmod 2k}$, we have

$$\rho_A = (x_1\ x_2)(x_3\ x_4)(x_5\ x_6)\ \cdots\ (x_{2k-3}\ x_{2k-2})(x_{2k-1}\ x_0).$$

Applying ϕ_A to the respective x_i here yields

$$(x_{2k-2}\ x_{2k-3})(x_{2k-4}\ x_{2k-5})(x_{2k-6}\ x_{2k-7})\ \cdots\ (x_2\ x_1)(x_0\ x_{2k-1}) = \rho_A.$$

Therefore, we see that ϕ induces a cycle map on ρ, so ϕ and ρ are commuting elements, by 2.2. And ϕ is obviously a product of disjoint transpositions, fixing no element of $\{1, 2, \ldots, n\}$. Thus the proof is complete.

We have now realized our goal of a more constructive route to Proposition 4.6. But furthermore, Proposition 4.9 yields a new proof of a result from [1, p.139].

Theorem 4.10. Suppose that n is even, $n > 4$. Let X be the set of all products of n/2 disjoint transpositions in S_n. Then the commuting graph Δ_X has diameter 2.

Proof. The diameter of Δ_X is ≤ 2 by 4.9. Define

$$\pi = (1\ 2)(3\ 4)(5\ 6)\ \cdots\ (n-1\ n),$$
$$\rho = (2\ 3)(4\ 5)(6\ 7)\ \cdots\ (n-2\ n-1)(1\ n),$$

a particular pair of elements of X. We observe that (ρπ)(1) = ρ(2) = 3, while (πρ)(1) = π(n) = n − 1. Therefore (ρπ)(1) 6≠ (πρ)(1), because n > 4. Hence π and ρ are not commuting elements. We conclude that the diameter of Δ_X is > 1.

We remark that in the terminology of [1], Δ_X is referred to as a commuting involution graph.

AN UPPER BOUND ON THE DIAMETER OF Δ_{SN} FOR COMPOSITE N AND N − 1

Theorem 5.1. Suppose that each of n and n−1 is a composite number. Then the diameter Δ_{Sn} is ≤ 5.

We remark that n ≥ 9 here, implicitly. We obtain the theorem through two propositions, that shall accompany Theorem 3.1.

Proposition 5.2. Suppose that n > 4, and l ∈ {n − 1, n} is a composite number. Let π ∈ R_n and $\gamma \in C_n^l$. Then the distance between π and γ in $\Delta_{R_n \cup C_n^l}$ is ≤ 5.

Proof. By 3.3, there exists a cycle δ ∈ R_n of length ≤ n/2 that commutes with π. We have |F ix(δ)| ≥ n/2, thus |F ix(δ)| ≥ 3 because n > 4. Choose a, b ∈ F ix(δ), and define τ = (a b) ∈ R_n. Then τ is disjoint from δ; hence τ and δ are commuting elements, by 2.8.

Now assume that $\gamma = (c_1\ c_2\ c_3\ \cdots\ c_l)$. Since l is composite, there exists a positive integer i ∈ (1, l) that divides l. We observe that γ commutes with γ^i, because a group element commutes with each of its powers. In decomposed form, we have

$$\gamma^i = \prod_{j=1}^{i} \left(a_j\ a_{j+i}\ a_{j+2i}\ \cdots\ a_{j+[(l/i)-1]i}\right).$$

In particular, γ^i is a product of i ≥ 2 disjoint cycles, each of length l/i ≥ 2. Hence γ^i ∈ R_n. Moreover since τ is a cycle of length ≤ i, there exists an element ρ ∈ R_n commuting with γ i and τ , by 4.3. Therefore, (π, δ, τ, ρ, γ^i, γ) is a commuting path in $\Delta_{R_n \cup C_n^l}$. The proposition follows

Proposition 5.3. Let l and m be elements of the set {n − 1, n}. Assume that l ≤ m, and that each of l and m is a composite number. Let $\gamma \in C_n^l$, and $\delta \in C_n^m$. Then the distance between γ and δ in $\Delta_{R_n \cup C_n^l \cup C_n^m}$ *is* ≤ 5.

Proof. Let $D_l \subseteq \{2, 3, \ldots, l-1\}$ be the set of all proper nontrivial divisors of l. Since l is composite, D_l is nonemtpy. Let $i = \max(D_l)$. Since $i \in D_l$, we have $l/i \in D_l$ as well. Therefore $l/i \leq i$, because i is maximal; so $\sqrt{l} \leq i$. Analogously, we let $D_m \subseteq \{2, 3, \ldots, m-1\}$ be the collection of all proper nontrivial divisors of m, nonempty because m is composite, and we let $j = \max(D_m)$. We then note that $m/j \in D_m$, and deduce that $\sqrt{m} \leq j$. Moreover we have $\sqrt{l} \leq j$, because $l \leq m$. Hence $l \leq ij$, and so $l/i \leq j$

Now assume tha

$$\gamma = (a_1\ a_2\ a_3\ \cdots\ a_l), \quad \delta = (b_1\ b_2\ b_3\ \cdots\ b_m).$$

Then we have, in decomposed form,

$$\gamma^i = \prod_{k=1}^{i} \left(a_k\ a_{k+i}\ a_{k+2i}\ \cdots\ a_{k+[(l/i)-1]i}\right);$$

$$\delta^j = \prod_{k=1}^{j} \left(b_k\ b_{k+j}\ b_{k+2j}\ \cdots\ b_{k+[(m/j)-1]j}\right).$$

We observe that γ^i consists of i nontrivial cycles, each of length l/i, and δ^j consists of j nontrivial cycles, each of length m/j. So obviously, $\gamma^i, \delta^j \in R_n$. Let σ be any nontrivial cycle in the decomposition of γ^i. Then $\sigma \in R_n$, because γ^i has multiple nontrivial cycles. By 2.3, σ commutes with γ^i. Furthermore since $l/i \leq j$, there exists an element $\rho \in R_n$ that commutes with σ and δ^j, by 4.3. Hence $(\gamma, \gamma^i, \sigma, \rho, \delta^j, \delta)$ is a commuting path in $\Delta_{R_n \cup C_n^l \cup C_n^m}$, because γ and δ commute with γ^i and δ^j, respectively. Thus we have the proposition

We finish the section with an argument for Theorem 5.

Proof. As noted earlier, we have $n \geq 9$, because n and n − 1 are composite numbers. The theorem is realized immediately by combining the results of Theorem 3.1, and Propositions 5.2 and 5.

THE EXISTENCE OF ELEMENTS AT DISTANCE 5 IN Δ_{SN}

We exhibit two pairs of elements at distance ≥ 5 in the commuting graph Δ_{S_n}. In each of our constructions, we shall require the following standard result.

Lemma 6.1. Let G be a group. Suppose that $g \in G$ has finite order i. Then for all positive integers j, $\langle g^j \rangle = \langle g^{\gcd(i,j)} \rangle$.

Proof. Assume that j = k · gcd(i, j), where k is a positive integer that is relatively prime to i. On the one hand, we observe that $\left(\gamma^{\gcd(i,j)}\right)^k = \gamma^j$. Thus $\gamma^j \in \left\langle \gamma^{\gcd(i,j)} \right\rangle$, and so $\langle \gamma^j \rangle \subseteq \left\langle \gamma^{\gcd(i,j)} \right\rangle$. On the other hand, by a well-known result of number theory, there exist integers x and y such that ix + ky = gcd(i, k) = 1. (See [3, p.11].) Therefore, we have

$$(\gamma^j)^y = \left(\gamma^{\gcd(i,j)}\right)^{ky} = \left(\gamma^{\gcd(i,j)}\right)^{1-ix} = \gamma^{\gcd(i,j)}\left(\gamma^i\right)^{-(x \cdot \gcd(i,j))} = \gamma^{\gcd(i,j)}.$$

Hence $\gamma^{\gcd(i,j)} \in \langle \gamma^j \rangle$, and thus $\left\langle \gamma^{\gcd(i,j)} \right\rangle \subseteq \langle \gamma^j \rangle$.

Proposition 6.2. Suppose that n is a positive integer, $n \geq 3$. Let

$$\gamma = (1\,2\,3\,\cdots\,n-1) \in C_n^{n-1}, \quad \delta = (1\,2\,3\,\cdots\,n) \in C_n^n.$$

Then the distance between γ and δ in Δ_{S_n} is at least 5.

Proof. We note that γ and δ are in fact elements of $S_n \setminus \{1_n\}$, because $n \geq 3$. Suppose that $d_{\Delta_{S_n}}(\gamma, \delta) \leq 4$. In particular, assume that (γ, ϕ, χ, ψ, δ) is a commuting path in Δ_{S_n}. Since ϕ and ψ commute with γ and δ, respectively, there exist positive integers s and t such that $\phi = \gamma^s$ and $\psi = \delta^t$, by 2.4. Let u = gcd(s, n − 1) and v = gcd(t, n); and note that u and v are proper divisors of n − 1 and n, respectively, because ϕ and ψ are nontrivial elements. Let $\pi = \gamma^u$, $\rho = \delta^v$, and H = hπ, ρi. By 6.1, we have <ϕ> = <π> and <ψ> = <ρ>. Hence each subgroup of S_n that contains ϕ and ψ will also contain π and ρ, and vice-versa. Therefore, H = <ϕ, ψ>.

We observe that precisely one of the integers n − 1 and n is divisible by 2; so u + v < (n − 1)/2 + n/2. Thus u + v, itself an integer, must be ≤ n − 1. Let m = min(u, v), and let a ∈ {n − m, n − m + 1, n − m + 2, . . . , n − 1}. We observe that n−1 is strictly less than a+u and a+v, but a+u+v ≤ 2(n−1). Therefore,

$$\begin{aligned} (\rho\pi)(a) &= \rho[a+u-(n-1)] = a+u+v-(n-1), \\ (\pi\rho)(a+1) &= \pi[(a+1)+v-n] = \pi[a+v-(n-1)] \\ &= a+u+v-(n-1). \end{aligned}$$

Thus (ρπ)(a) = (πρ)(a+1), and so $(\rho^{-1}\pi^{-1}\rho\pi)(a) = a+1$. Hence $a+1 \in [a]_H$, because $\rho^{-1}\pi^{-1}\rho\pi \in H$. Moreover $[a]H = [a+1]_H$, by 2.10. Therefore, we conclude that

$$[n-m]_H = [n-m+1]_H = \cdots = [n-1]_H = [n]_H. \tag{6.1}$$

Suppose that b ∈ {1, 2, . . . , n − m − 1}. We observe that i = 0 is a solution to b + im < n − m; thus there exists a maximum nonnegative integer i for which the inequality holds. For this i, we have

$$b + (i+1)m \in \{n-m, n-m+1, n-m+2, \ldots, n-1\}.$$

We observe that if m = u, then $\pi^{i+1}(b) = b + (i+1)m$. And if m = v, then $\rho^{i+1}(b) = b + (i+1)m$. Either way, we have [b]H = [b + (i + 1)m]H, because each of π^{i+1} and ρ^{i+1} is an element of H. Together with (6.1), this implies that

$$[1]_H = [2]_H = \cdots = [n-1]_H = [n]_H.$$

Hence $[n]_H = \{1, 2, \ldots, n\}$.

Define $K = \langle \varphi, \chi, \psi \rangle$. Then K may be explicitly described as the set of all products of the form $\eta_1\eta_2 \cdots \eta_w$, where w is a positive integer, and each η_j is an element of $\{\phi, \phi^{-1}, \chi, \chi^{-1}, \psi, \psi^{-1}\}$. (See [2, p.62].) We recall that χ sits between ϕ and ψ in our commuting path; hence χ commutes with ϕ and ψ. Therefore χ commutes with ϕ^{-1} and ψ^{-1} as well. And of course χ commutes with itself and its inverse. Thus χ commutes with all products $\eta_1\eta_2 \cdots \eta_w$. In other words, χ is a member of the center of K.

Since $\phi \in h\gamma i \setminus \{id\}$, we see that F ix($\phi$) = {n}. Therefore by 2.6, n ∈ F ix(χ), because ϕ and χ are commuting elements. Moreover since χ is an element of the center of K, we have [n]K ⊆ F ix(χ), by 2.11. But H is a subgroup of K, because $H = \langle \varphi, \psi \rangle$ and $\phi, \psi \in K$. Thus $[n]_H \subseteq [n]_K$, and so [n]K = {1, 2, . . . , n}. We conclude that F ix(χ) = {1, 2, . . . , n}, which implies that χ is the identity element of S_n. This is a contradiction. Hence we have the proposition.

To prove the main result of the paper, we must still demonstrate the existence of a pair of elements at distance 5 in $\Delta_{R_n \cup C_n^{n-1}}$, when n−1 is composite.

For the remainder of the section, the following setup shall apply.

- Let n be a positive integer such that n − 1 is a composite number.
- Let M be the maximum proper divisor of n − 1.
- Define the following elements of C_n^{n-1}:

$$\gamma = (1\,2 \cdots n-1); \tag{6.2}$$

$$\delta = (1\,2 \cdots M\;\; n\;\; M+1\;\; M+2 \cdots n-2). \tag{6.3}$$

- Let p and q be arbitrary prime divisors of n − 1, possibly alike.
- Let r = (n − 1)/p, s = (n − 1)/q, and m = min(r, s).
- Define $H = \langle \gamma^r, \delta^s \rangle$

We note that $n \geq 5$, $M > 1$, and $m > 1$, because $n - 1$ is composite. We also point out that r and s are proper divisors of $n - 1$, hence $r, s \leq M$.

Lemma 6.3. Suppose that a is an integer such that $M + 1 - m \leq a \leq M - 2$. Then $[a]_H = [a + 2]_H$.

Proof. First assume that $r \leq s$. We observe that

$$M + 1 \leq a + r \leq 2M - 2 \leq n - 3.$$

Let i be the maximum positive integer such that $M + 1 \leq a + ir \leq n - 2$.

Then since $r \leq s$, we have $n - 2 < a + ir + s \leq (n - 2) + M$. Therefore

$$(\delta^s \gamma^{ir})(a) = \delta^s(a + ir) = a + ir + s - (n - 2).$$

But we have $M + 1 \leq a + s \leq n - 3$ as well, so

$$(\gamma^{ir} \delta^s)(a + 2) = \gamma^{ir}(a + s + 1) = a + ir + s + 1 - (n - 1).$$

Hence $(\delta^s \gamma^{ir})(a) = (\gamma^{ir} \delta^s)(a + 2)$, and thus $(\delta^{-s} \gamma^{-ir} \delta^s \gamma^{ir})(a) = a + 2$. It follows that $[a]_H = [a + 2]_H$, because $\delta^{-s} \gamma^{-ir} \delta^s \gamma^{ir} \in H$.

Now let us assume that $s < r$. Let j be the maximum positive integer such that $M + 1 \leq a + js \leq n - 3$. Then $n - 2 < a + r + js \leq (n - 3) + M$, because s is strictly less than r. Thus

$$(\gamma^r \delta^{js})(a + 2) = \gamma^r(a + js + 1) = a + r + js + 1 - (n - 1).$$

But on the other hand

$$(\delta^{js} \gamma^r)(a) = \delta^{js}(a + r) = a + r + js - (n - 2).$$

Thus $(\gamma^r \delta^{js})(a + 2) = (\delta^{js} \gamma^r)(a)$. So once again, $[a]_H = [a + 2]_H$.

Lemma 6.4. Suppose that a and b are integers such that $M-m+1 \leq a, b \leq M$. Assume that a and b have opposite parity. Then $[a]_H \cup [b]_H = \{1, 2, \ldots, n$

Proof. Let i and j be the odd and even elements of the set $\{m - 1, m\}$, respectively. Define

$$\begin{aligned} C &= \{M - m + 1, M - m + 3, M - m + 5, \ldots, M - m + i\}, \\ D &= \{M - m + 2, M - m + 4, M - m + 6, \ldots, M - m + j\}. \end{aligned}$$

We observe that $C \cup D = \{M-m+1, M-m+2, \ldots, M-1, M\}$, so $a, b \in C \cup D$. Also, either the elements of C are strictly even and those of D are strictly odd, or vice versa. Thus one of the elements a and b is a member of C, and the other is a member of D. But by 6.

$$\begin{aligned} &[M - m + 1]_H = [M - m + 3]_H = [M - m + 5]_H = \cdots = [M - m + i]_H, \\ &[M - m + 2]_H = [M - m + 4]_H = [M - m + 6]_H = \cdots = [M - m + j]_H. \end{aligned}$$

Therefore $C \cup D \subseteq [a]_H \cup [b]_H$.

Suppose $1 \leq x \leq M - m$. Let k be the maximum nonnegative integer such that $x + km \leq M - m$, and let $y = x + (k + 1)m$. Then $y \in C \cup D$, because $C \cup D$ consists of m consecutive integers. We observe that $\gamma^{(k+1)m}(x) = \delta^{(k+1)m}(x) = y$. Also, if m = r then $\gamma^{(k+1)m} = (\gamma^r)^{k+1} \in H$, and if m = s then $\delta^{(k+1)m} = (\delta^s)^{k+1} \in H$. Therefore, $[x]_H = [y]_H$. But $y \in [a]H$ or $y \in [b]H$, because $y \in C \cup D$. Hence $x \in [a]_H$ or $x \in [b]_H$. In other words, $x \in [a]_H \cup [b]_H$.

Now assume that $M + 1 \leq z \leq n - 1$. Let l be the maximum nonnegative integer such that $z-lr \geq M+1$. Then $z-(l+1)r \in \{1, 2, \ldots, M\}$, since $r \leq M$. And we have $\gamma^{(l+1)r}(z) = z-(l+1)r$. Therefore $[z]_H = [z-(l+1)r]_H$, because $\gamma^{-(l+1)r} = (\gamma^r)^{-(l+1)} \in H$. But we have already shown that $\{1, 2, \ldots, M\} \subseteq [a]H \cup [b]H$. Thus $z \in [a]H \cup [b]$

Finally, we observe that $\delta^{-s}(n) = M - s + 1$. Hence $[n]_H = [M - s + 1]H$. Therefore $n \in [a]_H \cup [b]_H$, because $M - s+1 \in \{1, \ldots, M-1\} \subseteq [a]_H \cup [b]_H$.

Proposition 6.5. Suppose p = 2 or q = 2, but p 6≠ q. Then

$$[n-1]_H \cup [n]_H = \{1, 2, \ldots, n\}.$$

Proof. Since $2 \in \{p, q\}$, and p and q are divisors of $n - 1$, we realize that $n - 1$ is even. Therefore $M = (n - 1)/2$.

Assume that p = 2. Then we have r = M and s = m; thus $\gamma^r(n - 1) = M$ and $\delta^{-s}(n) = M - m+1$. Therefore $[n-1]_H = [M]_H$, and $[n]_H = [M - m+1]_H$. Since $p \neq q$, q is an odd prime. Hence s is even, because $n - 1$ is even, and so M and $M - m + 1$ have opposite parity, because s = m. Thus by 6.4, we have $[M]_H \cup [M - m + 1]_H = \{1, 2, \ldots, n\}$, and therefore $[n - 1]_H \cup [n]_H = \{1, 2, \ldots, n\}$.

Now suppose that q = 2. Then by analogy to the above case, r = m and s = M, and r is even. We observe that the set $\{M - m+1, M - m+2, \ldots, M\}$ consists of r consecutive integers. Therefore the set includes elements a and b such that $a \equiv 0 \pmod r$ and $b \equiv 1 \pmod r$. Let i and j be integers such that $a = ir$ and $b = jr + 1$. Then $\gamma^{ir}(n - 1) = a$, and $(\gamma^{jr}\delta^{-s})(n) = \gamma^{jr}(1) = b$, because s = M. Hence $[n - 1]_H = [a]_H$, and $[n]_H = [b]_H$. But a and b have opposite parity, because r is even. Thus by 6.4 once again, we have $[n - 1]_H \cup [n]_H = \{1, 2, \ldots, n\}$.

We point out that $[n - 1]_H \neq [n]_H$ is a possibility under the hypotheses of 6.5. For example if $n - 1 = 6$, then

$$\gamma = (1\,2\,3\,4\,5\,6), \quad \delta = (1\,2\,3\,7\,4\,5).$$

(6.4)

If p = 2 and q = 3, we have r = 3 and s = 2. Therefore γ^{r} = (1 4)(2 5)(3 6), and δ^{s} = (1 3 4)(2 7 5). Thus

$$[6]_H = \{1, 3, 4, 6\}, \quad [7]_H = \{2, 5, 7\}.$$

Definition 6.6. If the natural action of H on {1, 2, . . . , n} has precisely one orbit, then we shall say that H is transitive.

Proposition 6.7. If p = q = 2, then H is transitive

Proof. We observe that r = s = M = (n − 1)/2. Thus for 1 ≤ a ≤ M − 1, we have

$$\left(\delta^{-s}\gamma^{r}\right)(a) = \left(\delta^{-M}\gamma^{M}\right)(a) = \delta^{-M}(a+M) = a+1.$$

Therefore [a]H = [a + 1]H, and furthermore, [1]H = [2]H = · · · = $[M]_H$.

Now, for M + 2 ≤ a ≤ n −1,

$$\left(\delta^{s}\gamma^{-r}\right)(a) = \delta^{M}(a-M) = a-1.$$

Hence [a]H = $[a-1]_H$, and so $[M+1]_H = [M+2]_H = \cdots = [n-1]_H$.

Finally, we notice that γ r (n^{-1}) = M, and δ^{-s} (n) = 1. Therefore we have $[n-1]_H = [M]_H$, and $[n]_H = [1]_H$. So we conclude that

$$[1]_H = [2]_H = \cdots = [n-1]_H = [n]_H.$$

Lemma 6.8. If each of r and s is ≤ n − M − 3, then H is transitive.

Proof. Let a = M −m+1. Since m ≥ 2, the set {M −m+1, M −m+2, . . . , M} contains at least two elements; thus {a, a + 1} is a subset. We point out that each of a + r and a + s is ≥ M + 1, but that

$$a+r+s = M+\max(r,s)+1 \leq M+(n-M-3)+1 = n-2.$$

Therefore,

$$\begin{aligned} \left(\delta^{s}\gamma^{r}\right)(a) &= \delta^{s}(a+r) = a+r+s, \\ \left(\gamma^{r}\delta^{s}\right)(a+1) &= \gamma^{r}(a+s) = a+r+s. \end{aligned}$$

Hence $(\delta^{-s}\gamma^{-r}\delta^{s}\gamma^{r})$ (a) = a + 1, and so $[a]_H = [a+1]_H$. But by 6.4, we have $[a]_H \cup [a+1]_H$ = {1, 2, . . . , n}. Thus the lemma follows.

Proposition 6.9. If p ≠ 2 and q ≠ 2, then H is transitive.

Proof. We observe that n − 1, being divisible by an odd prime, is not a power of 2. In the case of n − 1 = 6 and p = q = 3, γ and δ are as in equation (6.4), and r = s = 2. Therefore γ^{r} = (1 3 5)(2 4 6), and δ^{s} = (1 3 4)(2 7 5). Thus we obviously have

$$[1]_H = [3]_H = [5]_H,\ [2]_H = [4]_H = [6]_H;$$
$$[1]_H = [3]_H = [4]_H,\ [2]_H = [7]_H = [5]_H.$$

Hence we see that $[1]_H = [2]H = \cdots = [7]_H$, and so H is transitive.

For n − 1 = 9, we have p = q = r = s = M = 3. And for n − 1 = 10, p = q = M = 5 and r = s = 2. But in either case, each of r and s is less than n − M − 3. Therefore H is transitive, by 6.8.

Now assume that n − 1 ≥ 12. Since p and q are both odd primes, each of r and s is ≤ (n − 1)/3. Therefore

$$M + \max(r, s) \ \leq\ \frac{n-1}{2} + \frac{n-1}{3} \ =\ n - \frac{n+5}{6} \ \leq\ n - 3.$$

Thus by 6.8, H is transitive once again.

In the proof of the culminating result of the current section, as follows, the prime numbers p and q that have been under consideration, and thus the group H, shall arise. In stating the proposition, we keep our assumptions that n − 1 is composite, and M is the maximum proper divisor of n − 1. The definitions of γ and δ, as in (6.2) and (6.3), remain as well. Our argument here revisits many of the techniques that we applied in the proof of 6.2.

Proposition 6.10. The distance between γ and δ in Δ_{Sn} is at least 5.

Proof. Suppose that the distance between γ and δ in Δ_{Sn} is ≤ 4. In particular, assume that (γ, ϕ, χ, ψ, δ) is a commuting path in Δ_{Sn}. Since ϕ and ψ commute with γ and δ, respectively, there exist positive integers t and u such that $\phi = \gamma^t$ and $\psi = \delta^u$, by 2.4. We observe that $\gamma^{\gcd(t,n-1)} \in \langle\varphi\rangle$, and $\delta^{\gcd(u,n-1)} \in \langle\psi\rangle$, by 6.1. Also, gcd(t, n−1) and gcd(u, n−1) are proper divisors of n−1, because ϕ and ψ are nontrivial elements. Suppose that n − 1 = p · j · gcd(t, n − 1) = q · k · gcd(u, n − 1), where p and q are prime numbers, and j and k are positive integers. Since $\gamma^{\gcd(t,n-1)} \in \langle\varphi\rangle$, we have $\left(\gamma^{\gcd(t,n-1)}\right)^j = \gamma^{(n-1)/p} \in \langle\varphi\rangle$ as well. Similarly, $\delta^{(n-1)/q} \in \langle\psi\rangle$. And we note that $\gamma^{(n-1)/p}$ and $\delta^{(n-1)/q}$ are nontrivial elements, because each of (n − 1)/p and (n − 1)/q is strictly less than n − 1

Now given the structure of our commuting path, we see that χ commutes with ϕ and ψ. So furthermore, χ commutes with each element of $\langle\varphi\rangle$, and each of $\langle\psi\rangle$. Hence χ commutes with $\gamma^{(n-1)/p}$ and $\delta^{(n-1)/q}$

$$K = \left\langle \gamma^{(n-1)/p},\ \chi,\ \delta^{(n-1)/q} \right\rangle.$$

We observe that χ is a member of the center of K, as in the proof of 6.2. Also, by 2.6, n−1 and n are fixed by χ, because we obviously have F ix $\gamma^{(n-1)/p}) =$ {n} and $Fix(\delta^{(n-1)/q}) = \{n-1\}$. Moreover we have [n−1]K ∪[n]K ⊆ F ix(χ), by 2.11. But letting $H = \langle\gamma^{(n-1)/p}, \delta^{(n-1)/q}\rangle$, $[n-1]_H \cup [n]_H = \{1, 2, \ldots, n\}$, in view of Propositions 6.5, 6.7, and 6.9. Hence $[n-1]_K \cup [n]_K = \{1, 2, \ldots,$ $n\}$, because H is a subgroup of K. We conclude that F ix(χ) = {1, 2, . . . , n}. In other words, χ is the identity element of S_n. This is a contradiction, so the proof is complete.

PROOF OF MAIN RES

In order to realize the main result of the paper, we prove one further proposition

Proposition 7.1. Suppose that p ∈ {n − 1, n} is a prime number. Let γ be an element of C^p_n. Then Δhγi is a connected component of Δ_{Sn} . Furthermore, the diameter of $\Delta_{\langle\gamma\rangle}$ is 0 or 1, according as p = 2 or p >2

Proof. We observe that $\langle\gamma\rangle$ is an abelian group, containing p − 1 nontrivial elements. Thus we see that the diameter of $\Delta_{\langle\gamma\rangle}$ is equal to 0 if p = 2, but equal to 1 if p >2.

Suppose that the connected component of γ in Δ_{Sn} strictly contains $\Delta_{\langle\gamma\rangle}$. Then there exists a commuting path (γ, π, ρ) in Δ_{Sn} such that γ and ρ are noncommuting elements. Since π commutes with γ, we have $\pi \in \langle\gamma\rangle$, by 2.4. Therefore the order of π is a divisor of p, the order of γ, by Lagrange's Theorem. (See [2, p.89].) Thus π has order p, because p is a prime and π is nontrivial. It follows that $\langle\pi\rangle = \langle\gamma\rangle$. Furthermore, we claim that π is itself a cycle of length p. We observe that if l is a positive integer, and $l \notin \{1, p\}$, then the decomposition of π cannot contain a cycle of length l, because the order of π is not divisible by l. Also, the decomposition of π cannot have two disjoint cycles of length p, since 2p > n. Hence we have our claim. Therefore by 2.4, $\rho \in \langle\pi\rangle$, because ρ and π, being adjacent in our commuting path, are commuting elements. Thus $\rho \in \langle\gamma\rangle$. In particular, we conclude that ρ and γ are commuting elements, which is a contradiction. Thus we have the proposition.

We may now give arguments to obtain Theorem 1.1.

Proof. We consider the three cases separately.

a. This follows at once from Theorem 5.1 and Proposition 6.2.

b. First consider the case of n = 3. We observe that R3 = Ø, and

$$C_3^3 = \{(1\ 2\ 3), (1\ 3\ 2)\} = \langle(1\ 2\ 3)\rangle \setminus \{1_3\}.$$

Thus $\Delta_{R_3 \cup C_3^3}$ is a connected component of Δ_{S3}, of diameter 1, by 7.1.

Also, for $\gamma \in C_3^2$, $\Delta_{\langle\gamma\rangle}$ is a component of Δ_{Sn} of diameter 0, by 7.1.

Regarding the n = 4 case, we observe that for each $\gamma \in C_4^3$, $\Delta_{\langle\gamma\rangle}$ is a connected component of Δ_{S4}, of diameter 1, by 7.1. And by 4.1, $\Delta_{R_4 \cup C_4^4}$ is connected of diameter 3. Thus we see that ΔR4 ∪ C4 4 is a component of Δ_{S4}, in particular.

For n > 4, we note that n is even, because n − 1 is a prime. Therefore by 4.1 and 7.1, once again, we realize that the connected components of Δ_{S_n} are $\Delta_{R_n \cup C_n^n}$, of diameter 4, and $\Delta_{\langle\gamma\rangle}$ for $\gamma \in C_n^{n-1}$, each of diameter 1.

c. By 7.1, $\Delta_{\langle\gamma\rangle}$ is a connected component of Δ_{Sn}, of diameter 1, for each $\gamma \in C_n^n$. We observe that n − 1 is even and > 2, because n is a prime and > 3. Hence n − 1 is composite. Therefore by 3.4, 5.2, 5.3, and 6.10, $\Delta_{R_n \cup C_n^{n-1}}$ is connected of diameter 5. Thus we see that $\Delta_{R_n \cup C_n^{n-1}}$ is a component of Δ_{Sn}.

REFERENCES

1. C. Bates, D. Bondy, S. Perkins, P. Rowley, Commuting involution graphs for symmetric groups, J. Algebra, 266 (2003), 133-153. http://dx.doi.org/10.1016/s0021-8693(03)00302-8
2. D. Dummit and R. Foote, Abstract Algebra, Prentice-Hall, Englewood Cliffs, New Jersey, 1991.
3. T. Hungerford, Algebra, Springer-Verlag, New York, 1974.
4. A. Iranmanesh and A. Jafarzadeh, On the Commuting Graph Associated with the Symmetric and Alternating Groups, J. Algebra Appl., 7 (2008), 129-146. http://dx.doi.org/10.1142/s0219498808002710

17

Modeling Quantum Behavior in the Framework of Permutation Groups

Vladimir Kornyak[1]

[1]Laboratory of Information Technologies, Joint Institute for Nuclear Research, 141980 Dubna, Moscow Region, Russia

ABSTRACT

Quantum-mechanical concepts can be formulated in constructive finite terms without loss of their empirical content if we replace a general unitary group by a unitary representation of a finite group. Any linear representation of a finite group can be realized as a subrepresentation of a permutation representation. Thus, quantum-mechanical problems can be expressed in terms of permutation groups. This approach allows us to clarify the meaning of a number of physical concepts. Combining methods of computational group theory with Monte Carlo simulation we study a model based on representations of permutation groups.

Citation: Vladimir Kornyak "Modeling Quantum Behavior in the Framework of Permutation Groups" EPJ Web Conf. 173 01007 (2018). https://doi.org/10.1051/epjconf/201817301007

INTRODUCTION

Since the time of Newton, differential calculus demonstrates high efficiency in describing physical phenomena. However, the infinitesimal analysis introduces infinities in the physical theories. This is often considered as a serious conceptual flaw: recall, for example, Dirac's frequently quoted claim that the most important challenge in physics is "to get rid of infinity". Moreover, differential calculus, being, in fact, a kind of approximation, may lead to descriptive losses in some problems – an illustrative example is given below in Sec. 3.1. In the paper, we describe a constructive version of the quantum formalism that does not involve any concepts associated with actual infinities.

The main part of the paper starts with Sec. 2, which contains a summary of the basic concepts of the standard quantum mechanics with emphasis on the aspects important for our purposes.

Sec. 3 describes a constructive modification of the quantum formalism. We start with replacing a continuous group of symmetries of quantum states by a finite group. The natural consequence of this replacement is unitarity, since any linear representation of a finite group is unitary. Further, any finite group is naturally associated with some cyclotomic field. Generally, a cyclotomic field is a dense subfield of the field of complex numbers. This can be regarded as an explanation of the presence of complex numbers in the quantum formalism. Any linear representation of a finite group over the associated cyclotomic field can be obtained from a permutation action of the group on vectors with natural components by projecting into a suitable invariant subspace. All this allows us to reproduce all the elements of the quantum formalism in invariant subspaces of the permutation representations.

In Sec. 4 we consider a model of quantum evolution inspired by the quantum Zeno effect – the most convincing manifestation of the role of observation in the dynamics of quantum systems. The model represents the quantum evolution as a sequence of observations with unitary transitions between them. The standard quantum mechanics assumes a single deterministic unitary transition between observations. In our model we generalize this assumption. We treat a unitary transition as a kind of gauge connection – a way of identifying indistinguishable entities at different times. A priori, any unitary transformation can be used as a data identification rule. So, we assume that all unitary transformations participate in transitions between observations with appropriate weights. We call a unitary evolution dominant if it provides the maximum transition probability. The Monte

Carlo simulation shows a sharp dominance of such evolutions over other evolutions. To compare with a continuous description, we present also the Lagrangian of the continuum approximation of the model.

FORMALISM OF QUANTUM MECHANICS

Here is a brief outline of the basic concepts of the quantum mechanics. We divide these concepts into three categories: states, observations and measurements, and time evolution.

States

A pure quantum state is a ray in a Hilbert space $\mathcal{H}$ over the complex field $\mathbb{C}$, i.e. an equivalence class of vectors $|\psi\rangle \in \mathcal{H}$ with respect to the equivalence relation $|\psi\rangle \sim a|\psi\rangle$, where $a \in \mathbb{C},\ a \neq 0$. We can reduce the equivalence classes by normalization: $|\psi\rangle \sim e^{i\alpha}|\psi\rangle, \||\psi\|| = 1, \alpha \in \mathbb{R}$. Finally, we can eliminate the phase "degree of freedom" α by transition to the rank one projector $\Pi_\psi = |\psi\rangle\langle\psi|$, which is a special case of a density matrix.

A mixed quantum state is described by a general density matrix ρ characterized by the properties: (a) $\rho = \rho^\dagger$, (b) $\langle\psi|\rho|\psi\rangle \geq 0$ for any $|\psi\rangle \in \mathcal{H}$, (c) $\operatorname{tr}\rho = 1$. In fact, any mixed state is a weighted mixture of pure states, i.e. its density matrix can be represented as a weighted sum of the rank one projectors. We will denote the set of all density matrices by D(H).

The Hilbert space of a composite system, XY = X × Y, is the tensor product of the Hilbert spaces for the constituents: $\mathcal{H}_{XY} = \mathcal{H}_X \otimes \mathcal{H}_Y$. The states of composite system $\mathcal{D}(\mathcal{H}_{XY})$, are classified into two types: separable and entangled states. The set of separable states $\mathcal{D}_S(\mathcal{H}_{XY})$, consists of the states $\rho_{XY} \in \mathcal{D}(\mathcal{H}_{XY})$ that can be represented as weighted sums of the tensor products of states of the constituents: $\rho_{XY} = \sum_k w_k \rho_X^k \otimes \rho_Y^k,\ w_k \geq 0,\ \sum_k w_k = 1$. The set of entangled states, $\mathcal{D}_E(\mathcal{H}_{XY})$, is by definition the complement of $\mathcal{D}_S(\mathcal{H}_{XY})$ in the set of all states: $\mathcal{D}_E(\mathcal{H}_{XY}) = \mathcal{D}(\mathcal{H}_{XY}) \setminus \mathcal{D}_S(\mathcal{H}_{XY})$.

Observations and measurements

The terms 'observation' and 'measurement' are often used as synonyms. However, it makes sense to separate these concepts: we treat observation as a more general concept which does not imply, in contrast to measurement, obtaining numerical information. Observation is the detection ("click of detector") of a system, that is in the state ρ, in the subspace S ≤ H. The mathematical abstraction of the "detector in the subspace" S of a Hilbert space is the operator of projection, Π_S, into this subspace. The result of quantum observation is random and its statistics is described by a probability measure defined on subspaces of the Hilbert space. Any such measure μ (·) must be additive on any set of mutually orthogonal subspaces of a Hilbert space: if, e.g., A and B are mutually orthogonal subspaces, then μ span (A, B) = μ (A) + μ (B). Gleason proved [1] that, excepting the case dim H = 2, the only such measures have the form μ_ρ (S) = t_r ($\rho\Pi_S$), where ρ is an arbitrary density matrix. If, in particular, ρ describes a pure state, $\rho = |\psi\rangle\langle\psi|$, and S is one-dimensional, S = span $(|\varphi\rangle)$, we come to the familiar Born rule: $\mathrm{tr}\,(\rho\Pi_S) = \mathbf{P}_{\mathrm{Born}} = |\langle\varphi \mid \psi\rangle|^2$.

Measurement is a special case of observation, when the partition of a Hilbert space into mutually orthogonal subspaces is provided by a Hermitian operator A. Any such operator can be written as $A = \sum_k a_k \Pi_{e_k}$, where $a_1, a_2, \ldots \in \mathbb{R}$ is the spectrum of A, and $e_1, e_2, \ldots$ is an orthonormal basis of eigenvectors of A. "Click of the detector" Π_{ek} is interpreted as that the eigenvalue a_k is the result of the measurement. The mean for multiple measurements tends to the expectation value of A in the state ρ: $\langle A\rangle_\rho = \mathrm{tr}\,(\rho A)$.

Time evolution

The time evolution of a quantum system is a unitary transformation of data between observations. For a density matrix, unitary evolution takes the form

$$\rho_{t'} = U_{t't}\rho_t U_{t't}^\dagger, \qquad (1)$$

where ρ_t is the state after observation at the time t, ρ_t' is the state before observation at the time t', and $U_{t't}$ is the unitary transition between the observation times t and t'. In standard quantum formalism, time is considered as a continuous parameter, and relation (1) becomes the von Neumann equation in the infinitesimal limit. The evolution of a pure state can be written as $|\psi_{t'}\rangle = U_{t't}|\psi_t\rangle$, and the corresponding infinitesimal limit is the Schrödinger equation. To emphasize the role of observation in quantum physics, we note that unitary evolution is simply a change of coordinates in Hilbert space and is not sufficient to describe observable physical phenomena.

EMERGENCE OF GEOMETRY WITHIN LARGE HILBERT SPACE VIA ENTANGLEMENT

Quantum-mechanical theory does not need a geometric space as a fundamental concept — everything can be formulated using only the

Hilbert space formalism. In this view, the observed geometry must emerge as an approximation. The currently popular idea [2–4] of the emergence of geometry within a Hilbert space is based on the notion of entanglement. Briefly, the scheme of extracting geometric manifold from the entanglement structure of a quantum state ρ in a Hilbert space H is as follows:

- The Hilbert space decomposes into a large number of tensor factors: $\mathcal{H} = \otimes_x \mathcal{H}_x,\ x \in X$. Each factor is treated as a point (or bulk) of geometric space to be built. A graph G — called tensor network — with vertices x ∈ X and edges {x, y} ∈ X × X is introduced.
- The edges of G are assigned weights based on a measure of entanglement, a function that vanishes on separable states and is positive on entangled states. A typical such measure is the mutual information: $I(\rho_{xy}) = S(\rho_x) + S(\rho_y) - S(\rho_{xy})$, where ρ_x denotes the result of taking traces of ρ over all tensor factors excepting the x-th (and similarly for ρ_x, ρ_{xy}); S (a) = − t_r a log a is the von Neumann entropy. The graph G is supplied with a metric derived from the weights of the edges
- Finally, the graph G is approximately isometrically embedded in a smooth metric manifold of as small as possible dimension using algorithms like multidimensional scaling (MDS)

CONSTRUCTIVE MODIFICATION OF QUANTUM FORMALISM

David Hilbert, a prominent advocate of the free use of the concept of infinity in mathematics, wrote the following about the relation of the infinite to the reality: "Our principal result is that the infinite is nowhere to be found in reality. It neither exists in nature nor provides a legitimate basis for rational thought — a remarkable harmony between being and thought." Adopting this view, we reformulate the quantum formalism in constructive finite terms without distorting its empirical content [5–7]

Losses due to continuum and differential calculus

Differential calculus (including differential equations, differential geometry, etc.) forms the basis of mathematical methods in physics. The applicability of differential calculus is based on the assumption that any relevant function can be approximated by linear relations at small scales. This assumption simplifies many problems in physics and mathematics, but at the cost of loss of completeness.

As an example, consider the problem of classifying simple groups. The concept of a group is an abstraction of the properties of permutations (also called one-to-one mappings or bijections) of a set. Namely, an abstract group is a set with an associative operation, an identity element, and an invertibility for each element. There are two most common additional assumptions that make the notion of a group more meaningful: (a) the group is a differentiable manifold — such a group is called Lie group; (b) the group is finite. It is clear that empirical physics is insensitive to assumption (b) — ultimately, any empirical description is reduced to a finite set of data. On the contrary, assumption (a) implies severe constraints on possible physical models.

The problem of classification of simple groups under assumption (a) turned out to be rather easy and was solved by two people (Killing and Cartan) in a few years. The result is four infinite series: A_n, B_n, C_n, D_n; and five exceptional groups: E_6, E_7, E_8, F_4, G_2.

The solution of the classification problem under assumption (b) required the efforts of about a hundred people for over a hundred years [8]. But the result — "the enormous theorem" — turned out to be much richer. The list of finite simple groups contains 16 + 1 + 1 infinite series:

- groups of Lie type:
  $A_n(q), B_n(q), C_n(q), D_n(q), E_6(q), E_7(q), E_8(q), F_4(q), G_2(q),$
  ${}^2A_n\left(q^2\right), {}^2B_n\left(2^{2n+1}\right), {}^2D_n\left(q^2\right), {}^3D_4\left(q^3\right), {}^2E_6\left(q^2\right), {}^2F_4\left(2^{2n+1}\right), {}^2G_2\left(3^{2n+1}\right);$
- cyclic groups of prime order, $\mathbb{Z}_p$;
- alternating groups, A_n, $n \geq 5$;

and 26 sporadic groups: M_{11}, M_{12}, M_{22}, M_{23}, M_{24}, J_1, J_2, J_3, J_4, Co_1, Co_2, Co_3, Fi_{22}, Fi_{23}, Fi_{24}, HS , McL, He, Ru, S uz, O 0N, HN, Ly, T h, B, M

Note that finite groups have an advantage over Lie groups in the sense that in empirical applications any Lie group can be modeled by some finite group, but not vice versa.

Replacing unitary group by finite group

The main non-constructive element of the standard quantum formalism is the unitary group U(n), a set of cardinality of the continuum.

Formally, the group U(n) can be replaced by some finite group which is empirically equivalent to U(n) as follows. From the theory of quantum computing it is known that U(n) contains a dense finitely generated — and, hence, countable — matrix subgroup $U_*(n)$. The group $U_*(n)$ is residually finite, i.e. it has a reach set of non-trivial homomorphisms to finite groups.

In essence, it is more natural to assume that at the fundamental level there are finite symmetry groups, and U(n)'s are just continuum approximations of their unitary representations.

The following properties of finite groups are important for our purposes:

- any finite group is a subgroup of a symmetric group,
- any linear representation of a finite group is unitary,
- any linear representation is sub representation of some permutation representation.

"Physical" numbers

The basic number system in quantum formalism is the complex field $\mathbb{C}$. This non-constructive field can be obtained as a metric completion of many algebraic extensions of rational numbers. We consider here constructive numbers that are closely related to finite groups and are based on two primitives with a clear intuitive meaning:

1. natural numbers ("counters"): $\mathbb{N} = \{0, 1, \ldots\}$;
2. kth roots of unity3 ("algebraic form of the idea of k-periodicity"): $r_k \mid r_k^k = 1$.

These basic concepts are sufficient to represent all physically meaningful numbers.

We start by introducing $\mathbb{N}$ $[r^k]$, the extension of the semiring $\mathbb{N}$ by primitive kth root of unity. $\mathbb{N}$ $[r_k]$ is a ring if $k \geq 2$. This construction allows, in particular, to add negative numbers to the naturals: $\mathbb{Z} = \mathbb{N}$ $[r_2]$ is the extension of $\mathbb{N}$ by the primitive square root of unity. Further, by a standard mathematical procedure, we obtain the kth cyclotomic field $\mathbb{Q}(r_k)$ as the fraction field of the ring $\mathbb{N}$ $[r_k]$. If $k \geq 3$, then the field $\mathbb{Q}(r_k)$ is a dense subfield of $\mathbb{C}$, i.e. (constructive) cyclotomic fields are empirically indistinguishable from the (non-constructive) complex field. Note that $\mathbb{Q} \cong \mathbb{Q}(r_2)$.

The importance of cyclotomic numbers for constructive quantum mechanics is explained by the following. Let us recall some terms. The exponent of a group G is the least common multiple of the orders of its elements. A splitting field for a group G is a field that allows to split completely any linear representation of G into irreducible components. A minimal splitting field is a splitting field that does not contain proper splitting subfields. Although minimal splitting field for a given group G may be non-unique, any minimal splitting field is a subfield of some cyclotomic field $\mathbb{Q}(r_k)$, where k is a divisor of the exponent of G. Thus, to work with any unitary representation of G it is sufficient to use the kth cyclotomic field, where k is related to the structure of G.

Constructive representations of a finite group

Let a group G act by permutations on a set Ω, $|\Omega| = N$. If we assume that the elements of Ω are "types" of some discrete entities ("ontological entities", "elements of reality"), then the collections of these entities can be described as elements of the module $H = \mathbb{N}^N$ over the semiring N with the basis Ω. The decomposition of the action of G in the module H into irreducible components reflects the structure of the invariants of the action. In order for the decomposition to be complete, it is necessary to extend the semiring N to a splitting field, e.g., to a cyclotomic field $\mathbb{Q}(r_k)$, where k is a suitable divisor of the exponent of G. With such an extension of the scalars, the module H is transformed into the Hilbert space H over $\mathbb{Q}(r_k)$. This construction, with a suitable choice of the permutation domain Ω, allows us to obtain any representation of the group G in some invariant subspace of the Hilbert space H. We obtain "quantum mechanics" within an invariant subspace if, in addition to unitary evolutions, projective measurements are also restricted by this subspace

The above is illustrated in Figure 1 by the example of the natural action of the symmetric group S_N on the set $\Omega = \{e_1, \ldots, e_N\}$. Note, that any symmetric group is a rational-representation group, i.e. the field of rational numbers $\mathbb{Q}$ is a splitting field for S_N.

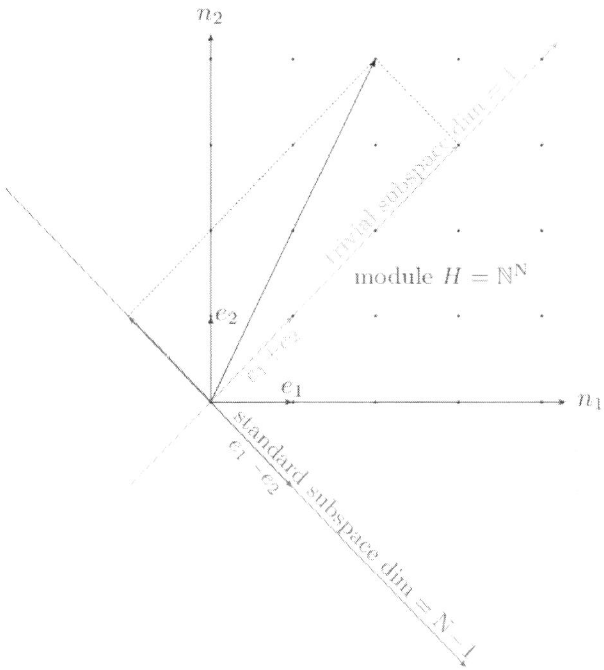

Figure 1. Natural representation of S_N decomposes into two irreducibles: 1D trivial and (N − 1)D standard representations.

Canonical bases: in trivial subspace

$e_1 + e_2 + \cdots + e_N$

in standard subspace

$e_1 - e_2$

$e_2 - e_3$

.

.

.

$e_{N-1} - e_N$

MODELING QUANTUM EVOLUTION

The fundamental discrete time $\mathcal{T}$ is represented by an ordered sequence of integers: $\mathcal{T} = \mathbb{N}$ or $\mathcal{T} = \mathbb{Z}$. We define a finite sequence of "instants of observations" as a subsequence of $\mathcal{T}$:

$$[t_0, t_1, \ldots, t_{k-1}, t_k, \ldots, t_n]. \quad (2)$$

The data of the model of quantum evolution include the sequence of the length n + 1 for states

$$[\rho_0, \rho_1, \ldots, \rho_{k-1}, \rho_k, \ldots, \rho_n] \quad (3)$$

and the sequence of the length n for unitary transitions between observations

$$[U_1, \ldots, U_k, \ldots, U_n]. \quad (4)$$

Standard quantum mechanics presupposes a single unitary evolution, U_k, between observations at times t_{k-1} and t_k. The single-step transition probability takes the form

$$\mathbf{P}_k = \mathrm{tr}\left(U_k \rho_{k-1} U_k^{\dagger} \rho_k\right). \quad (5)$$

The evolution can be expressed via the Hamiltonian: $U_k = e^{-iH(tk-tk-1)}$. In physical theories, Hamiltonians are usually derived from the principle of least action, which, like any extremal principle, implies the selection of a small subset of dominant elements in a large set of candidates. Thus it is natural to assume that, in fact, all unitary evolutions take part in the transition between observations with their weights, but only the dominant evolutions are manifested in observations. Therefore, in our model, we use the following modification of the single-step transition probability

$$\mathbf{P}_k = \sum_{m=1}^{M} w_{km} \,\mathrm{tr}\left(U_{k,m} \rho_{k-1} U_{k,m}^{\dagger} \rho_k\right), \quad (6)$$

where $U_{k,m} = U(g_m)$, $g_m \in G$; $G = \{g_1, \ldots, g_M\}$ is a finite group; U is a unitary representation of G; wkm is the weight of mth group element at kth transition.

The operators $U_{k,m}$, that maximize $\mathrm{tr}\left(U_{k,m} \rho_{k-1} U_{k,m}^{\dagger} \rho_k\right)$, will be called dominant evolutions.

The *single-step entropy* is defined as $\Delta \mathbf{S}_k = -\log \mathbf{P}_k$. (7)

Continuum approximation of (7) leads to the Lagrangian $\mathcal{L}$. Taking the logarithm of the probability of the whole trajectory, $\mathbf{P}_{0\to n} = \prod_{k=1}^{n} \mathbf{P}_k$, we arrive at the entropy of trajectory $\mathbf{S}_{0\to n} = \sum_{k=1}^{n} \Delta \mathbf{S}_k$, the continuum approximation of which is the action $S = \int \mathcal{L} dt$.

Continuum approximation of discrete model

Continuum approximation of the above model requires the following simplifying assumptions:

- Sequence (2) should be replaced by a continuous time interval $[t_0, t_n] \subseteq \mathbb{R}$.
- Sequences (3) and (4) are to be replaced by continuous functions of time, ρ = ρ (t) and U = U (t).
- The relation $\mathrm{tr}(\rho^2) = 1$ is necessary to ensure the continuity of probability. This relation holds only for pure states $\rho = |\psi\rangle\langle\psi|$. So, we will consider ψ instead of ρ.
- Assuming that U belongs to a unitary representation of a Lie group, we use the Lie algebra approximation, U ≈ 1 +iA, where A = A (t) is a function whose values are Hermitian matrices.
- We introduce derivatives and use the linear approximations $\Delta A \approx \dot{A}\Delta t$ and $\Delta\psi \approx \dot{\psi}\Delta t$

Applying these assumptions and approximations to the single-step entropy (7) and taking the infinitesimal limit we obtain the Lagrangian:

$$\mathcal{L} = \underbrace{\langle\psi|\dot{A}^2|\psi\rangle - \langle\psi|\dot{A}|\psi\rangle^2}_{\text{dispersion of } \dot{A} \text{ in state } \psi} - i\left(\langle\dot{\psi}|\dot{A}|\psi\rangle - \langle\psi|\dot{A}|\dot{\psi}\rangle + 2\langle\psi|\dot{A}|\psi\rangle\langle\psi|\dot{\psi}\rangle\right) - \langle\psi|\dot{\psi}\rangle^2.$$

Dominant unitary evolutions in symmetric group

The dominant evolutions between the states represented by the vectors from the module $H = \mathbb{N}^N$ for the group S_N can be computed as follows.

Let $|n\rangle = \begin{pmatrix} n_1 \\ \vdots \\ n_N \end{pmatrix}, |m\rangle = \begin{pmatrix} m_1 \\ \vdots \\ m_N \end{pmatrix}, |1\rangle = \begin{pmatrix} 1 \\ \vdots \\ 1 \end{pmatrix}$ be N-dimensional vectors with natural components.

The Born probabilities for the pair $|n\rangle$ and $|m\rangle$ are

$$\mathbf{P}_{\text{nat}}(|n\rangle, |m\rangle) = \frac{\langle n|m\rangle^2}{\langle n|n\rangle\langle m|m\rangle} \text{ — natural representation,}$$

$$\mathbf{P}_{\text{std}}(|n\rangle, |m\rangle) = \frac{\left(\langle n|m\rangle - \frac{1}{N}\langle n|1\rangle\langle 1|m\rangle\right)^2}{\left(\langle n|n\rangle - \frac{1}{N}\langle n|1\rangle\langle 1|n\rangle\right)\left(\langle m|m\rangle - \frac{1}{N}\langle m|1\rangle\langle 1|m\rangle\right)} \text{ — standard representation.} \quad (8)$$

Let R_a denote the permutation (as well as its representation), that sorts the components of vector $|a\rangle$ in some order. It is not hard to show that the unitary operator $U = R_m^{-1}R_n$ maximizes the probability $\mathbf{P}_*(U|n\rangle, |m\rangle)$, where the permutations R_n and R_m sort the vectors $|n\rangle$ and $|m\rangle$ identically in the case of natural representation, and either identically or oppositely — depending on the value of the numerator in (8) — in the case of standard representation.

Energy of permutation

Planck's formula, E = hν, relates energy to frequency. This relation is reproduced by the quantum mechanical definition of energy as an eigenvalue of the Hamiltonian, H = i~ ln U, associated with a unitary transformation. Consider the energy spectrum of a unitary operator defined by a permutation. Let p be a permutation of the cycle type $\{\ell_1^{m_1},\ldots,\ell_k^{m_k},\ldots,\ell_K^{m_K}\}$, where ℓ_k and m_k represent lengths and multiplicities of cycles in the decomposition of p into disjoint cycles. A short calculation shows that the Hamiltonian of the permutation p has the following diagonal form

$$H_p = \begin{pmatrix} \mathbb{1}_{m_1} \otimes H_{\ell_1} & & \\ & \ddots & \\ & & \mathbb{1}_{m_K} \otimes H_{\ell_K} \end{pmatrix}, \quad \text{where} \quad H_{\ell_k} = \frac{1}{\ell_k} \begin{pmatrix} 0 & & & \\ & 1 & & \\ & & \ddots & \\ & & & \ell_k - 1 \end{pmatrix}.$$

We shall call the least nonzero energy of a permutation the base energy:

$$\varepsilon = \frac{1}{\max(\ell_1, \ldots, \ell_K)}. \tag{9}$$

Simulation shows that the base ("ground state", "zero-point", "vacuum") energy is statistically more significant than other energy levels.

Monte Carlo simulation of dominant evolutions

Figure 2 shows several dominant evolutions for the standard representation of the groups S_{100} and S_{2000}. Each graph represents the time dependencies of Born's probabilities for the dominant evolutions between four randomly generated pairs of natural vectors. The dominant evolutions are marked by labeling their peaks with their base energies: $\varepsilon \in \left\{\frac{1}{61}, \frac{1}{69}, \frac{1}{72}, \frac{1}{98}\right\}$ and $\varepsilon \in \left\{\frac{1}{1416}, \frac{1}{1789}, \frac{1}{1939}, \frac{1}{1972}\right\}$ for S_{100} and S_{2000}, respectively. We see that with increasing the group size, non-dominant evolutions become almost invisible against the sharp peaks of dominant evolutions.

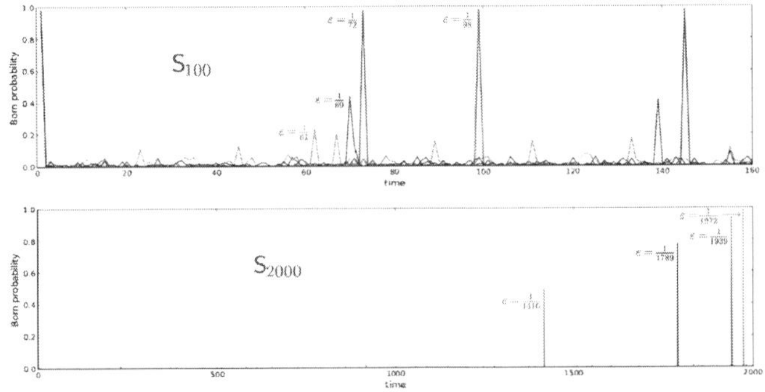

Figure 2. Dominant evolutions between randomly generated states. Born probability vs time

SUMMARY

1. A constructive version of quantum formalism can be formulated in terms of projections of permutations of finite sets into invariant subspaces.
2. Quantum randomness is a consequence of the fundamental impossibility of tracing the individuality of indistinguishable entities in their evolution.
3. The natural number systems for quantum formalism are cyclotomic fields, and the field of complex numbers is just their non-constructive metric completion.
4. Observable behavior of quantum system is determined by the dominants among all possible quantum evolutions.
5. The principle of least action is a continuum approximation of the principle of selection of the most probable trajectories.

REFERENCES

1. A.M. Gleason, Indiana Univ. Math. J. 6, 885–893 (1957)
2. M. Van Raamsdonk, Gen. Rel. Grav. 42, 2323–2329 (2010)
3. J. Maldacena, L. Susskind, Fortschr. Phys. 61 781–811 (2013)
4. C. Cao, S.M. Carroll, S. Michalakis, Phys. Rev. D 95, 024031 (2017)
5. V.V. Kornyak, EPJ Web of Conferences 108, 01007 (2016)
6. V.V. Kornyak, Mathematical Modelling and Geometry, 3, No 1, 1–24 (2015)
7. V.V. Kornyak, Phys. Part. Nucl. 44, No 1, 47–91 (2013)
8. R. Solomon, Bull. Amer. Math. Soc. (N.S.) 38, No 3, 315–352 (2001)

INDEX

T

U

V

W

Z